BICYCLING SCIENCE

BICYCLING SCIENCE
Third edition

David Gordon Wilson
with contributions by Jim Papadopoulos

The MIT Press
Cambridge, Massachusetts
London, England

This book was set in Stone Sans and Stone Serif on 3B2 by Asco Typesetters, Hong Kong.
Printed and bound in the United States of America.

Library of Congress Cataloging-in-Publication Data

Wilson, David Gordon, 1928–
Bicycling science / David Gordon Wilson with contributions by Jim Papadopoulos. — 3rd ed.
    p. cm.
Rev. ed. of: Bicycling science / Frank Rowland Whitt. 2nd ed. c1982.
Includes bibliographical references and index.
ISBN-13: 978-0-262-23237-1 (hc. : alk. paper)—978-0-262-73154-6 (pbk. : alk. paper)
ISBN-10: 0-262-23237-5 (hc. : alk. paper)—0-262-73154-1 (pbk. : alk. paper)
1. Bicycles—Dynamics. 2. Bicycles—History. I. Papadopoulos, Jim. II. Whitt, Frank Rowland. Bicycling science. III. Title.

TL410.W546   2004
629.227′2′015313—dc21   2003056132

10 9 8 7 6

Contents

Preface

The preface to the second edition of this book was written in 1981, and it gave the background of how the book came to be written. The book started with Frank Rowland Whitt, a chemical engineer working for a branch of Britain's war office but otherwise consumed with an enthusiasm for bicycles. He had been technical editor for the publications of the Cyclists' Touring Club. I had come to the United States from Britain in 1961 to work for a company designing components for jet engines. The Bank of England would not let me take my small savings out of the country, and, feeling guilty at abandoning my native land, I decided to offer some of these funds as a prize for improvements in bicycles and other human-powered vehicles. The U.K. journal *Engineering* was gracious enough to give the competition a great deal of publicity in 1967, Liberty Mutual added to the prize, and by the close of the competition in 1969, seventy-three entries were received from six countries. One entry came from Frank Whitt. He wrote that he wanted to meet me, and the next time I went to London we had a meal together. As we were parting, he thrust into my hands an old envelope stuffed with dog-eared papers with a cover sheet labeled "Bicycle Motion" and asked if I would just get it published in the United States, because he had been unsuccessful in the United Kingdom. (The reason he gave was that at that time, British publishers insisted on SI units' being used, and Frank had a strong distaste for them.)

After I had experienced a year or two of rejections by U.S. publishers, the MIT Press offered to publish the book, but only if I rewrote much of it and contributed up-to-date material. It was published in 1974 and became an unexpected success. Around 1980, the press asked for a second edition, but we had hardly started when Frank Whitt suffered a massive stroke that took him out of contributing to the book and resulted in his death some time later. I greatly missed his wisdom and experience and collegiality-by-correspondence. The second edition was in many ways a new book, because it was considerably expanded and changed. I also allowed some embarrassing errors to be incorporated. It was therefore a pleasure when the press asked me for a third edition in 1998, because by that time there were many corrections and enhancements that I wanted to incorporate.

Above all I wanted the chapter on steering and stability to be re-written completely by someone with a far better understanding of the subject matter than had I. I asked Jim Papadopoulos (another transplant from the United Kingdom), a graduate of MIT with a Ph.D. in mechanical engineering, someone who has devoted his life to the improvement of scientific and engineering knowledge of bicycles and bicycling, and a recognized genius. He agreed and also asked if he could be a coauthor of the third edition. It was easy to agree to such an offer from such a person. Jim took responsibility for five chapters. Alas! He was almost immediately dogged by problems in his family life that have prevented him finishing his work. I have reluctantly taken over full responsibility for the book, while acknowledging substantial contributions from him.

HUMAN POWER

1 A short history of bicycling

Introduction

Those who are ignorant of history are not, in truth, condemned to repeat it, as George Santayana claimed. However, people do spend a great deal of time reinventing types of bicycles and of components, and one purpose of this necessarily brief history is to give would-be inventors a glimpse of some of their predecessors. Sir Isaac Newton said that we make advances by standing on the shoulders of giants, but we must first know that there were giants and what they accomplished. Another purpose is to kill the many-headed Hydra of bicycling myths. People invent these myths—for instance, that Leonardo da Vinci or one of his pupils invented the chain-driven bicycle—for nefarious or self-serving or humorous purposes, and the myths are immediately picked up by journalists and enthusiasts and almost instantly become lore, however false. Historians repeatedly denounce the fakes, but the amateur historians continue to report them as if they were true. These people seem to practice a crude form of democracy: if they read something in ten publications and the contrary in one, then the one reported most often is, they believe, correct.

We have become the disciples of a group of cycle-historians that has become a powerful international movement having scholarly proceedings and meetings. Derek Roberts, the founder of the group, has written correction sheets for every new book incorporating cycling history, pointing out inaccuracies in detail. John Pinkerton encouraged Roberts to gather these together and published *Cycling history—myths and queries* (1991) in a further attempt to stem the tide of inaccurate versions of history. We are embarrassed to confess that Roberts had to write a correction sheet for the second edition of this book. In this present brief history we will endeavor to lay to rest previous myths, and we will do our utmost not to create more. We have been graciously guided by Roberts, by the late John Pinkerton, prominent member of the group and a publisher of cycling-history books, and by Hans-Erhard Lessing, a leading cycle-historian, former curator, and university professor. He himself has documented several major bicycling myths (some quoted below) previously regarded as historical facts. Others in this group who have been of particular help to the author are Nick Clayton and David Herlihy. Cycle historians themselves are far from agreement on many aspects of their profession: cycle history is a field in which views are strongly held and defended, and amateurs must tread with great care; the author has greatly appreciated this group's advice, which has not always been unanimous.

There have been three significant periods in cycling history, each covered in more detail below. Despite the myths of supposed earlier two-wheelers, the first bicycle (a "running machine" that the rider straddled and propelled with his feet on the ground) was invented in Germany in 1817, and this is when the history of the bicycle and the motorcycle begins. It led to a promising acceptance in several countries but was suppressed by the authorities in several places, so that by 1821 it had virtually died out. (Others, including Pinkerton [see below] believed that it was simply a fad of the rich and that fashions come and go in such a period.) It was not until the early 1860s that someone in France added cranks and pedals to the front wheel of a running machine, and another international rush developed. If we define a modern bicycle as a vehicle having two wheels in line connected by a frame on which a rider can sit, pedal, and steer so as to maintain balance, then this is the start of its history. This rush lasted much longer than that of 1817–1821. The front wheel was made progressively larger, and the high bicycle or "ordinary" was born. It was fun but it was dangerous,[1] and designers and inventors tried for many years to arrive at a safer machine. Success came with the so-called safety, first in 1878 with the Xtraordinary and the Facile, and reaching significant commercial success with John Kemp Starley's safeties of 1885 which, with Dunlop's pneumatic tires reinvented in 1888, became by 1890 very similar to the safety bicycle of today.

These, then, are the three principal developments that we shall discuss below in this short history. We shall also mention the tricycle period, the repeated enthusiasm for recumbent bicycles, and the enormous popularity of the modern all-terrain (or mountain) bicycle (the ATB).

Early history

It was through the use of tools that human beings raised themselves above the animals. In the broadest sense of the term, a tool might be something as simple as a stone used as a hammer or as complex as a computer controlling a spacecraft. We are concerned with the historical and mechanical range of tools that led to the bicycle, which—almost alone among major human-powered machines—came to use human muscles in a near-optimum way. A short review of the misuse of human muscle power throughout history (Wilson 1977) shows the bicycle to be a brilliant culmination of the efforts of many people to end such drudgery.

Many boats, even large ones, were muscle-powered until the seventeenth century. Roman galleys had hundreds of "sweeps" in up to three banks. Figure 1.1 shows a large seventeenth-century galley having fifty-four sweeps, with five men on each. The men were likely to be criminals, chained to their benches. A central gangway was patrolled by overseers

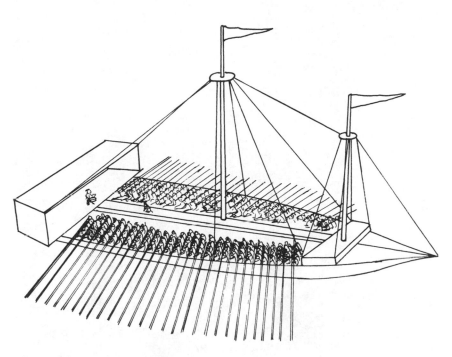

Figure 1.1
Early-seventeenth-century galley, with drummer in the stern and a whip-bearing overseer on the central gangway. (From a drawing in the British Museum reproduced in the *Encyclopedia Britannica*, sketched by Dave Wilson.)

equipped with whips to provide persuasion for anyone considered to be taking life too easily. The muscle actions used by these unfortunate oarsmen were typical of those considered appropriate in the ancient world. The hand, arm, and back muscles were used the most, while the largest muscles in the body—those in the legs—were used merely to provide props or reaction forces. (They didn't have the sliding seat of today's competitive rowers.) The motion was generally one of straining mightily against a slowly yielding resistance. With five men on the inboard end of a sweep, the one at the extreme end would have a more rapid motion than the one nearest to the pivot, but even the end man would probably be working at well below his optimum speed. Most farm work and forestry fell into the same general category. Hoeing, digging, sawing, chopping, pitchforking, and shoveling all used predominantly the arm and back muscles, with little useful output from the leg muscles. In many cases, the muscles had to strain against stiff resistances; it is now known that muscles develop maximum power when they are contracting quickly against a small resistance,

Figure 1.2
Engraving showing use of capstans in the erection of an obelisk at the Vatican in 1586. (The penalty for disrupting work was death.) (From N. Zabaglia, *Castelli e Ponti* [Rome, 1743].)

in what is termed a good "impedance match." We would call this good impedance match an optimum gear ratio.

One medieval example of the use of appropriate muscles in a good impedance match is the capstan (figure 1.2). Several people walked in a circle, pushing on radial arms, to winch in a rope. The capstan's diameter was chosen to give comfortable working conditions, and each pusher could choose a preferred radial position on the bar.

Other relatively satisfactory uses of muscle power were the inclined treadmill (figure 1.3) and Leonardo da Vinci's drum or cage for armaments (rotated by people climbing on the outside) (Reti 1974, 178–179), and treadmill-driven pumps (figure 1.4). This type of work may not have been pleasant, but per unit of output it was far more congenial than that of a galley slave.

The path of development, in this as in most other areas, was not a steady upward climb. Even though relatively efficient mechanisms using

Figure 1.3
Inclined treadmill powering a mill. (From Gnudi and Ferguson 1987.)

Figure 1.4
Leonardo's human-powered drum. (From Reti 1974.)

leg muscles at good impedance matches (figure 1.5) had been developed, sometimes hundreds of years earlier, some designers and manufacturers persisted in requiring heavy hand cranking for everything from drill presses to pneumatic diving apparatus to church-organ blowers, even though in all these cases pedaling seems clearly advantageous.

The first clearly human-powered vehicles known to history (if we exclude classes like wheelbarrows and carts pulled or pushed by men) were carriages supposedly propelled by footmen, in France in the 1690s (Ritchie 1975, 16). (An alleged earlier effort by a pupil of Leonardo da Vinci has been convincingly shown by Lessing [1998a] to be a fake.)

The first bicycle

It seems likely that the most important discovery in the development of the bicycle was made by chance. Baron Karl von Drais, a resident of Mann-

Figure 1.5
Treadmill geared winch (the first recumbent exerciser?). (From Gnudi and Ferguson 1987.)

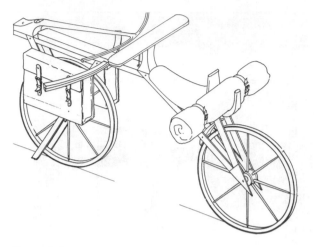

Figure 1.6
Draisienne. (Drawn from Drais's plans by Joachim Lessing; the cloak and side panniers are reconstructed. The wheel diameter chosen by Drais was 690 mm, 27 inches. Courtesy of Hans-Erhard Lessing.)

heim, studied mathematics and mechanics at Heidelberg and was an inventor of a binary digit system, a paper-strip piano-music recorder, a typewriter, and—during a series of bad harvests since 1812—two human-powered "driving machines" on four wheels. In 1815 the Indonesian volcano Tambora exploded, expelling the greatest known mass of dust in the atmosphere (estimated at seven times the amount from Krakatoa in 1883) and making 1816 "the year without a summer" in central Europe and the New England states. Starvation was widespread, and horses were killed for lack of fodder, the price of oats then playing the same role as the oil price today. Lessing believes that the consequent shortage of horses led von Drais to develop his two-wheeled "running machine" with front-wheel steering from the outset (figure 1.6). Our earlier assumption was that he had no preconception that the steering would enable him to balance but simply thought that it would be a convenience. However, Lessing (1995, 130) has made a powerful argument that ice skating, which "had long been a means of travel and transport in the Netherlands with its many canals" led to roller skating. Lessing quotes sources describing "a pair of skates contrived to run on small metallic wheels" to imitate ice skating on theater stages between 1761 and 1772. A preserved flyer for an outdoor demonstration between The Hague and Scheveningen in 1790 shows what appear to be the earliest in-line roller skates. These did not appear in technological magazines of the time, therefore it is hard to tell if von Drais had knowledge of them. But von Drais was an ice skater himself, so balancing on one foot on a skate could have started him thinking about something larger, necessarily with steering. (Roller skates that could be steered were patented by James Plimpton later, in 1863; he became a multimillionaire as a result [Lessing 1995].) A better-documented influence was the rediscovery of the Chinese wheelbarrow (using even a sail) with its central wheel under the load, since this was a topic at the University of Heidelberg.

However it was attained, the major discovery in bicycle history had been made, and it was scarcely recorded. Von Drais's vehicle was, however, noted in the German newspapers in 1817 and those of the United Kingdom in 1818 and the United States in 1819. In Paris, where von Drais obtained a five-year patent (Wolf 1890) it was called *le vélocipède* or the Draisienne, misspelled "Tracena" in the United States initially. In Britain it became known as the Pedestrian Accelerater and was nicknamed Hobby Horse (Street 1998). (Live horses needed constant care. These mechanical "horses" could be used or left at will and were thus treated as a hobby.)

Despite some initial skepticism and ridicule, von Drais was soon demonstrating that he could exceed the speed of runners and that of the horse-pulled "posts," even over journeys of two or three hours. His ability to balance when going down inclines and to steer at speed must have been important in this, but it awed the unathletic majority of the population. He indeed must have the principal claim to being the originator of balance on two wheels by steering.

Von Drais had many imitators. One was the London coachmaker Denis Johnson, who made a seemingly more elegant conveyance having a mainly iron instead of a wooden frame (it was therefore probably a little heavier). It was soon called the "dandy-horse." He set up a school in which young gentlemen could learn to ride. In the next year or so, use of the vehicle could be considered to have spread to clergymen, mailmen, and tradesmen, if contemporary cartoonists are to be taken seriously. However, its cost was too high for it to be used by any but the rich. In 1821, Lewis Gompertz fitted a swinging-arc ratchet drive to the front wheel (figure 1.7) so that the rider could pull on the steering handles to assist his feet. However, by this time so many restrictions had been put upon velocipedes that they lost their usefulness: "[F]or they gave orders that those who rode velocipedes should be stopped in the streets and highways and their money taken from them. This they called putting down the velocipede by fines" (Davies 1837/1986). (Pinkerton [2001] believed that Davis was exaggerating: velocipede users were almost exclusively the very rich and therefore unlikely to be harassed.)

Von Drais's premier place in what might be regarded as the three-step history of the development of the safety bicycle is assured, and it is

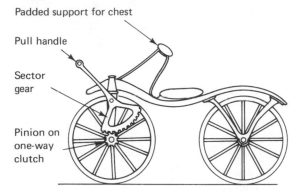

Figure 1.7
Gompertz's hand drive. (Sketched by Dave Wilson.)

relatively free from controversy. In contrast, the second and third steps (and "steps" seems an appropriate name, for they each resulted in "step-changes" in bicycle performance) are shrouded in some mystery and arguments among present-day proponents of one claimant or another.

In the previous edition of this book, and in many other reputable books of bicycle history including Ritchie 1975, credence has been given to a second step being taken in Dumfriesshire, Scotland, in 1839 or 1840 by Kirkpatrick Macmillan, who had been thought to have fitted cranks to the (large) rear wheel of a bicycle, with connecting rods going to swinging arms near the front-wheel pivot point (figure 1.8). Alas! Bicycle inventors seldom leave behind much incontrovertible evidence, and this is certainly true of Macmillan. His claimed development is reckoned by Nicholas Oddy (1990), Hans-Erhard Lessing (1991), and Alastair Dodds (1992) to be another myth. Lessing points out that in the chauvinistic atmosphere of that period (and later!), unscrupulous people repeatedly manufactured "proofs" that someone from their own countries were the first to invent some notable device. (The velocipede credited to Macmillan by a relative was actually the McCall velocipede of 1869, i.e., from step 2.) However, others believe with conviction that Macmillan did in fact produce a rideable pedaled bicycle much earlier than this.

As implied above, the hobbyhorse-velocipede "boom" died down substantially by 1821. The second step in bicycle development had to wait from then until the 1860s (see below). Why so long? One can speculate that the countries in which two-wheeled vehicles had been developed and received with such enthusiasm—principally Germany, France, Netherlands, the United States, and Britain—were now in the grip of railway

Figure 1.8
A copy of the velocipede attributed by some to Kirkpatrick Macmillan, made around 1869 by Thomas McCall of Kilmarnock. (Reproduced, with permission, from Ritchie 1975.)

mania. There was a new, fast way to travel, and this technology lured the creative dreams and efforts of inventors and mechanics away from the more mundane human-powered transportation. The parallels with what was to happen eighty years later, when the enthusiasm for the safety bicycle was to evaporate before the flaming passion for the automobile, are striking. Lessing (1995) points out that roller skating had lost its popularity on the arrival of the safety bicycle, with the rinks closing down in Europe, but not in the United States.

It would be an exaggeration to claim that all development in human-powered vehicles stopped during this time. From 1817 to 1870 the term "velocipede" was used for any foot-propelled vehicle. Such vehicles were used by some enthusiasts (including Prince Albert, husband of Queen Victoria), but not extensively. The machines' size, weight, and cost and the poor roads deterred walkers from changing their mode of travel. Willard Sawyer, a coachmaker in Kent, England, made increasingly sophisticated four-wheeled velocipedes, such as that shown in figure 1.9, and exported them around the world, from about 1840 to 1870 (McGurn 1999, 24–26). They were used by a few enthusiasts, but no movement developed. Undoubtedly there were lone mechanics and inventors in various countries making what seemed to be improvements to the Draisienne.

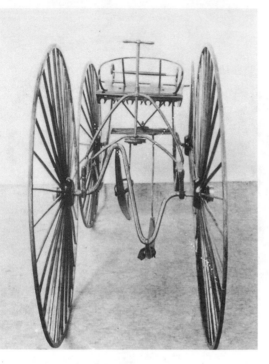

Figure 1.9
A Sawyer four-wheeled velocipede. (Reproduced, with permission, from Ritchie 1975.)

The second step: pedaling propulsion

The next (second) step in bicycle development has become highly contro-versial. We added a chapter on bicycle history to the second edition of this book. We credited Pierre Michaux with the significant step of adding pedals and cranks to the front wheel of a Draisienne, thus starting the astonishing period that lasted from the 1860s to the turn of the century when at least some parts of the earth appeared to have gone "bicycle-crazy." We were following what we thought were established historical facts. We were quite wrong in perpetuating a myth about the supposed existence of unsteerable hobbyhorses before the advent of the steerable machines of von Drais. (The senior author of the second edition, Frank Whitt, should be absolved with respect to this error. He suffered a severe and eventually fatal stroke early in the work and was, alas, able to contribute only marginally.) We might have been wrong in giving Michaux credit for the pedaled velocipede. Historian David Herlihy (1997) has been researching the contributions of Pierre

Figure 1.10
The first commercial Michaux velocipede. (From Clayton 1998.)

Lallement, who arrived in Brooklyn in 1865 (possibly with a crude bicycle with cranks and pedals) after serving an apprenticeship in Nancy, France. He impressed James Carroll, who provided funds for U.S. patent no. 59,915 (1866; viewable at ⟨www.uspto.gov⟩), the first for such a machine. The Michaux family later claimed that Lallement copied Pierre Michaux's ideas, and many believe that this is true. Herlihy believes that the opposite occurred, and that he can show the relationships among early French pioneers of the pedaled bicycle (H. Cadot, Michaux, Lallement, and the Olivier brothers), who played a major part. Pierre Michaux certainly produced pedaled velocipedes in increasing numbers in 1867–1869 (figure 1.10).

Whoever deserves the credit, there is no doubt about the results. A wild enthusiasm for *le vélocipède bicycle* (the bicycular velocipede) started in Paris in 1868 and spread to Belgium, the Netherlands, Germany, the United States, and Britain. The first true bicycle boom was underway.

Why, and why then? Lessing claims that having learned to ride a bicycle during childhood, we are unable to understand the fear of balancing of former times (unless we try to teach cycling to an unknowing adult). This fear of balancing hindered the earlier mechanics in thinking of two-wheelers with the feet permanently off safe ground. After Meyerbeer's opera *Le prophète* with roller skaters on stage had promoted roller skating throughout the Continent in the 1840s, ice skaters developed the new art of figure skating. Trying to imitate this on roller skates created the need for the "rocking" roller skates with rubber-block steering invented by a Bostonian, James Plimpton, in 1863. His empire of covered roller-skating rinks where the roller skates were rented, never sold, spanned the United States, Europe, and the whole Commonwealth. Roller skating became all the rage in the 1860s, and a large percentage of the rich learned to balance with both feet on wheels. Only on the basis of this broad balancing experience could someone on a two-wheeler ask: why not take the feet off the ground permanently and put them on cranks? Moreover, Paris during this time got new macadamized boulevards that eased the use of the new machine that had double the weight of the Draisiennes. But above all the machine was fun to ride, and thousands did so unimpeded by the authorities.

We might not find their experience so entrancing nowadays. The wooden wheels of the machines they rode had thick compression spokes and iron rims. It was only in the late 1860s that rubber was fastened onto the rims to cushion the harsh ride and ball bearings were first used on bicycles to give easier running (although Davies [1837] mentions that some Draisiennes were fitted with "friction rollers" to lessen the friction). Then the French leadership was lost when, in the Franco-Prussian war of 1870–1871, the French bicycle factories were required to turn to armaments (Ritchie 1975, 61).

What of the apparent lack of American contributions to the mainstream of bicycle development? What happened to the Yankee genius in engineering and mechanics? The U.S. patent office was in fact flooded with applications to patent improvements to velocipedes from 1868 on. The French and British makers found it necessary to follow the developments taking place across the Atlantic (Ritchie 1975, 61 et seq.). In 1869 Pickering's Improved Velocipedes were exported from New York to Liverpool. But the American craze, which *Scientific American* stated had made the art of walking obsolete, suddenly petered out in 1871 as quickly as it had started, leaving new businesses bankrupt and inventors with nowhere to go (Ritchie 1975, 66). There was then a lull until 1877, when the high-wheel bicycle was imported. Colonel Albert Pope started manufacturing them a year later. But conditions in the United States were less conducive for bicycles than those in Europe. In Europe, the high bicycle enabled people to travel much farther than was comfortably possible on a velocipede, and in Britain the roads were good enough for the country to be traversed from Lands End in southwest Cornwall to John O'Groats in northeast Scotland (924 miles; 1,490 km) in seven days (Ritchie 1975, 126–127). In the United States the distances between towns were (except perhaps in New England) enormous, and the roads were poor (Ritchie 1975, 82–83). Accordingly, the bicycle did not have, and did not convey, as much freedom, and the market was therefore smaller and far more dispersed than in Europe. It is doubtful that bicycles were used anywhere in the United States for long-distance travel except by a few enthusiasts and people who wanted to set records.

Development was fast in Britain, however, where production had been started more to fill the unsatiated French demand than to supply any domestic market. Technical leadership in the area was repeatedly taken by James Starley. The suspension or tension wheel had already been tried early in the century and was developed in Paris by Eugene Meyer in 1869 (Clayton 1997) and Grout in 1870. Around 1870 Starley and William Hillman introduced the "lever-tension" wheel, with radial spokes and a lever for tensioning and torque transmission (figure 1.11), and in 1874 Starley patented the logical extension of this idea, the tangent-tension method of spoking (figure 1.12). This has remained the standard spoking method to this day.

The high-wheeler or "ordinary"

With the advent of tension spoking, front wheels could be and were being made larger and larger to give a longer distance per pedal revolution, and therefore greater speed. The Ariel bicycle was patented by Starley on August 11, 1870, having already a larger-than-normal driving wheel. (For a while, some French race organizers tried to restrict the diameter to about a meter (Dodge 1996, 58)—perhaps a harbinger of the restrictions later imposed by the Union Cycliste Internationale (UCI)?)—Starley and others recognized the advantages of using a geared step-up transmission, but experimenters found that the available chains quickly froze up in the grit and gravel of contemporary roads. Soon front wheels were made as large as comfortable pedaling would allow. One bought one's bicycle to fit one's inside leg length. The largest production "high-wheeler" or "ordinary" would have a driving wheel about 60 inches (about 1.5 m) in diameter (figure 1.13). In the English-speaking world we still translate gear ratios into equivalent driving-wheel diameters, and this size corresponds to the middle gear of a typical modern bicycle. (The French and others in Europe use la developpement, the wheel's circumference, the distance traveled in one full turn of the cranks.) The 1870s were the years of the dominance of the high-wheeler. By the end of the decade, top-level bicycles were made with ball bearings in both wheels and in the steering head, the rims and forks were formed from hollow tubing, the steer axis had been tilted to create a castering effect, the tire rubber was greatly improved over the crude type used in 1870, and racing bicycles had been reduced to under 30 lb (13.6 kg). A ridable James ordinary weighing only 11 lb (5 kg) was produced in 1889.

Figure 1.11
The Starley-Hillman lever-tension wheel, 1870, shown by the late John Pinkerton in 2001. (Photo: Dave Wilson.)

Figure 1.12
Tangent-tension spoking. (From Sharp 1896.)

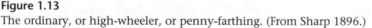

Figure 1.13
The ordinary, or high-wheeler, or penny-farthing. (From Sharp 1896.)

The ordinary was responsible for the third two-wheeler passion, which was concentrated among the young upper-class men of France, Britain, and the United States and was fostered by military-style clubs with uniforms and even buglers (Dodge 1996, 82–84). The ordinary conferred unimagined freedom on its devotees; it also engendered antipathy on the part of the majority who didn't or couldn't bicycle. Part of the antipathy was envy. The new freedom and style were restricted to rich young men. Strict dress codes prevented all but the most iconoclastic of women from riding high-wheelers. Family men, even if they were still athletic, hesitated to ride because of the reported frequent severe injuries to riders who fell (some feel that these reports were exaggerated). Unathletic or short men were excluded automatically. These prospective riders took to tricycles (Sharp 1896, 165–182), which for a time were produced in as many models as the ordinaries.

There were two technological responses to the need to serve the "extra-ordinary" market. James Starley played a prominent role in the first, and his nephew, John Kemp Starley, in the second.

Tricycles and quadricycles

The first of these responses was the development of practical machines of three or four wheels in which the need to balance was gone and the rider could be seated in a comfortable, reasonably safe, and perhaps more digni- fied position. Such vehicles had been made at different times since at least the start of the century, but the old heavy construction made propelling them a formidable task. In fact, the motive power was allegedly often pro- vided by one or more servants, who in effect substituted for horses (there is considerable doubt about the truth of these reports). Starley's Coventry Tricycle, patented by Starley's son and nephew in 1876, could be used with comparative ease by women in conventional dress and by relatively staid males. The Starleys produced this vehicle for several years from 1877. Early in the production run it was also made with more-conventional cranks with circular foot motion (figure 1.14). (The early version was then called the Coventry Lever tricycle, and the latter the Coventry Rotary.) Starley had found a chain that worked, at least in the possibly more protected conditions of a tricycle. The Coventry Lever and its successors had one large driving wheel on the left of the seat and two steering wheels, one in front and one behind, on the right. Starley saw the advantage of two large driving wheels on either side of the rider(s) and a single steering wheel in front. For this arrangement to work, power had to be transmitted to two

Figure 1.14
Starley's Coventry rotary tricycle. (From Sharp 1896.)

wheels, which, in a turn, would be going at different speeds. Starley reinvented the "balance gear" (Sharp 1896, 240–241), which is now known as the differential. Starley's Royal Salvo tricycle became the predominant form—for single riders, for two sitting side-by-side, and even for one behind the other (figure 1.15). This is not to say that there were no other forms; the reverse of this arrangement, for instance, with the steering wheel trailing the large driving wheels, was used for tradesmen's carrier machines (Pinkerton 1983). But the front-steerer was perceived as giving better control (one did not have to steer the rear wheel toward a pedestrian or a pothole to take avoiding action, as is necessary with rear-steerers).

Gradually the front wheel was made larger and the driving wheels smaller, as could be done with chain drives of increasing efficiency and reliability. By 1884 or 1885 the front wheel was connected directly to the handlebars (figure 1.16). This was a simpler and more reliable arrangement than the rack-and-pinion and other indirect systems that had been used. The modern tricycle had evolved, with the modern riding position in which one sits or stands almost over the cranks and splits the body weight among handlebars, pedals, and saddle.

This modern tricycle of late 1884/early 1885 was also very similar to the emerging form of the modern bicycle. In fact, the second response to the exclusion of so many from the high-wheeler movement was the

Figure 1.15
Starley's Royal Salvo tricycle. (From Sharp 1896.)

Figure 1.16
An early modern tricycle. (From Sharp 1896.)

development of a configuration that would make less likely a headfirst fall from a considerable height, that could be ridden in conventional dress, and that did not require gymnastic abilities.

Some improvements to the high-wheeler fulfilled only the first of these desiderata. Whatton bars (figure 1.17) were handlebars that came under the legs from behind, so that in the event of a pitch forward the rider could land feet first. (Cycle clubs—but not the police—recommended that riders of standard high-wheelers put their legs over the handlebars when going fast downhill, as in figure 1.18, for the same reason.) Some modern recumbent bicycles have similar handlebar arrangements. The designer of the American Star took the approach of making over-the-handlebars spills much less likely by putting the small wheel in front, giving it the steering function, and reducing the wheel size by using a lever-and-strap drive to the large wheel through one-way clutches (figure 1.19). Unfortunately, this arrived too late (1881) to have much impact, because the true "safety" bicycle was evolving rapidly by that date. Another type of bicycle that was safer to ride than the high ordinary was the "dwarf" front-driver, such as Hillman's 1884 Kangaroo (figure 1.20 shows an 1886 Kangaroo Dwarf Roadster) with a geared-up drive to a smaller front wheel (Sharp 1896, 152, 158). Such machines were offered because riders accustomed to front-drive machines did not always take kindly to the rear-drive safeties. Small-wheeled Bantam bicycles with an epicyclic hub gear (figure 1.21) were marketed as late as 1900.

The third step: the arrival of the modern "safety" bicycle

It had long been recognized that it would be most desirable from the viewpoint of safety to have the rider sitting between two wheels of moderate

Figure 1.17
Whatton bars. (From *Cycling* [1887].)

Figure 1.18
"Coasting—Safe and Reckless." (From *Cycling* [1887].)

Figure 1.19
The American Star, a treadle-action bicycle of 1880. (From Baudry de Saunier 1892.)

Figure 1.20
1886 Kangaroo Dwarf Roadster. (From Sharp 1896.)

Figure 1.21
Bantam geared front-drive safety bicycle. (From Sharp 1896.)

Figure 1.22
Starley safety bicycle. (From Sharp 1896.)

size. Many attempts were made over the years. The first Paris velocipede show, at which rubber tires, variable gears, free-wheels, tubular frames, sprung wheels, and band brakes were shown, was held in 1869. But the direct ancestors of today's bicycles evolved rapidly in the one or two years before 1885, when several were shown in Britain's annual Stanley Bicycle Show. James Starley had died in 1881, but his nephew John Kemp Starley, working with William Sutton, produced a series of Rover safety bicycles (Pinkerton and Roberts 1998) in 1885 that, by the end of that year, had direct steering and something very close to the diamond frame used in most bicycles today (figure 1.22).

One major development in the mainstream flowing to the modern bicycle remained: the pneumatic tire. This was patented in 1888 by John

Boyd Dunlop, a Scottish veterinarian in Belfast, although another Scot, R. W. Thomson, had patented pneumatic tires for horse-drawn vehicles in 1845 (Thomson 1845), and some were still in use in the 1880s (Du Cros 1938). Dunlop's early tires (made to smooth the ride of his son's tricycle) were crude, but by May 1889 they were used by W. Hume in bicycle races in Belfast—and he won four out of four. Success in racing in those days gave a clear signal to a public confused by many diverse developments. Cyclists saw that, as in the case of the safety versus the high-wheeled bicycle, a development had arrived that promised not only greater speed, or the same speed with less effort, but greater comfort and, especially, greater safety. Within eight years, solid tires had virtually disappeared from new bicycles, and Dunlop was a millionaire in pounds sterling.

With the arrival of the pneumatic-tired direct-steering safety bicycle, only refinements in components remained to be accomplished before the modern-day bicycle could be said to have been fully developed. Various types of epicyclic spur-gear variable-ratio transmissions for the brackets and rear hubs of chain-driven safety bicycles came on the market in Britain in the 1890s. Some heavier devices were available earlier for tricycles. The Sturmey-Archer three-speed hub (1902) was the predominant type, as it still is in many parts of the world (Hadland 1987), but there were many competitors at the turn of the century. The derailleur or shifting-chain gear was developed in France and Britain in 1895 but was not popular. It was developed by degrees in Europe and was eventually accepted for racing in the 1920s (Berto, Shepherd, and Henry 1999).

Undoubtedly, much more will be discovered about the history of the modern traditional single-rider bicycle, and unrecognized inventors will receive the honor due them. Inquiring readers can find much more history than we have space for here in the excellent books referenced and those listed at the end of the chapter.

Waxing and waning enthusiasm

Although the enormous enthusiasm for the bicycle that was found in most "Western, developed" countries in the 1890s waned sharply toward the end of the decade, that is not to state that the bicycle fell into wide disuse. Not many workers could afford bicycles, but they were used by well-to-do people for commuting and shopping, and later, in Europe at least, for sport and for weekend and vacation travel mainly by the "cloth-cap" (i.e., working) class. The hapless author was not allowed to ride a bicycle until he was nine (and then he was allocated an old single-speed "clunker"), and he was given an old three-speed "sports" bike when he was eleven, in 1939, the year war was declared in Europe. Petrol ("gas") was first rationed and then made unavailable for private use in Britain during World War II, and the bicycle was used widely. Riding with my elder brothers and mother and father was an important part of growing up. Going with my schoolboy friends to see local bomb damage and downed planes, to visit local towns for attractions such as swimming holes, and to plan increasingly longer trips ending with a 1,000-mile (1,600-km) tour into Scotland in 1944 were all liberating and, one hopes, character-forming activities. The camaraderie of European bicyclists everywhere made trips of any length very enjoyable.

When motor fuel and cars became available and affordable again (well after the Second World War ended), the bicycle in many Western countries was reduced to being used by children and by what were seen as fringe groups. In the third world, the bicycle was a necessity for anyone who could afford one. In most of these countries and especially in China, the proportion of person-trips and even of freight moved by bicycle were and possibly still are far higher than that taken by the railroads and road traffic.

A modern bicycle boom started in the United States in around 1970, for reasons difficult to discern. (It followed rather closely the end of a two-year competition in the design of human-powered vehicles organized by the author that created considerable public interest at the time, so that he is tempted to puff himself up to take credit, just as the cock crows at the dawn he has obviously caused.) Sales of bicycles rose rapidly to exceed comfortably the annual sales of automobiles. The buyers were overwhelmingly middle-class, college and professional people, U.S. bicycling thereby contrasting with the center of gravity of the sport in Britain. At the start, the popular style was the "English bicycle," predominantly Raleigh three-speed models, but soon "English racers" (an increasing proportion being actually French and Japanese), nowadays called "ten-speeds," became fashionable.

All-terrain bicycles

Most of these "road" bicycles enthusiastically purchased in the United States were used for a few kilometers and then left unridden, so that the bicycle boom began to peter out. But in 1970, at the time the enthusiasm for lightweight road bikes in the United States was increasing, a few enthusiasts in Marin County, California, began experimenting with old Schwinn clunkers for downhill off-road racing (Berto 1998). Others had done so in different countries before this, but they had not started a movement. Berto interviewed nine then-young men who, in this small area of California, continued experimenting throughout the 1970s with configurations of bicycles that gave advantages first for fast purely downhill travel and later for cross-country and uphill riding. Several started companies to produce the designs they developed. Rather suddenly, "beginning around

1982, a sea change affected the sales of bicycles in America and Europe. The buyers switched from road bikes to all-terrain or 'mountain' bikes. Tires went from skinny to fat, and riders went from a crouched position on dropped handlebars to a more erect position on flat handlebars" (Berto 1998, 25). This second boom in popularity of bicycles has been different in character from the road-bike boom, because a far higher proportion of the bikes purchased has been used to a significant extent. Perhaps most have not in fact been used for off-road recreation but have been seen as an extremely practical bike for negotiating rough urban streets in commuting or shopping use. They have left far behind their original heavy clunker image and have become high-tech lightweights. They have reached extraordinary levels of sophistication, many having front and rear suspension, wide-range twenty-seven-speed gears, hydraulic disk brakes, and frames made from aluminum, titanium, or carbon fiber. The technology developed for so-called mountain bikes is leading the bicycle industry generally. However, at the time of writing (2003) the sales of ATBs have peaked and are falling somewhat. Enthusiasts for "recumbent" bicycles wonder if there will be another bike boom featuring their configuration.

Recumbents

One reason for discussing recumbents rather than tandems, folding bicycles, pedicabs, or goods transporters is that most modern record-breaking machines are recumbents. Another is that greater safety can result from the use of the recumbent riding position in highway bicycles. In addition, what we know of the history of this variant form might help to illustrate the past and present character of the cycle industry.

Many early cycles (particularly tricycles) used the semirecumbent position. The "boneshaker" was often ridden with the saddle well back on the backbone spring and the feet at an angle considerably higher than that for the modern upright safety. In contrast with the riders of the high-wheeler and of the safety, who were told to position the center of gravity vertically over the center of the crank, the semirecumbent rider sits in something like a chair and puts her/his feet out forward on the pedals. The pedal-force reaction is taken not by the weight of the body (or, when that is exceeded, by pulling down on the handlebars), but by the backrest.

The first known semirecumbent bicycle (by which we mean one in which the rider's center of gravity was low enough relative to the front-wheel road contact point for there to be a negligibly low possibility of his being thrown over the front wheel in an accident) was built in Geneva by Challand (von Salvisberg 1897, 47) sometime before 1895 (figure 1.23). Challand called it the Normal Bicyclette. The rider sat rather high, directly

Figure 1.23
Challand's recumbent bicycle, 1896. (From von Salvisberg 1897, p. 47.)

over the rear wheel. In 1897 U.S. patent no. 577,895 was awarded to I. F. Wales for a somewhat strange-looking recumbent bicycle with hand and foot drive (figure 1.24) (Barrett 1972). A much more modern-looking recumbent bicycle was constructed by an American named Brown and taken to Britain in 1901 (figure 1.25) (Dolnar 1902). By this time orthodoxy rested firmly with the traditional safety bicycle, and the derision that had successively greeted the Draisienne, the velocipede, and the safety had been forgotten. Dolnar's review of the Brown recumbent in *The Cyclist* of January 8, 1902, was derisive to the point of sarcasm:

The curiously unsuitable monstrosity in the way of a novel bicycle shown in the single existing example of Mr. Brown's idea of the cycle of the future here illustrated.... The illustration(s) fully show(s) the rider's position and the general construction of this crazy effort.... The weight (30 lb.) and cost of the machine are greatly increased.... The mounting and dismounting are easy, and this is a fine coasting machine, the great wheel-base making very smooth riding ... and turns in a small circle. The machine runs light and is a good hill-climber, and it is only fair to say that the general action of this queerest of all attempts at cycle improvement is easy and good—far better than its appearance indicates.... The surprising fact is that any man in his sober senses could believe that there was a market for this long and heavy monstrosity at the price of a hundred dollars (£20).

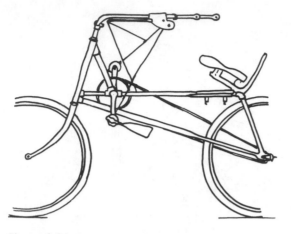

Figure 1.24
Design for hand-and-foot-powered recumbent patented by I. F. Wales in 1897.
(Sketched by Frank Whitt.)

Figure 1.25
Brown's 1900 recumbent bicycle. (From a sketch of the Sofa Bicycle in *The
Cyclist* (U.K.), November 13, 1901, p. 785.)

Figure 1.26
The Velocar. (From the advertisement of a licensor.)

Recumbents were more successful in Europe. After the First World War, the Austrian Zeppelin engineer (and, later, car designer) Paul Jaray built recumbents in Stuttgart in 1921 (Lessing 1998b).

A racing recumbent called the Velocar (figure 1.26) was developed in France in 1931–1932, from four-wheeled pedaled vehicles of that name (Schmitz 1994). With a Velocar, a relatively unknown racing cyclist, Francis Faure, defeated the world champion, Henri Lemoine, in a 4-km pursuit race and broke track records that had been established on conventional machines ("The Loiterer" 1934). A genuine orthodoxy pervaded the bicycle industry and the UCI, which controlled world bicycle racing. Instead of setting up a procedure and special category for machines such as the Velocar, the UCI, at the urging of the cycle trade, banned unconventional types from organized competition. This decision denied novel ideas the opportunity of being tested and publicized through racing and thereby deterred experimentation and development.

Only with the open-rule human-powered-vehicle competitions, started in California in 1974 (and resulting in the International Human Powered Vehicle Association, or IHPVA) has the inventiveness of human-powered-vehicle designers been given an incentive. With all classes of "open" races now being won by recumbent machines of a large variety of types, the technological history of this vehicle, and of bicycles in general, is again being written. The single-rider 200-m flying-start record for a streamlined bicycle is 130.3 km/h in 2002 and is likely to be faster by the time this book is published. These are exciting times. We wonder (and this is just speculation on the part of the author) if there may not also be a parallel in this new period of development with the period that started around 1866. The excitement over railway travel had seemed to drain away either the excess energies of inventors or the support for their activities, so that bicycle development languished. Occasional inventions like those of Gompertz were not followed up. But perhaps by the mid-1860s the railway was accepted, and it was apparent that it was not going to solve all transportation problems. Similarly, in the 1890s the motorcar arrived, and suddenly it was fashionable not only to travel in one, but to be involved in developing them. And two bicycle mechanics produced the first successful powered airplane only a little later. From then almost until the present day there has been a widely acknowledged love affair with the automobile, and with the airplane, first in the developed countries and later in the undeveloped countries. Only when disenchantment set in over the damage that these methods of transportation were inflicting on our cities did widespread enthusiasm for bicycle development surface once more.

May future histories record that new developments led to a new wave of popularity for human-powered travel, one that will last longer than some of the booms of the past.

Bicycle technology

For partisans of the bicycle, it is a matter of pride that the bicycle has frequently led to new technologies, or even fertilized new industries, such as

- mass production and use of ball bearings;
- production and use of steel tubes;
- use of metal stamping in production;
- differential gearing;
- tangent-spoked wheels (later used in cars, motorcycles, airplanes);
- bushed power-transmission chain;
- mass production and use of pneumatic tires;
- good-roads movement;
- Harley and Davidson, bicycle racers;
- Wright brothers, bicycle manufacturers; and
- the underpinnings of the automobile age.

Note

1.   John Pinkerton, a long-time rider of high bicycles, believed that the supposed dangers are highly exaggerated.

References

Barrett, Roy. (1972). "Recumbent cycles." *The Boneshaker* (Southern Veteran-Cycle Club, U.K.):227–243.

Baudry de Saunier, L. (1892). *Le cyclisme, théoretique et pratique*. Paris: Libraierie Illustre.

Berto, Frank J. (1998). "Who invented the mountain bike?" In *Cycle History: Proceedings of the 8th International Cycle History Conference, Glasgow, Scotland, August 1997*. San Francisco: Van der Plas.

Berto, Frank, Ron Shepherd, and Raymond Henry. (1999). *The Dancing Chain*. San Francisco: Van de Plas.

Clayton, Nick. (1997). "Who invented the penny-farthing?" In *Cycle History: Proceedings of the 7th International Cycle History Conference, Buffalo, NY, 1996*. San Francisco: Van der Plas.

Clayton, Nick. (1998). *Early Bicycles*. Princes Risborough, U.K.: Shire.

Cycling. (1887). Badminton Library. London: Longman's, Green.

Davies, Thomas Stephens. (1837). "On the velocipede." Address to the Royal Military College, Woolwich, U.K. Reported in *The Boneshaker*, nos. 108 and 111 (1986), with notes by Hans-Erhard Lessing.

Dodds, Alastair. (1992). "Kirkpatrick MacMillan—Inventor of the bicycle: Fact or fiction." In *Proceedings of the 3rd International Cycle History Conference*. Saint Etienne, France: Ville de Saint Etienne.

Dodge, Pryor. (1996). *The Bicycle*. Paris: Flammarion.

Dolnar, H. (1902). "An American stroke for novelty." *The Cyclist* (London) (January 8):20.

Du Cros, Arthur. (1938). *Wheels of Fortune*. London: Chapman & Hall.

Gnudi, Martha Teach (translator), and Eugene S. Ferguson (annotator). (1987). *The Various and Ingenious Mechanisms of Agostino Ramelli (1588)*. New York: Dover Publications, and Aldershot, U.K.: Scolar.

Hadland, Tony. (1987). *The Sturmey-Archer Story*. Self-published (U.K.).

Herlihy, David V. (1997). "H. Cadot and his relevance to early bicycle history." In *Cycle History: Proceedings of the 7th International Cycle History Conference, Buffalo, NY, September 1996*. San Francisco: Van der Plas.

Lessing, Hans-Erhard. (1991). "Around Michaux: Myths and realities." In *Actes de la deuxieme conference internationale sur l'histoire du cycle, St. Etienne, France*, vol. 2. Saint-Etienne, France: Ville de Saint-Etienne.

Lessing, Hans-Erhard. (1995). "Cycling or roller skating: The resistible rise of personal mobility." In *Cycle History: Proceedings of the 5th International Cycle History Conference*. San Francisco: Van der Plas.

Lessing, Hans-Erhard. (1998a). "The evidence against 'Leonardo's bicycle.'" In *Cycle History: Proceedings of the 8th International Cycle History Conference, Glasgow, August 1997*. San Francisco: Van der Plas.

Lessing, Hans-Erhard. (1998b). "The J Wheel—Streamline pioneer Paul Jaray's recumbent." In *Cycle History: Proceedings of the 9th International Cycle History Conference, Ottawa, Canada, August 1998*. San Francisco: Van der Plas.

"The Loiterer." (1934). In "Velocar versus normal," *Cycling* (March 2):202.

McGurn, Jim. (1999). *On Your Bicycle: The Illustrated Story of Cycling*. York, U.K.: Open Road.

Oddy, Nicholas. (1990). "Kirkpatrick MacMillan, the inventor of the pedal cycle, or the invention of cycle history." In *Proceedings of the 1st International Cycle History Conference, Glasgow*. Cheltenham, U.K.: Quorum Press.

Pinkerton, John. (1983). At Your Service: A look at Carrier Cycles. Birmingham, U.K.: Pinkerton Press.

Pinkerton, John. (2001). Personal communication with D. G. Wilson, September.

Pinkerton, John, and Derek Roberts. (1998). *A History of Rover Cycles*. Birmingham, U.K.: Pinkerton Press.

Reti, L., ed. (1974). *The Unknown Leonardo*. New York: McGraw-Hill.

Ritchie, A. (1975). *King of the Road*. Berkeley, Calif.: Ten Speed Press.

Roberts, Derek. (1991). *Cycling History—Myths and Queries*. Birmingham, U.K.: Pinkerton Press.

Schmitz, Arnfried. (1994). "Why your bicycle hasn't changed for 106 years." *Human Power* 13, no. 3 (1994):4–9; reprint of article originally published in *Cycling Science* (June 1990).

Sharp, Archibald. (1896). *Bicycles and Tricycles*. London: Longmans, Green; reprint, Cambridge: MIT Press, 1977.

Street, Roger. (1998). *The Pedestrian Hobby-Horse at the Dawn of Cycling*. Christchurch, U.K.: Artesius Publications.

Thomson, R. W. (1845). "Carriage wheels." U.K. patent no. 10,990.

Von Salvisberg, P. (1897). *Der Radfahrsport in Bild und Wort*. Munich, 1897; reprint, Hildesheim, Germany: Olms Presse, 1980.

Wilson, David Gordon. (1977). "Human muscle power in history." In *Pedal Power*, ed. James C. McCullagh. Emmaus, Penn.: Rodale.

Wolf, W. (1890). "Fahrrad und Radfahrer." Spamer, Leipzig; reprint Dortmund, Germany: Hitzegrad, 1979.

Zabaglia, N. (1743). *Castelli e Ponti*. Rome.

Recommended reading

Cycling History, vols. 1–16 (1991–2002). Volumes 1–3 are out of print and available only in some libraries. Volumes 4 onward are published by Van der Plas Publications, San Francisco.

2 Human power generation

Introduction

As a power producer, the human body has similarities and dissimilarities to the engine of an automobile. Energy is taken in through fuel (food and drink, in the case of humans). "Useful" energy is put out in the form of torque on a rotating crankshaft (in the case of cars) or in a variety of muscular movements (in the case of humans); and "waste" energy is dissipated as heat, which may be beneficial (for both) in cold weather. The peak efficiencies of the two systems (in cars, the energy transmitted to the crankshaft divided by the energy in the fuel; in humans, the extra food used in working) are remarkably close to one another, in the region of 20 to 30 percent. But automobile engines seldom work at peak efficiency, and in any case, peak efficiency in a car engine is attained only close to full power, whereas the rider of a multispeed bicycle can operate much closer to peak efficiency at all times. And whereas the automobile is powered by a "heat engine," the human body is similar to a fuel cell, a device that converts chemical energy in fuel directly to work. Also, human output, unlike that of the automobile engine, changes over time because of fatigue, possibly hunger, and eventually the need for sleep. A human can draw on body reserves (i.e., stores of several different fuels); the piston engine can work steadily until the fuel runs out and then delivers nothing. Humans also vary greatly from one to another, and from one day to another, and from one life stage to another, in terms of the power output they can produce.

The intention of the author and contributor in this chapter is to provide a basic understanding of how energy gets to the muscles of the rider of a bicycle and of how muscles produce power at the pedals on the bicycle he is riding. Readers should then be qualified to absorb the main conclusions of research papers in this area. We shall also comment on some bicycle configurations and mechanisms as they relate to the generation of human power. We take the philosophical position that athletes do sophisticated things to maximize performance, many of which are not yet understood. Timing and direction of foot force, choice of crank length and gear ratio, when to stand up or "bounce" the upper body—all seem to diverge from simple logic. We are reminded of the agreement of the thermodynamicist and the practical engineer in stating that "science has learned more from the steam engine than the steam engine has learned from science." (The second law of thermodynamics was formulated long after the first successful steam engines had been developed.)

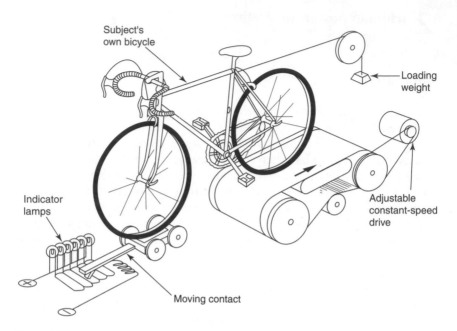

Subject's
own bicycle

Loading
weight

Indicator
lamps

Adjustable
constant-speed
drive

Moving contact

Figure 2.1
Muller ergometer. Load and speed are set; rider tries to keep center lamp lit. Run
stops when rearmost lamp lights up.

Measuring human power output

Exercise bicycles and ergometers of the pattern depicted in figure 2.1 have
been employed long and successfully. The flywheel's inertia minimizes
crank-speed variations during brief variations in pedaling torque. For accurate work the wheel speed and the average braking torque must be measured precisely. One effective preelectronic technique involves a band brake
whose drag is set by a weight. Rider power at a given belt speed (figure 2.2)
is controlled by the slope (or any rearward pull force, if used), the rolling
resistance due to compression of the pneumatic tires and the rubber belt,
and bicycle drive-train inefficiency.

Much of the information referred to in this chapter has been
obtained through careful experiments, typically with ergometers. Most
ergometers are pedaled in the same way as bicycles; other types are
"rowed," "skied," or "swum." All are capable of precise energy measurements. However, we must keep in mind some reservations about such
human-performance research:

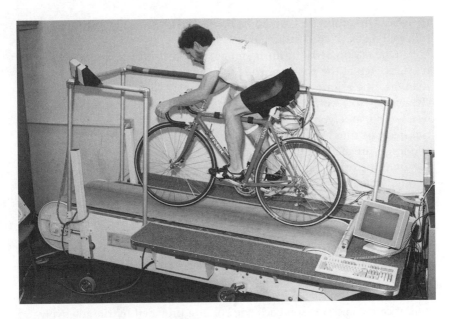

Figure 2.2
Treadmill bicycle ergometer. (Courtesy Maury Hull, University of California.)

• People vary widely in performance, and unless very many are tested (as has seldom been the case), the data obtained through testing cannot be generalized to the whole of humanity. There has also been a bias toward testing athletes (already self-selected for physical capability) and college students, predominantly male, in Western countries, and this population is not representative of humanity everywhere.

• Pedaling or rowing an ergometer usually feels stranger than riding a novel type of bicycle. It may take a month of regular riding before one becomes accustomed to a novel bicycle, as one's muscle actions gradually adapt to a new motion, body position, or restraints. Muscle adaptation to full oxygen-using capability can take years of extensive training. Subjects are seldom given the opportunity to adapt for more than a few minutes (occasionally hours) to working an ergometer before tests are performed and measurements are taken.

• Quite apart from imperfect adaptation to an ergometer, a person's response to years of exercise is rarely, in the extant research in this area, followed from start to finish. Comparing a group of exercisers to a group of nonexercisers may suggest that exercise confers physical vigor, but the logic is weak: already-vigorous people may simply be the ones who tend to exercise. The proper test would be to track two equivalent groups as they followed different specified regimens.

·   One reason pedaling an ergometer may feel strange is that the inertial resistance felt at the pedal (provided by the flywheel) is often much smaller than (as little as one-tenth of) the inertial resistance of the rider and bicycle, leading to a bothersome variation in pedal speed at substantial power levels. Also, an ergometer is usually fixed to the ground, whereas a bicycle can freely be tilted and moved relative to the pedaler, so that body motions and forces are affected.

·   On the other hand, a competitive bicyclist must crouch to minimize aerodynamic drag, possibly restricting breathing. Crouching is unnecessary on an ergometer but should possibly be enforced in research studies if accurate comparison to road racing is desired.

·   Subjects pedaling ergometers may not be given adequate cooling, and their long-term output can be limited by heat stress, as revealed by copious sweating. (There are exercisers on the market in which most of the power is dissipated in fans, thus simulating the square-law effect of wind resistance, but the air flow on such exercisers is not usually directed at the pedaler and in any case could not approach the cooling provided by the relative wind in bicycling.)

·   The motivation of competition (for maintaining a painful effort) can far exceed the stimulation of a laboratory setting.

Therefore, power output on ergometers (especially in the long term) is likely to be lower than could be achieved by the same subjects pedaling or rowing their own familiar machines through cooling air in a race that they want to win.

Some of the available test data on human power output are, however, taken from subjects bicycling on pavement, with various ingenious means used to measure work output (and/or oxygen consumption, which in steady state can be roughly related to fuel used and also to work output, if the subject's work-oxygen relation has been calibrated in the lab—figure 2.3). Such measurements may be more realistic than ergometer data. Even in such measurement schemes, however, someone wearing various sensors, possibly including a breathing mask, is likely to find that there is a noticeable resistance to movement and/or to breathing created by the measurement apparatus and that this will reduce performance somewhat (Davies 1961).

Modern on-bicycle power-measuring systems such as Schoberer Rad Messtechnik (SRM) and PowerTap (see chapter 4) are free from the foregoing objections, and we anticipate a very substantial rise in reported performances as more riders are sampled, using these systems, on their own bicycles, and especially in the heat of competition. For the shortest

Figure 2.3
Cyclist using breathing-rate-measuring equipment. (Courtesy Nijmegen University.)

times, simply using fast accurate, ergometer electronics that sense speed will also detect heretofore unexpectedly high peak power. (For example, Nuescheler's 2400 W for five seconds ⟨http://www.recordholders.org/de/records/roller1.html⟩ is almost double the peak power indicated in the second edition of this book [BSII]. Other Nuescheler records can be found at links such as ⟨http://www.recordholders.org/de/records/roller3.html⟩.)

Most ergometers have frames, saddles, handlebars, and cranks similar to those of ordinary bicycles. The crank drives some form of resistance or brake, and the whole device is fastened to a stand, which remains stationary during use. Other ergometers can measure the output from hand cranking in addition to that from pedaling. Some permit various types of foot motion and body reaction, including rowing (sliding-seat) actions. The methods employed for power measurement range from the crude to the sophisticated. One problem of ergometry is that human leg-power output varies cyclically (as does that of a piston engine) rather than being smooth (as with a turbine). Even in steady pedaling, a device indicating instantaneous power (pedal force in the direction of pedal motion, multiplied by pedal speed) would show peak values of perhaps 375–625 W, with an average of perhaps 250 W. Therefore, some form of averaging is usually employed. In some cases the subject is supposed to keep pedaling at a constant rate over a minute or two to obtain accurate results; in other systems the power can be integrated and averaged electronically over any desired number of crank revolutions (Von Döbeln 1954; Lanooy and Bonjer 1956).

There are additional problems associated with the determination of very-short-duration extreme power levels (from 1 kW to 2 kW or even greater). It is very hard to hold power constant; the usual dramatic increase in rpm reduces the pedaler's ability to produce power; and for the very shortest times it is important to measure the work done over completed crank rotations only. The best-accepted high-power ergometer test is known as the Wingate anaerobic test, in which a high resistance is suddenly applied, and the pedaler immediately strives to pedal at maximum speed for thirty seconds, initially accelerating the flywheel dramatically (even above 150 rpm if the pedaler is powerful), then allowing its speed to drop as fatigue sets in. Timing equipment determines the interval of each successive flywheel rotation, allowing average power during that rotation to be determined.[1]

Sturdy old exercise bicycles with heavy, braked flywheels are very similar in function to laboratory-grade ergometers. They can be adapted for accurate power measurement, if the problem of controlling and measuring torque can be solved. Load devices based on the dissipation of a small, tire-driven roller heat up the tire, which reduces the rolling resistance substantially. Magnetic (eddy-current) load units also heat up their conductive elements, increasing the electrical resistance and more than halving the initial magnetic torque. Air-blowing units require calibration and are moderately affected by the proximity of objects that alter airflow. Frictional brake drag also tends to be affected by temperature rise, so the unit must be designed using negative feedback to impose a torque that is essentially independent of the friction coefficient.

Describing pedaling performance quantitatively

The usual way to describe pedaling performance quantitatively is to fix a power level (usually by asking the subject to maintain a fixed pedaling speed at a known braking torque) and to determine the time to exhaustion. Different power levels can be sustained for durations anywhere between a few seconds and several hours. The results are plotted as a power-duration curve. As of this writing, such a curve seems to provide the best overall picture of the person's power-producing strengths and weaknesses.

The advantage of testing pedaling performance indoors, on an ergometer, is that the resistance is likely to be steady. Outdoors, even "level-road" riding may involve periods of double or triple the intended power, because of slight grades in the road, wind gusts, or accelerations.

Because each individual has different muscle mass, muscle makeup, inherited abilities, and state of conditioning, he or she will have a unique power-duration curve. When it comes to good athletic performance, some people are relatively stronger over particular durations and thus are better suited for events of those durations. This is partly why sprinters are not also climbers. (Another aspect of bicycling performance, of course, is that different body types may have more or less aerodynamic drag—important in level riding—and more or less weight—important when riding uphill.)

Power-duration data for "first-class athletes" and "healthy men" (NASA designations in the original graph) and for good cyclists are shown in figure 2.4. These data will be referred to repeatedly throughout the book. They are derived from ergometer tests, from tests of bicyclists on bicycles, and from estimates based on the results of time-trial races. Each data point given is the maximum duration of pedaling at a fixed power level: the curves do not reflect human power drop-off with time.

The top performances at different power levels are typically achieved by different types of individuals. The outer envelope reflects outstanding performances by rather large, strong men, with sprinters producing the short-time data and distance racers the longer-time results. However, the performance of any particular individual, in a given state of training and feeding, can be described by a curve of roughly similar form. (See the following section on critical-power curve fitting to power-duration data.) The chain-dotted line represents the author's estimate of the maximum performance of the best athletes with an optimum mechanism, perhaps one using hands and feet.

Critical power: curve fitting to power-duration pedaling data
The power-duration curves of individuals have been subjected to a variety of curve-fitting efforts, now commonly known by the term "critical power" (defined as the greatest power level that short-term tests suggest could be sustained "forever"). Such efforts are interesting because they encapsulate data efficiently and permit mathematical approaches to pedaling optimization and because they may reveal aspects of the physiological mechanisms governing endurance.

The simple regression used originally for such curve fitting of individual power-duration data appeared as a linear relationship between total

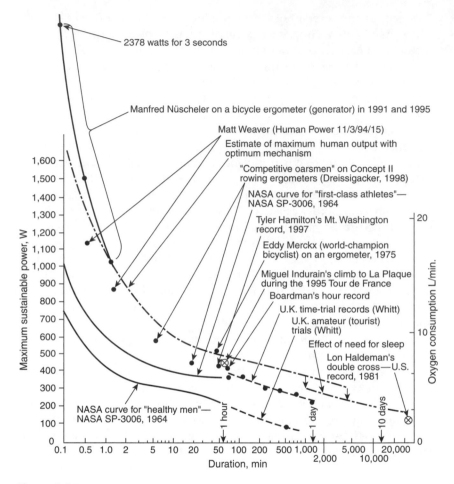

Figure 2.4
Human power output, principally by pedaling. Curves connect the termina-
tions through exhaustion of *constant-power* tests. (Data collected by Dave Wilson added to an original NASA chart.)

work performed (that is, the selected power level times duration) and the duration, in the form (see Hill 1993):

total work = anaerobic work capacity + (critical power × duration).

(Anaerobic work capacity refers to an amount of stored energy that can be released very quickly.) This equation embodies the simplified idea that any power beyond the pedaler's steady-state capacity is drawn from a finite energy reserve. Alternatively, this equation can be expressed as a linear relation between sustained power and 1/*duration*:

power = (anaerobic work capacity × 1/duration) + critical power.

From these two expressions it is also possible to eliminate duration and find a straight-line relation between 1/*total work* and 1/*power*, whose intercepts are 1/*anaerobic work capacity* and 1/*critical power*, respectively.

A variety of papers using such equations exhibit nice curve fits over ranges between two and twelve minutes, at power levels typically in the range from 200 W to 400 W (obviously not championship power levels, which would be two-and-a-half or even three times as great). In principle, two data points suffice to construct the line, but of course further trials will demonstrate the variability and quality of fit. An initial guess at short-duration power settings, based on rider mass, might be 2 and 4 W/kg for an unfit person; 4 and 6 W/kg for a fit recreational cyclist; and 6 and 10 W/kg for a cycling champion. (Each of these power levels can be equated to a given vertical velocity pedaling up a steep hill or running up flights of stairs.)

Some criticisms of this simple correlation are outlined in Gaesser et al. 1995; for example, the erroneous implication that the entire anaerobic work capacity can be depleted in a relatively short time span. (In fact, some anaerobic work capacity will be held back; and the shortest-term maximum power will fall well below predictions.) Other researchers have determined that the "critical power" determined by a series of relatively short tests is well above the lactate threshold (described below) and that very few riders can sustain that intensity for even 30 minutes (see, for example, Jenkins and Quigley 1990). Morton and Hodgson (1996) provide a comprehensive review of various proposed equations and conclude that the above model "has a simple appeal, its parameters are well understood, and it has always been found to be a good fit to data over the 2- to 15-minute range. Extensions ... incorporate a more realistic representation of the human bioenergetic system, and fit data over a wider range of power and duration, from 5 s to 2 h." Morton (1994) observed that long tests near critical power suffer from variability in motivation and proposed substituting a series of ramp tests (i.e., pedaling tests wherein the required power level rises at a rate of between 15 W each minute and 90 W each minute, until exhaustion of the rider). In 1997 he demonstrated the equivalence of ramp and constant-power determinations of critical power (Morton et al. 1997); if this finding is borne out by other researchers, the ramp method will undoubtedly prove to be very valuable.

In principle, specialized power-duration curves could be developed for any particular conditions of interest, for example, with two different cadences or body positions, or before and after a preliminary fatiguing effort similar to a hill climb, or perhaps following a change in diet. And performance research should focus on changes to the entire power-duration curve, not just the duration at one single power level. As an example of this, Jenkins and Quigley (1992) subjected twelve untrained male college students to eight weeks of ergometer endurance training (three sessions per week, forty minutes per session). On average, critical power (CP) increased from 196 W to 255 W over the course of the training, with no significant effect on anaerobic work capacity (AWC).

When it comes to characterizing human performance improvements, actual power-duration data (or directly derived performance parameters such as CP and AWC) seem more directly relevant than physiological measurements such as lactate threshold, maximal oxygen uptake, or fuel efficiency. And they are more easily measurable, requiring only a known-resistance exercise bicycle, or an on-bicycle power-monitoring system with a wind trainer to supply the load (note that slight speed adjustments will be needed as the drive tire warms up).

Anaerobic power: the Wingate test

Anaerobic power is revealed in a person's ability to leap or to sprint up a few flights of stairs. As described below, it is governed by immediate[2] and anaerobic (i.e., liberated through rapid partial metabolism of glycogen without oxygen) energy stores in the specific muscles being used. Because of the special problems of short-term, high-power ergometry, anaerobic power is not often assessed.

Currently the best accepted anaerobic power measurement is the Wingate anaerobic test (Inbar, Bar-Or, and Skinner 1996), which commonly uses a simple flywheel-style ergometer, accurately braked by a weight-loaded friction band. A typical protocol is for the rider to stay seated, pedaling at 60 rpm with no resistance. A large resistance equal to 8.5 percent of body weight is suddenly applied to the friction band at time 0,[3] and the rider strives to produce maximum power (while remaining seated) for thirty seconds. Flywheel speed is measured every five seconds (or better yet, the time of every completed crank revolution is logged). A powerful sprinter may bring the pedal rpm up to 160 within the first few seconds of a test, only to have it drop to about 60 rpm by the end.

Apart from energy used to accelerate the flywheel and to cover transmission losses (which should be small), pedaling power output is the wheel's peripheral speed times braking force. Three numbers are determined: the average speed (based on the total number of flywheel revolutions for the entire test); the highest speed (i.e., highest average over five seconds); and the lowest speed (i.e., lowest average over five seconds). From these and the known resistance are calculated average power (AP), peak power (PP), and minimum power (MP). Finally, the fatiguability index (FI) is defined as the percentage drop from PP to MP. Roughly speaking, the PP corresponds to immediate fuel sources, whereas the MP tends to approximate the maximum glycolytic power (see below).

As an example of Wingate research applied to cycling, Passfield and Doust (1998) investigated the effects of seventy minutes of medium-power pedaling (at 65 percent of VO_2max where VO_2 is the flow rate of oxygen absorbed in the lungs, and VO_2max is the maximum that can be absorbed by an individual working at maximum intensity) versus seventy minutes of rest on the results of two Wingate tests, one before and one after the seventy-minute interval. No significant difference was observed in the resting case. However, in the pedaling case, PP was found to be reduced by 2.6 percent, and AP was found to be reduced by 5.3 percent.

The Wingate test has commonly been used on noncyclist subjects to evaluate effects of diet or exercise. Naturally it is also used in evaluating elite competitors. However, it is not likely to ascertain the true five-second peak power directed to the flywheel, as would be revealed by on-bicycle power instrumentation such as PowerTap[4] or SRM. One reason for this is the inertia of the flywheel: during the violent initial acceleration, actual power may briefly reach twice the brake power or even more, and PP will be underestimated. (MP and AP are not affected by initial acceleration as much as the PP.) Reiser, Broker, and Peterson 2000 give an example in which inertial power correction yields 20 percent higher PP. Such a correction requires knowledge of the flywheel moment of inertia.

Another hindrance to true peak-power determinations is the relatively low value of the resisting force felt at the pedals (48 percent of body weight for the ergometer used by Reiser, Broker, and Peterson [2000]),[5] since a high power output is possible only at an extremely high pedaling rpm. This issue was addressed by Hermina (1999), who tested fifteen elite road cyclists at brake resistances from 7.5 to 14.5 percent of body weight. At the lowest resistance the mean PP was 951 W, whereas at the greatest it was 1450 W.

One way to restrict cadence for a peak-power determination, used by Martin, Wagner, and Coyle (1997), is to consider just one pedal revolution, in which case flywheel acceleration alone (no brake) provides the power determination. Of course this power level is sustained for less than a second. Thirteen subjects in Martin, Wagner, and Coyle's study averaged a one-revolution maximal power of 1317 W. (The peak torque over the pedal revolution was also identified at about 62 percent higher than average torque.)

Perhaps the best way to determine maximal five-second power is on a fixed-speed (isokinetic), motor-driven ergometer. Electronic instrumentation would be needed to average the measured torque of such an ergometer, but the cadence would be controlled perfectly. Such an ergometer was used by Beelen and Sargeant (1991) to show that peak power is commonly produced at 120–130 rpm (however, the spinning champion Manfred Nüscheler produces his peak power, above 2200 W, at 150 rpm).

Physiology of high-power pedaling: a primer

The physiology of exercise is a complex subject, evolving substantially from decade to decade as research progresses. Neither the book's author nor its contributing author is a researcher in this general field, so our attempt (following) to reconcile and summarize material published mostly during the 1980s risks criticism by experts in the field. Nevertheless it seems worth presenting, because the subject is complicated, and the field remains awash in mythology from still earlier decades. We hope that the simple presentation below will prepare readers to gain insight from current and future exercise research.

For the big picture we have relied heavily on comprehensive texts by Astrand and Rodahl (1977), by Brooks, Fahey, and White (1996), and by McArdle, Katch, and Katch (1996), all of which merit repeated study. Many specialized details about muscle-fiber behavior are engagingly presented by McMahon (1984).

Overview of how muscles work

Human muscle cells convert chemical potential energy into mechanical work, using a variety of fuels, originally derived from foodstuffs, that are stored in the body. The energy released by oxidizing the foods is commonly measured in kilocalories (confusingly abbreviated as "Calories," with a capital C, when shown on food packaging). The proportion of the energy content actually delivered as muscular work is 20–30 percent (the remainder being released as heat and as energy in body wastes). This percentage is the food's energy efficiency previously mentioned.

Every muscle is composed of a large number of fibers (or cells) of three more or less distinct types. A platoon of fibers, known as a motor unit, is assigned to each of the hundreds of nerves (motor neurons, or motoneurons) controlling a given muscle (figure 2.5). Fibers may be visual-

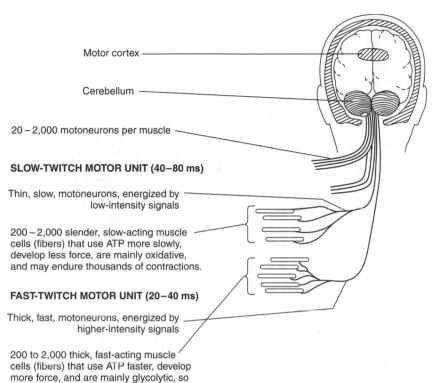

Motor cortex

Cerebellum

20 – 2,000 motoneurons per muscle

SLOW-TWITCH MOTOR UNIT (40–80 ms)

Thin, slow, motoneurons, energized by
low-intensity signals

200 – 2,000 slender, slow-acting muscle
cells (fibers) that use ATP more slowly,
develop less force, are mainly oxidative,
and may endure thousands of contractions.

FAST-TWITCH MOTOR UNIT (20–40 ms)

Thick, fast, motoneurons, energized by
higher-intensity signals

200 to 2,000 thick, fast-acting muscle
cells (fibers) that use ATP faster, develop
more force, and are mainly glycolytic, so
that they fatigue after 50–100 contractions.

A motor unit is a single nerve (motoneuron) controlling a platoon of similar fibers.
In any one muscle the fibers of hundreds of different motor units are indiscriminately
interlaced, leading to a spotted or checkered appearance when the fast fibers are
stained. Fast fibers make up between 10 and 90 percent of the total fibers
in a given muscle.

Figure 2.5
Muscle fibers and motor units.

ized as extending from one muscle endpoint to the other, but this is not
always accurate. A tilted fiber arrangement called *pennation* involves more
fibers of shorter length, effectively creating a shorter, wider muscle. A pennated muscle exerts more force than one with direct fibers, but can't
shorten to the same degree.

The term "pennated" means that the fibers of the muscle are angled
to the direction of muscle contraction, rather than straight. This arrangement permits the connection of two long, overlapping tendons with many
short fibers. This increases the force a muscle can exert compared to that
from a smaller number of long fibers, but reduces its range of motion.

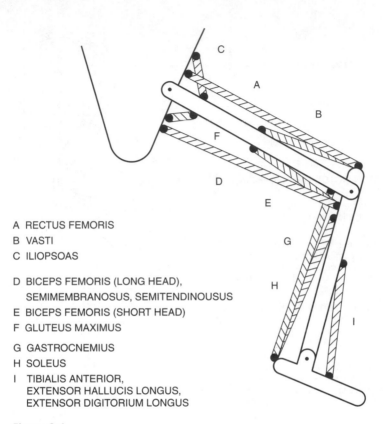

A RECTUS FEMORIS
B VASTI
C ILIOPSOAS

D BICEPS FEMORIS (LONG HEAD),
   SEMIMEMBRANOSUS, SEMITENDINOUSUS
E BICEPS FEMORIS (SHORT HEAD)
F GLUTEUS MAXIMUS

G GASTROCNEMIUS
H SOLEUS
I TIBIALIS ANTERIOR,
   EXTENSOR HALLUCIS LONGUS,
   EXTENSOR DIGITORIUM LONGUS

Figure 2.6
Stylized functional representation of the major muscles acting at the hip, knee, and ankle. (From Papadopoulos 1987.)

Muscles exert tension only (this physiological condition is termed *contraction*), and therefore can perform mechanical work only as they shorten, drawing together their attachment points on two different bones. A limb or hand "pushes" only because the body has a system of levers (composed of bones), pivots (the joints), tensioning cords (the tendons), and antagonistic muscles, so that one set of muscles produces a pull of the limb or of its extremity, and the other set produces a push. A schematic representation of the main leg muscles is presented in figure 2.6.

If a muscle actually lengthens while contracted (as occurs when one is lowering a barbell, for example, or slowly squatting), it is absorbing and dissipating work, rather than producing it. Such behavior is known as *eccentric contraction* or *negative work* and must be minimized if power or endurance is to be maximized. It is a matter of faith (for the contributing author) that humans, no less than animals, instinctively adjust their behavior to prevent energy from being lost in negative work.

Muscle fibers are caused to contract by nerve stimuli in the presence of a fuel. No less than six types of fuel are used individually or in combination in muscles. The choice of fuels to be used is not under conscious control. Instead, the power level elected by the muscle user effectively "calls upon" the appropriate fuel choice or choices, at least until depletion. In cycling specifically, at the very highest power levels for a given individual (generally above 1500 W, 2 hp, for strong men), exhaustion occurs in just a few seconds. At a considerably lower power level, say 500 W, a strong rider may last a few minutes; at 350 W, an hour or longer; at 250 W, it may be possible to pedal all day. All these durations have analogues in cycling events: short match sprints (about ten seconds), track time trials (a few minutes), road time trials, and long road races, which can last even for days (e.g., the Race across America).

After a brief contraction, a fiber will again relax. However, if a muscle is required to exert force for a longer time—for example, while supporting a weight—the nerve stimulating the motor unit involved will "fire" repeatedly, and if the firing period is shorter than the fiber relaxation time, the motor unit will exert a steady, maximal tension. During such "isometric" contraction, there is no shortening of the muscle.[6] Even though the weight is not being lifted during this time, and so in the thermodynamic sense no external work is being done, the muscle still requires energy either from its stores or from the bloodstream. To maximize external work and to minimize fatigue when bicycling, isometric contractions should therefore be avoided as much as possible.

Beyond the elementary picture of the muscle presented here lies the entire complex subject of exercise physiology, which must be explored to understand human bicycling performance.

The six muscle fuels

Muscles make use of six different types of fuels, some of which are short-acting and others of which are long-duration fuels. Figure 2.7 charts the movement and transformation of these six muscle fuels.

Two fast-acting fuels The so-called immediate fuels are adenosine triphosphate (ATP), and phosphocreatine (or creatine phosphate) (PCr). These are created within the muscle fibers from other fuels and do not release any harmful waste products requiring processing or removal other than heat. ATP is the only fuel used directly by a cell's contractile proteins; all other fuels are useful only insofar as they can regenerate ATP within the muscle fiber. ATP can be used as fuel without delay (no oxygen required) and replenished just as rapidly by conversion of one of many other fuels.

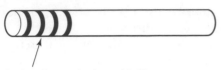

Contractile proteins in muscle fibers are
powered solely by high-energy ATP

1. IMMEDIATE FUELS   ATP stores in fibers (3 s) (*These ATP sources require no
oxygen and produce no waste. There is therefore no transport delay.*)
                                      PCr stores in fibers (10 s)

2. FATS (LIPIDS)   Energy-rich compounds composed primarily of carbon and
hydrogen that are stored throughout the body, enough for days of continuous effort. (*These
are mobilized and transported slowly to fibers. Oxygen is also required. They are used
primarily by the weaker SO fibers. Fats are the main fuel while resting or gently exercising.
High power levels suppress fat usage. Endurance training mobilizes fat and increases its
usage at medium and high power, improves oxygen delivery and increases oxidizing
structures, e.g., mitochondria.*)

3. CARBOHYDRATES   of the (CH2O)n variety, primarily glucose, its long-chain form
glycogen, and its partially reacted forms pyruvate/lactate. (*Sufficient glycogen is stored in the
muscles for immediate use, giving two hours of high power, or a few minutes of extreme power.
Glycogen does not circulate. However, glucose circulates in the blood for the use of muscles
and the brain, delivered from the intestines or from breakdown of liver glycogen.*)

*There are two stages of carbohydrate use in muscles. Stage 1 is the fast nonoxidative
(anaerobic) glycolysis/glycogenolysis to pyruvate, which has the greatest potential in FG or
FOG fibers. It delivers only 7 percent of the available energy in carbohydrate. No oxygen is
required. Glycogen is in place. Glucose is brought in. This stage is potentially very powerful
for one or two minutes. The excess unoxidized pyruvate accumulates as lactate. It must be
cleared from working fibers. It enters neighboring fibers or is transported by blood to far fibers
or the liver. May oxidize or reconstitute glycogen later. Inadequate transport or take-up limits
high-power activity.*

*Stage 2 takes the oxidizable pyruvate at a limited rate to produce slow complete oxidation
within the mitochondria of SO or fOG fibers to deliver the remaining 93 percent of energy.
Oxygen supply is essential but believed not to be limiting.*

Figure 2.7
Movement and transformation of the six muscle fuels.

Each muscle fiber stores enough ATP for about 2 to 5 seconds of all-out
effort and enough PCr (which can be metabolized very rapidly without
oxygen to form ATP) for about a further 10 seconds of ATP effort.

Because it can be utilized without any need for oxygen, the stored
reserve of ATP is an anaerobic energy source. It is the key resource in leaping, or in accelerating from rest in a 100-m dash, or in lifting a maximal
weight. At much lower power levels lasting minutes or longer, ATP is still
the only fuel powering the contractile proteins; however, at such power
levels it is generated at a steady rate (for example, by oxidation of other
fuels), and the reserve is not drawn down.

In a shortening contracted muscle, the transformation of ATP releases approximately equal amounts of heat and contractile work. That is, this final stage of the work-producing process has up to 50 percent efficiency.

Three longer-duration carbohydrate fuels   A preliminary warning is perhaps in order before even beginning the discussion of carbohydrate fuels: although of surpassing interest in respect to intense efforts lasting twenty seconds through two hours, the various routes for carbohydrate usage are hard to grasp. The interested reader is advised to consult other literature for more details.

The carbohydrate fuels are the simple sugar glucose (essentially six carbon atoms combined with six water molecules); its stored form, the long-chain polysaccharide (starch) compound, glycogen; and its partially metabolized form, lactate. Glucose and glycogen can be used either aerobically (with oxygen, and slowly) or anaerobically (without oxygen, and far more quickly, but extremely incompletely). Anaerobic carbohydrate metabolism leaves behind high-energy lactate, either to be used immediately elsewhere or later (when oxygen is available) or to be reconsituted to glucose or glycogen. When used aerobically, the body's glucose and glycogen can provide power for a couple of hours. Alternatively, the glycogen in a muscle can be depleted anaerobically by conversion to lactate in just a few minutes.

Glucose reaches the muscle from the bloodstream. It might enter the blood from the digestive system or be released from the liver, where it either is stored as glycogen or has been resynthesized from lactate. Glucose can be delivered to the muscles only fast enough to supply up to one-third of the energy needs of intense steady-state exercise, so incoming glucose alone is not sufficient to produce high power levels. However, an adequate blood-glucose level is essential, because glucose is also the primary fuel for the brain. If exercise depletes the body's supply of glucose, allowing blood levels to drop, the rider will feel weak and dizzy (hypoglycemic). Periodic intake of carbohydrates (for example, in a sugar drink) is effective in preventing this condition and is also somewhat beneficial for long-time exercise performance.

Once glucose enters a muscle cell (fiber), it can release energy in one or two steps. The first (anaerobic) step is to split in half to form two lactate molecules, with each half also giving up hydrogen to form pyruvate.[7] This decomposition releases only about 7 percent of the energy available in the glucose, but it can occur rapidly without using oxygen. This splitting is called "glycolysis" (a term also commonly misapplied to the splitting of muscle glycogen, which is more properly called "glycogenolysis").

The second step proceeds in either of two ways. If the pyruvate is taken into the muscle cell's oxidative structure (mitochondria) and enough oxygen is also taken in, the other 93 percent of the energy is released aerobically in its conversion to water and carbon dioxide. In this aerobic process of generating ATP, the amount of heat produced roughly equals the energy available from the synthesized ATP, so that the steady-state formation of ATP is about 50 percent efficient. (The 50 percent efficiency of forming ATP aerobically and the 50 percent efficiency of using ATP to power the muscle lead to an overall aerobic "fuel efficiency of working" of around 25 percent.)

On the other hand, if the pyruvate is not oxidized in this way because there are too few mitochondria in the muscle cell to process the amount of pyruvate being produced, it simply regains its hydrogen to become lactate. Lactate created in anaerobic glycolysis (more usually, glycogenolysis, since glucose cannot be delivered very quickly, and glycogen stored in the muscle is readily available) must quickly be cleared from the muscle fiber if it is to continue functioning. The accumulation of too much lactate in the blood will also put an end to exercise through the increasing pain that results as it accumulates.

In exercise at very high power levels, lactate concentrations in the blood may become unendurable within thirty seconds. However, at somewhat lower power levels, quite a few minutes may pass before that condition is reached, both because less lactate is being produced, and because the body's lactate-removal system is able to handle most of it.

The point to remember is that glucose can be used either slowly and completely, achieving a high yield and medium power level, or rapidly but incompletely, achieving a low yield and high power for a short time only. Although an excessive accumulation of lactate prevents further work, lactate is far from a worthless poison: it is a highly significant fuel, since most of the energy of the precursor carbohydrate remains within it to be used. Apparently, lactate is reconverted to pyruvate, either in the liver, where it is further reconstituted to glucose, or in a muscle fiber, where it can be oxidized to perform work, or can even be restored to glycogen. (The specific outcome apparently depends on the state of fatigue, hunger, and whether exercise continues.) The transport around the body of this highly mobile energy form has been termed "the lactate shuttle" (Brooks, Fahey, and White 1996). However, the literature is not very definite on many issues collectively referred to as "the fate of lactate."

As will be discussed below, some lactate is produced even at low and medium aerobic power levels. In exercise at a constant rate, the concentration of lactate in the blood will climb to a fixed level, usually under 5 milliMole/liter (mM/l), related to exercise intensity and removal rate. If lactate is produced at a rate greater than it can be cleared (stored, oxidized, or reconverted), then the concentration of lactate in the blood begins an upward trend that will eventually terminate working. The critical exercise intensity that produces lactate at this rate is termed the "onset of blood-lactate accumulation" (OBLA), described by McArdle, Katch, and Katch (1996).

In recent years it has been generally accepted that elevated lactate concentration defines the highest tolerable steady-state (i.e., over the range of 20 to 120 minutes) exercise intensity. However, a welter of terms and proposed definitions have somewhat muddied matters. Such concepts as the lactate threshold (LT) and anaerobic threshold (AT) (now considered a misnomer because lactate elevation is not usually due to an inadequate oxygen supply), have also been defined, either when lactate reaches a specific concentration (4 mM/l) or at the point at which the plotted relation between steady-state concentration and exercise intensity increases slope. (The ventilatory threshold, or onset of panting, was originally believed to mirror the lactate threshold; however, Brooks, Fahey, and White [1996] have clarified that the near simultaneous onset of panting as the lactate threshold is reached is sheer coincidence.) The blood lactate concentration of elite riders in a ten- to fifteen-minute race may reach 15 mM/l, whereas in a one-hour race the lactate level is below 8 mM/l because of the lower intensity of the power output.

Glucose has been discussed first in this section for reasons of simplicity, not of importance. Far more important to athletic muscle power than glucose itself is its starch, muscle glycogen, a long-chain polymer of glucose. Fuel for one and a half or even three hours of high aerobic (i.e., oxygen-using) power can be stored within the working muscles in the form of glycogen; unfortunately, this fuel is found to be incapable of moving from well-stocked fibers to others from which it has been depleted (Bergstrom and Hultman 1966). (Its energy can be transported to other fibers only in the form of lactate.) Muscle glycogen is typically 2 percent of a rider's muscle mass, if the rider is on a normal diet; it is one quarter of this, or 0.5 percent, if the rider is on a low-carbohydrate diet; and it can be up to 4 percent after the depletion/overfeeding scheme known as "carbohydrate loading" or glycogen supercompensation (Hermansen, Hultman, and Saltin 1967). Glycogen is stored in the muscle with three times its mass of water. Therefore, a person with 20 kg muscle mass engaging in carbohydrate loading may store up to $4\% \times 4 \times 20$ kg $= 3.2$ kg of glycogen with its water.

Muscle-stored glycogen can be degraded to pyruvate extremely rapidly compared to glucose, as glycogen does not have to travel through the bloodstream as glucose does. The pyruvate created through glycogen degradation can be used aerobically just as fast as the muscle mitochondria can process it (unless the oxygen supply is artificially restricted; see Coyle et al. 1983) and thus the combination of incoming glucose and muscle-stored glycogen produces higher aerobic power than incoming glucose alone.

To achieve power levels higher than the mitochondria and oxygen-delivery systems can support, "anaerobic glycogenolysis" (pyruvate generation) can be increased to far higher levels than in aerobic work. In producing two or three times the maximum power level available through aerobic glycogenolysis, while releasing only 7 percent of the fuel's energy, it evidently degrades glycogen thirty to forty times as fast as in complete oxidation. Thus a store of glycogen sufficient for a two-hour aerobic effort can be depleted in just a few minutes of intense effort, although rapid lactate accumulation may prevent this from occurring all at once. The immediate aftermath is a painfully high blood concentration of lactate. (The time required to achieve a given lactate level depends, as noted above, on how much the production rate exceeds the clearance rate.)

Fat, the fuel for very-long-duration effort   Fat is the final fuel in our list. There are many different fatty compounds, composed principally of numerous carbon atoms with up to twice as many hydrogen atoms, plus relatively few atoms of oxygen. Because it completely combines both carbon and hydrogen with oxygen, fat releases about twice as much energy per gram as a carbohydrate. Furthermore, unlike glycogen, it is not stored with additional water. Body fat, our major energy store, is principally triglyceride, a glycerol molecule joining three fatty-acid molecules. For fat to travel in the bloodstream, the fatty acids are joined to proteins to form lipoproteins, which are soluble in the blood.

Fat is used only aerobically and for most of us is solely a low-intensity fuel. It supplies most of the body's energy needs at rest and even during exercise up to medium intensity. However, it takes considerable time to reach the muscles and is taken up by the muscle cells relatively slowly. At its greatest delivery rate, it supplies oxidative energy more slowly than muscle glycogen. However, the body holds enough stores for many days: fat stores can fuel weeks of effort. We store 50,000 to 200,000 kilocalories as fat, because completely oxidized fat yields 9 kcal/g or 38 J/g, enough energy for 100–200 hours of hard work (or more realistically, 200–400 hours of moderate work interspersed with rest). Stored glucose and glycogen can furnish only 1–2 percent of that amount of energy.

Experts imply in their papers that low-power carbohydrate oxidation (as occurs in cycling during easy pedaling) facilitates fat oxidation, whereas high-power oxidation actually inhibits fat use. An important benefit of protracted endurance training is an increase of fat usage, and a decrease of glycogen usage, at higher power levels (Holloszy and Coyle 1984).

All of the fuels discussed are interconverted, transported, stored, and used differently. Short-lived chemical intermediaries have not been mentioned, but still play crucial roles in human power production. The fuel stores and the ability to transport fuel, oxygen, and waste products depend on genetics, training, and state of hunger or fatigue.

Exercising to reduce body fat

In addition to the base energy supply rate (i.e., basal metabolism) needed to sustain a resting person, about 100 kcal/h (equivalent to about 115 watts), extreme effort uses another 1200 kcal/h; a more common athletic effort is 900 above base level, and an easy workout uses 600 above base level.[8] It is possible to metabolize an entire kilogram of glycogen in a hard ride; presumably its associated 3 kg of water are released as sweat. However, as just mentioned, such high-intensity effort is reckoned to inhibit fat utilization, so exercise at that level may not even touch the fat stores. To reduce fat stores, it is more productive to bicycle at low to moderate levels for longer periods than to exercise vigorously for shorter periods and then become ravenous to replace depleted glycogen.

Types of muscle fibers

A muscle is typically controlled by 50–500 nerves (motoneurons). Each nerve controls a bundle, which is known as a motor unit, of several hundreds or even thousands of fibers, of which there are three types (discussed below). Each muscle fiber is a single hairlike cell (between 0.01 and 0.1 mm thick and sometimes as long as the muscle) containing a great many force-producing protein filaments known as *myofibrils*. The fibers of each unit are intertwined with fibers from nearby motor units within the muscle. The fibers in any one motor unit are all of the same fiber type. The proportion of each type of fiber, in a given muscle of a given person, is found to be relatively unalterable, at least over one season of training in adulthood. (However, recent research reported by Andersen, Schjerling, and Saltin [2000] shows that this is not always the case.) Furthermore, the total number of fibers in a muscle is considered fixed early in life: muscle dimensional changes are due primarily to hypertrophy (increase in size) of the constituent fibers.

Three fairly distinct types of muscle fiber can be distinguished by chemically staining a muscle cross section: slow oxidative, fast glycolytic, and fast oxidative glycolytic. Each type differs in how it uses fuel and produces force and work, although the differences among them in these areas may not always be marked, as cells adapt through training: their behaviors are actually placed along a continuum. Any one muscle is composed of a mixture of the three types, all more or less adapted through training to either endurance (aerobic) or force/power (immediate and glycogenolytic) activities.

At one end of the spectrum, slow oxidative (SO) fibers (also known as Type 1 fibers) are richly supplied with oxygen-using mitochondria. These fibers are reddish because of the oxygen-storing myoglobin they contain, as in the dark meat of a chicken. Both the mitochondrial density of and the number of capillaries supplying oxygen to these fibers can be increased substantially through endurance training. SO fibers are ideal for steady-state (endurance) activities, taking up oxygen at the highest rate to metabolize glucose, glycogen, or fat aerobically. SO fibers never grow very thick and can't exert much force. They respond slowly to nervous stimulation and so are referred to as "slow-twitch" fibers. They have little ability to support rapid, oxygen-free liberation of carbohydrate energy (anaerobic glycolysis or glycogenolysis). On the other hand, they are able to contract repeatedly without fatigue. Since steady muscle force is actually produced by repeated contractions of individual fibers, postural muscles tend to be SO.

At the other extreme, fast glycolytic (FG) fibers (also known as Type 2b fibers) respond faster and more forcefully to nervous impulses. They largely lack both mitochondria and myoglobin and hence are pale, like the white meat of a chicken. Their metabolic predilection is for rapid anaerobic conversion of glucose or glycogen into lactate, producing high force and power with little delay (fast twitch). They are frequently described as "fatiguable," presumably through glycogen depletion or lactate accumulation. By overload training, FG fibers can be enlarged in cross section, therefore increasing short-term muscle strength fueled either by carbohydrate or immediate sources (ATP and PCr). The extent to which they can also perform aerobically, especially after endurance training, is not clear in the literature. It has been suggested that glycogen stores can be higher in FG than in other types of fibers and that they are much better at using PCr.

A third type of fiber is the fast oxidative glycolytic (FOG, also known as Type 2a). It is believed that some FOG fibers may be converted from FG fibers as a result of endurance training. If so, they give up some glycolytic capacity for a substantial boost in aerobic ability. Textbooks describe FOG fibers as combining characteristics of SO and FG fibers.

We have suggested that each power level commanded of the muscles invokes some combination of fuel transport and conversion mechanisms. Presumably, duration at any given power level will be determined by exhaustion of one resource, or by saturation with one waste that is not being removed fast enough. A lesser rate of using fuel or producing waste will therefore permit longer duration. Differences in the physiological mechanisms operating at different power levels would be expected to alter the position or slope of the limiting power-duration curve. This is illustrated in figure 2.8, which plots record running speed versus the logarithm of duration over the range 10 to 10,000 seconds (Brooks, Fahey, and White 1996). (The unexpected positive slope for times below 20 seconds presumably reflects the energy cost of acceleration from rest.) Once the physiological

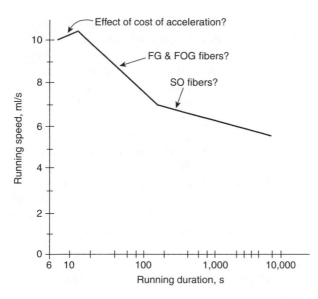

Figure 2.8
Log plot of duration versus average speed (presumed related to power) for running records. (Adapted from Brooks, Fahey, and White 1996.)

limits inherent in a person's power-duration curve are understood, more-rational training may be possible.

Fiber recruitment

The proper selection of muscle fibers to perform a given task is important. If glycolytic motor units were recruited first for endurance (low-force) activities, they would quickly become depleted of glycogen and would not have made use of much of the available oxygen. Although motor-unit recruitment is not directly under conscious control, it does seem to be a function of the central nervous system. (Short-term gains in weight-lifting ability are attributed to improved fiber recruitment, rather than actual muscle-strength gains.)

It is difficult to investigate the recruitment of different types of muscle fibers, not only because of the variability among individuals, but also because of the problems of discerning rapid chemical interactions in extremely small portions of living cells. Chemical tracers labeled using radioactive isotopes may be employed to this end, or certain reactions can be blocked chemically. For very short-lived chemical species, individual muscle cells can be studied in vitro and quick-frozen at various stages of performing work. In many cases, actual human muscle tissue or blood from a particular vein can be sampled during exercise. We have to conclude that there are yet many unanswered questions—or at least, questions for which there is no general agreement as to the correct answers—on many aspects of muscle physiology, control, and response to training regimens.

The recruitment of a fiber is a preferential energization: fibers are recruited via the stimulation of the motoneuron innervating the motor unit to which they belong. This stimulation appears to originate in the brain, and the fiber selection may be a simple matter of neuron properties (i.e., some nerves being triggered at a low level of stimulus), or there may be something more selective at work. This is the arena of Henneman's "size principle" (in single efforts) (Henneman, Somjen, and Carpenter 1965), and of Gollnick's (later Ahlquist's) fiber-specific fuel-depletion studies in repetitive pedaling efforts (see below). Once a fiber is recruited, the frequency of stimulation determines the time average of the force it exerts.

According to Henneman's size principle, the smaller (and weaker) SO fibers are recruited for contractions at low force levels, whereas the larger, stronger FG and FOG fibers are not recruited until contraction force levels reach a substantial fraction of the maximum. It is not obvious how this force-based principle, which apparently relates to just one or a few strong muscle contractions (and is effected by the greater stimulation thresholds of the fast motoneurons), might apply to thousands of repeated pedaling contractions in which the force is mostly in the range of between 5 and 20 percent of maximum. (Repeated contractions were studied by van Bolhuis, Medendorp, and Gielen (1997), who found that nerve signals were produced in advance of each actual contraction to compensate for the delay found in tension development.) Perhaps contractions of this nature cause muscle-shortening velocity to affect motor-unit recruitment. For example, riding faster up a steep hill at a lower gear ratio would decrease required foot force compared to that required in a higher gear, yet excessive lactate will be produced, and fatigue will occur rapidly.

Such issues are not commonly explored in the literature. Gollnick et al. (1973) showed that fast (glycolytic) fibers were preferentially recruited in high-power sprints. Subsequently, Gollnick, Piehl, and Saltin (1974) attempted to clarify the role of pedaling force on this phenomenon and found it had no effect. But a more recent paper by Ahlquist et al. (1992) showed that high-force (overgeared) submaximal pedaling recruits more fast fibers and depletes their glycogen (as proposed by Forester [1984]), whereas lower-force (high-cadence) pedaling relies more on slow fibers. Whether this conclusion is true for all riders, and what its implications are for each subpopulation in respect to fatigue, we do not know.

The subject of fiber types and recruitment seems to us to deserve greater attention than has been devoted to it up to this point. In principle, pedalers with different muscle types might be expected to have very different power-duration curves, choice of gearing, and potential to benefit from using additional muscles. We expect that research conducted on successful distance cyclists examines a biased population, that is, primarily those with well-developed oxidative fiber potential (SO and FOG). This is a select group, and results of research conducted on this group should not be generalized to the remainder of the population.

Here are three questions to ponder (and to which we have no answers!) with respect to muscle fibers and their recruitment:

• Can a person with a preponderance of FG fibers actually push the oxygen delivery system to its full capacity with leg work only? That is, can the maximum oxygen consumption (VO_2max) always be determined with a conventional pedaled ergometer? (Occasional reports in the literature suggest that it may not be able to be.)
• Is there a difference in overgeared (low-cadence) pedaling fatigue between FG fiber types and SO fiber types? Is this response altered due to training?
• Might occasional intervals of stand-up (high-force) pedaling actually serve a useful physiological function: mobilizing the glycogen stored in FG fibers by creating lactate, which can then migrate to fuel glycogen-depleted SO fibers?

High-power aerobic metabolism: lactate threshold and glycogen depletion
Not all of the physiological mechanisms defining an individual's power-duration curve have been studied to the same degree. Two that have received considerable attention, namely, highest steady-state power (aerobic) and highest power sustainable for a minute or two (anaerobic), relate to important types of cycling efforts. An authoritative review paper by Coyle (1995) comprehensively examines a variety of hypotheses about maximum steady-state pedaling power. High-power aerobic metabolism operates as follows: the highest power levels sustainable for about 30 minutes or longer are essentially steady state, not using up any rapidly depleted resources, nor increasingly accumulating painful lactate. From minute to minute, virtually all the power produced through such metabolism involves inhaled oxygen.

In such high-intensity steady efforts, the maximum duration of the effort is often set by the carbohydrate stores: depletion of muscle glycogen[9] or even blood glucose. Experiments by Bergstrom et al. (1967) and also described in textbooks such as Brooks, Fahey, and White 1996 comparing initial muscle glycogen to maximum possible duration of effort have amply confirmed that in well-trained endurance athletes, at least, muscle glycogen is the limiting resource that terminates hard steady-state effort (i.e., determines endurance). In addition, measurements of glycogen levels in both legs when only one leg is allowed to pedal (Bergstrom et al. 1967; see also Astrand and Rodahl 1977, chap. 14) have shown that glycogen is not mobile: the working leg depletes its stores and is exhausted, whereas the resting leg remains fully charged.

Consumption of muscle glycogen in cycling can be reduced by increasing the energy delivered by the fat system and also by increasing the pedaling rpm (reducing fast-twitch fiber recruitment, with its associated anaerobic glycogenolysis). Furthermore, muscle glycogen stores can be supercharged through the process commonly known as carbohydrate loading: depleting glycogen stores substantially over a couple of days, then eating a superabundance of carbohydrates. This two- to four-day process has proved to double the levels of endurance achieved by normal well-fed but "unloaded" persons. Since muscle glycogen is also useful in shorter, more powerful efforts, carbohydrate loading would seem to be a useful practice for all but the shortest events. Since glycogen regeneration after consumption and depletion is supposed to take more than a day, an important unanswered question concerns the size of the glycogen stores that can be maintained by athletes in multiday events.

In the past, it was widely believed that the maximum rate at which fuel could be oxidized was set by the rate at which the lungs and circulation could deliver oxygen to the working muscles, and that this rate could be determined in a test of VO_2max, as described below. However, at least for athletic endurance cyclists, this is no longer believed to be generally true. Instead, the rate of fuel oxidation seems to be limited by the somewhat lesser rate at which working muscles can oxidize fuels without excessive lactate production, which depends on their total mass, their fiber type, and their degree of adaptation (via mitochondria and capillaries) to oxidizing fats (Coyle et al. 1991). The most effective fibers for taking up oxygen are the relatively weaker and slower-acting SO fibers. Appropriate training can double the capacity of these fibers to use oxygen. Their weakness is not a problem for the cyclist, since the typical foot force produced in long cycling events is only about 10 percent of the maximum achievable pedaling force.

As there is no obvious connection between the maximum steady rate at which muscles can take up oxygen and VO_2max, the latter ought to be power-limiting in part of the population. (Coyle et al. 1983, discussing heart-disease patients, is an extreme example of this.) However, if this is true, individuals with limits set by VO_2max are not often found among successful competitors (Brooks, Fahey, and White 1996). It is now believed that the maximum work intensity tolerable by an individual is determined by the blood lactate level arising from the balance between lactate production of the working muscles and various lactate "clearing" mechanisms. A small amount of lactate is produced whenever pyruvate is available, that is, whenever carbohydrates are used as fuel. Much more is produced when SO fibers are required to produce more than a certain amount of energy from carbohydrates, or when FG fibers are recruited, or when a fiber has a poor oxygen supply. The rate of lactate production can then overwhelm the body's lactate-clearing capacity, which typically seems to occur quite independent of how much oxygen the circulation can deliver. Training will reduce the amount of lactate produced at any given workload and increase the rate at which it can be used (cleared), therefore reducing the level in the blood. In addition, training can increase the body's ability to tolerate elevated lactate levels.

The upper limit to long-term, steady-state power is closely associated with the level at which the slope of the curve of steady-state blood lactate versus power increases (Coyle et al. 1991). Up to 1990, no consensus had been reached about precisely how to define the body's lactate threshold, for example, as a concentration level, as a slope change, or as an increase of concentration above the baseline. A seemingly rational definition, though perhaps requiring more testing to establish, is the onset of blood lactate accumulation, or OBLA. This is the exercise intensity at which blood-lactate concentration can no longer remain steady: production exceeds clearance, and the blood-lactate concentration climbs inexorably until exercise ends.

There is good reason to expect that pedaling styles or devices that allow the use of a greater mass of muscles will increase a rider's maximum steady (i.e., aerobic) power level. Indeed, it is accepted that top Nordic skiers, who use arms as well as legs, tend to take up more oxygen than top cyclists and so probably produce greater steady-state power. Again, these observations leave unanswered questions to ponder:

- Are Nordic skiers, who use more of their major muscles than cyclists, limited by their ultimate cardiovascular capacity? When a racing cyclist's oxygen supply is curtailed by the crouched racing position, does oxygen delivery become a limiting factor? (This curtailed capacity is not the standard VO_2max measured in a laboratory.)
- Riders are known to exhibit reduced aerobic power as they age. Is it possible that they are maintaining muscle oxidative capacity but losing oxygen delivery capacity and thus eventually becoming limited by VO_2max? Figure 2.9 plots average speeds in 50-mile trials versus age and estimates of breathing capacity from the speeds.

Interesting physiological data from a maximum-power long-term effort are available from pilots evaluated for the MIT Daedalus flight; in these evaluations, the pilots were required to pedal for an estimated four hours. As shown in figure 2.10 (Bussolari 1987), two subjects were required to

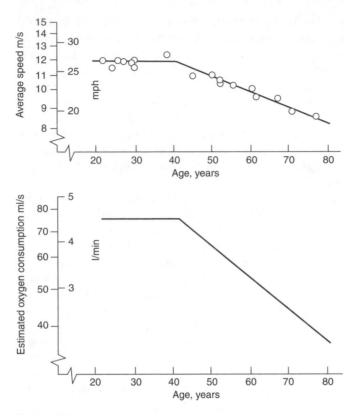

Figure 2.9
Speeds and estimated breathing capacities of 50-mile time trialists versus age.
(Plotted by Dave Wilson from data supplied by Frank Whitt.)

pedal at 70 percent of their maximum aerobic power (i.e., 70 percent of the
power eliciting maximal oxygen uptake) and were monitored through
measurements of their inspired and expired breathing and through blood
samples taken periodically throughout the four hours. They were allowed
to drink as much as they wished. The solid lines in the various panels of the
figure show the data for a female pilot who, according to Bussolari, who
conducted the study, had engaged in carbohydrate loading before the test
and drank periodically throughout. She finished the four hours in a condition good enough to have allowed her to continue for another thirty to
sixty minutes. The dashed lines show the data for a male pilot who did
not attempt carbohydrate loading, who drank less than half the quantity
of liquid that the female pilot consumed, and who had to quit after
three and a half hours because of leg soreness and cramping. It can be

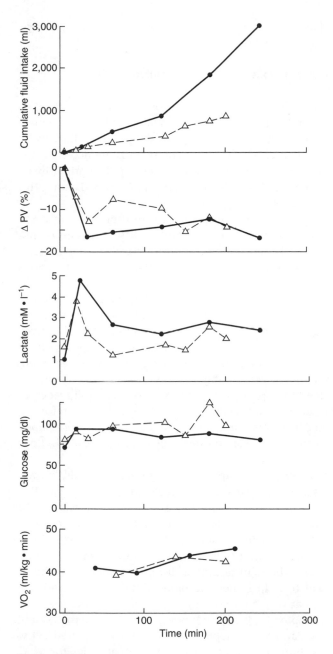

Figure 2.10
Measurements taken on two pilots of human-powered aircraft during a four-hour test at 70 percent of maximum aerobic power. (From Bussolari 1987.)

seen in the figure that these discomforts were not brought on by a high lactate level.

High-power anaerobic metabolism: lactate accumulation and fast-twitch-fiber population

In significant anaerobic efforts such as a sprint or climbing a short, steep hill, the power output of the muscles involved is far greater than the maximum aerobic power, by a factor of three to six, and is produced initially by the immediate fuels ATP and PCr. As those compounds are depleted, the power exerted by the muscles drops to a lower level, supplied primarily by the anaerobic glycogenolysis of muscle glycogen. As mentioned above, this results in a massive release of lactate.

When high-power work is performed for a very short time only, the glycogenolytic system is hardly engaged; the anaerobic fuel systems do not build up any lactate. This is one principle behind so-called interval training: great effort can be expended repeatedly, if the duration of such effort is kept short.

High-power anaerobic (both immediate and glycolytic) metabolism is predominant in the five- to twenty-second range of exercise duration. For lesser efforts, causing exhaustion in under two minutes, oxygen-derived power still represents less than 50 percent of the total energy expended (and some oxygen is already stored in muscle myoglobin), so an outstanding oxygen-delivery system is presumably of little value in such efforts.

Although lactate buildup ends high-power glycogenolysis, repeated intense efforts will actually deplete FG fibers of their glycogen. Therefore, to be able to perform many sprints, for example, one wishes to have a superabundance of glycogen, which can be developed through carbohydrate loading.

A muscle's "anaerobic work capacity," determined in critical-power curve fitting, suggests a rapidly deliverable "reserve quantity" of work. Presumably anaerobic work capacity can be approximated by measuring the stores of immediate fuels and adding an amount of muscle glycogen sufficient to raise blood lactate to intolerable levels when consumed. (Some riders can tolerate higher lactate levels than others, and some riders can clear lactate faster than others. In addition, drinking a solution of bicarbonate of soda is said to help buffer the blood lactate, thus permitting somewhat longer effort at very high power.)

The fibers best adapted for brief, high-power activities are the FG and maybe the FOG fibers. Part of the adaptation of these fibers to these activities is a growth in volume (cross-sectional area) that provides for more work-producing protein and greater force. This adaptation shows up as a larger muscle. In addition, there are enzymes that act to permit the conversion of glycogen to lactate, and their levels must be adequate for this conversion.

A high population of enlarged fast fibers is probably necessary to produce the maximum level of glycogenolytic power possible. However, it is repeatedly stressed in the literature that this is not the whole story, inasmuch as some very strong people, including bicycle sprinters, do not have particularly large muscles. It seems that the ability to recruit the proper fibers at precisely the right time is also important. Whether this facility is innate, as opposed to trainable, is little discussed.

Oxygen uptake

Oxygen usage is potentially a very powerful tool for revealing how much "fuel" is being metabolized to supply a person's energy needs and can be calculated through a process called indirect calorimetry. Two measurements are required: the amount of oxygen absorbed, and the amount of carbon dioxide emitted, in a given time. When carbohydrates are oxidized, every oxygen molecule is converted to a carbon dioxide molecule. Thus a 1:1 ratio between carbon dioxide and oxygen (the so-called respiratory quotient) indicates a state of pure carbohydrate usage. On the other hand when fats are oxidized, only about 70 percent of the oxygen forms carbon dioxide, whereas the rest creates water. Thus a carbon dioxide–to–oxygen ratio of 0.7:1 indicates a state of pure lipid usage. Ratios between 0.7:1 and 1:1 can be used to calculate the proportions of carbohydrate and fat usage. Fat oxidation yields 4.70 kcal per liter of oxygen, whereas carbohydrates deliver 5.05 kcal per liter of oxygen used. However, some caveats are in order: over the short term, not all energy is produced via oxidation. Brief, intense efforts rely on immediate fuels and anaerobic glycogenolysis, and their oxygen cost is deferred. (And after exercise ends, changes in, e.g., body temperature alter the basal metabolism, thus obscuring the total fuel usage because it is taken as being constant.) Furthermore, reservoirs for oxygen and carbon dioxide in the blood and lungs can significantly delay evidence of usage and production, leading to uninterpretable respiratory quotients well outside the steady-state range of 0.7–1.0. Thus, oxygen measurements at unsustainable exercise intensities must be interpreted cautiously.

An exercising person's rate of oxygen use can be determined from the volume per minute exhaled times the diminution in oxygen concentration in the exhaled air. The analysis can be performed either continuously or after the exhaled air has been collected in a Douglas bag. (Once calibrated, a given individual's heart rate provides a rough approximation of his oxygen usage.)

If oxygen usage rate is plotted for an increasing sequence of intensities of a particular exercise (allowing appropriate time for somewhat steady conditions to develop), the curve often shows a relatively sharp "knee" and levels out at an apparent maximum in oxygen uptake for that exercise. Even short, intense efforts can elicit the maximum uptake.

As mentioned above, VO_2max was long believed to represent a systemic (heart and lung) limitation on oxygen delivery. Although such a limitation surely does exist, it is likely that a bicyclist's VO_2max actually represents the working muscles' ability to take in oxygen. It has been found, for example, that different exercises lead to somewhat different values of VO_2max for the same person.

The maximum rate of oxygen delivery (VO_2max) may relate primarily to heart-stroke volume, which can be increased 10–15 percent with training, and blood hematocrit (red-cell concentration), which can be elevated by artificial means such as altitude training, blood doping, or use of the hormone erythropoietin (EPO). Even intense training cannot increase VO_2max much in those who are already reasonably fit. VO_2max decreases with age above some threshold (typically around age forty) (Carpenter et al. 1965).

Considerable experience has shown that VO_2max, in itself, does not normally define maximal steady-state pedaling performance: almost no one is able to sustain exercise while taking up oxygen at her/his maximum rate. A variety of studies (e.g., Coyle 1995) have found a far better correlation of long-term power to some kind of lactate threshold than to VO_2max.

A focus on VO_2max dominated exercise studies for decades. VO_2max was viewed as a fundamental determinant of performance, a capacity to be enhanced by training, and a natural measure of "endurance exercise intensity" (via percentage of maximum). Nowadays some version of the lactate threshold (e.g., OBLA) is seen as the trainable limit. VO_2max is frequently well above this limit (and in any event is not very trainable).[10] Indeed, for someone whose legs have a predominance of FG fibers, we wonder whether VO_2max can even be found on a pedaling ergometer.

This new perspective on performance determinants encourages a cautious optimism that employing more large muscles could permit riders to put out greater long-term power, perhaps even approaching their systemic limit to oxygen delivery. Bicycles with both hand and foot cranking are continually being reinvented to this end. The lack of notable racing-performance success hints at an array of difficult requirements, including a smooth energy-conserving pedaling motion and the ability to pedal hard without disturbing the steering. (This is one arena in which ergometer-based success should clearly precede construction of an on-road prototype!)

Although VO_2max does not obviously define maximum steady-state pedaling power, we cannot yet advocate ignoring it altogether as a determinant of athletic performance. It has been proposed that there is a mitochondrial stress associated with aerobic energy supply, such that a preferred process (perhaps fat oxidation) becomes impossible as mitochondria approach their maximum oxygen usage rate, causing a switch to carbohydrate usage and an overwhelming production of lactate (Holloszy and Coyle 1984). If this is borne out, it could be that VO_2max indeed influences maximal steady-state performance, through a linkage to OBLA. Furthermore, the very substantial differences between individual oxygen uptakes is a form of physiological destiny. Those whose VO_2max is only half that of an endurance champion can never match the champion's OBLA.

As already mentioned in the discussion of ergometers, the aerodynamic advantages of a bent body position can significantly reduce the power required to ride fast. But at the same time bending severely at the waist clearly restricts breathing. (To reduce this bend, specialist triathlon bicycles move the rider's body forward slightly in relation to the pedals.) For some pedalers we speculate that oxygen delivery is hampered by the crouched position, possibly dropping below the amount usable by the pedaling muscles and thereby clearly hindering performance. Riders of recumbent bicycles claim the advantages over standard positioning of lower aerodynamic drag and freer breathing.

Recovery from exertion

Cycling is often a sport of surges—short-term efforts above steady state that deplete ATP and PCr and may build up lactate. After each such effort, a recovery process is necessary. Continuous gentle pedaling is known to clear lactate faster than simply resting. The competitor who is best attuned moment-by-moment to her physiological reserves has a distinct advantage over those who are less aware of them.

Originally, anaerobic energy supply was considered a kind of deficit spending ("oxygen debt") that would have to be paid off by oxygen taken in after intense exercise was ended. The need for this postexercise makeup oxygen has widely been considered a stress and a limit to performance. But Brooks, Fahey, and White (1996) have identified substantial errors in this line of thinking and more accurately labeled the phenomenon as "excess postexercise oxygen consumption" (EPOC). Clearly more complex than has been supposed, EPOC also does not have a compelling role as a performance limiter; more properly, the focus in discussions of performance limitations should be on clearance of lactate.

Energetics in pedaling

In ideal circumstances, the (extra) energy cost of pedaling could be attributed directly to the work done on the pedals. However, in practice, some muscles (not necessarily in the leg) are used in pedaling in isometric or even in eccentric contraction. Furthermore, there are various immediate and delayed metabolic costs of using or replenishing the various fuels that feed the muscles in use. We do not know the relative contributions to fuel inefficiency of each such factor.

It is obvious that other, nonpedaling muscles are increasingly engaged at high-force or high-cadence pedaling. When force is high, the rider must use these muscles to prevent being lifted from (or slid along) the saddle. The same is true at high cadence, when the momentum of the descending thigh mass tends to straighten the leg fully and lift the rider from the saddle. (A nearly straight knee approximates a toggling lift mechanism.) But it doesn't seem that muscle use can be the only factor determining muscular fatigue. One of the main conundrums in studies of pedaling is why a lowered seat should so greatly harm performance, since the same work is being asked of the same muscles.

Quite a few muscles actuate the joints of the leg (see figure 2.6, or an anatomy book such as Wirhed 1984). Confusion about the functions of these muscles can be reduced by first concentrating on the one-joint muscles, namely, those that cross just one joint. It can be seen that each joint has one muscle situated to extend it and an opposing muscle situated to flex it. These opposed pairs would not normally co-contract (i.e., exert opposing tensions simultaneously) when the goal is power production, because one would then be performing negative work, which irreversibly absorbs useful energy. (Co-contraction is instead a strategy used to enhance structural stiffness or to resist injury.)

What remains are two-joint muscles such as the rectus femoris and the long head of the biceps femoris, which exert torque about two joints without touching the intervening bone. These can deviate from the simple logic of one-joint muscles. For example, the two muscles just mentioned are both simultaneously shortening in leg extensions such as jumping or pedaling. Therefore, in such motions, when these muscles are co-contracted, both perform positive work. (The initially surprising observation of these working muscles seeming to oppose each other is referred to as "Lombard's paradox.")

Two-joint muscles add a wrinkle to the detection of negative work. If all muscles were of the single-joint variety, that a joint was absorbing work (i.e., articulating in the direction it was being urged by the moment being supported)[11] would mean that one muscle was undeniably undergoing eccentric contraction. (However, the total amount of negative work would be unknown, because positive work is produced simultaneously whenever the opposing muscle is co-contracted.) In the two-joint case, however, the appearance of work absorption at one joint does not mean that any muscles are actually performing negative work.

Short of measuring individual muscle tensions and rates of lengthening, the only sure case of negative work by any leg muscles occurs when the increase in leg energy (kinetic plus potential), plus the work of the foot on the pedal, is negative. (Examples: the pedal pushes the foot upward without any increase in the leg's energy, or the leg's energy decreases by more than the amount of work done on the pedal, as when the thighs come to rest when turning the pedals without any chain.) Modest amounts of negative work are essentially undetectable by external measurements.

Personal energy requirements

The human engine has an additional characteristic not generally found in machines: some fuel must be consumed to keep it going even when it is at rest. (In this sense it is somewhat similar to a traditional steam plant, in which fuel must be burned continuously to keep steam pressure up even when no power is being delivered.) Human energy requirements are conventionally split into basal metabolism (used while resting and digesting) and work metabolism.

Basal metabolism is often expressed in terms of body surface area, for example, 39 kcal/h/sq m for young adult males, 37 for middle-aged males, and 36 for adult women (McArdle, Katch, and Katch 1996). Body surface area has been related to height and mass by a number of correlations, for example, the formula of Gehan and George (1970):

$$body\ area = 0.164(height^{0.422})(weight^{0.515}),$$

where body area is measured in square meters, height in meters, and weight in kilograms. (Unfortunately we are unable to supply error bars.) This leads to the estimate of 73 kcal/h, or 1,750 kcal/day, basal metabolism for a male of height 1.75 m and mass 75 kg. About 30 percent additional energy beyond the basal-metabolism level is needed to carry out ordinary daily activities.

If basal metabolism represents heat lost when the skin is maintained at a comfortable temperature, it should depend on environmental conditions. People in hot third-world countries who are condemned to a marginal existence can maintain life and health at a fraction of the previous paragraph's 73 kcal/h. It has been found that when insufficient food aid has been available during famines, a diet considered to be well below the starvation level for Westerners has resulted in such people's putting on weight.

Work metabolism is estimated as actual kilocalories or joules of work divided by an "efficiency factor" typically between 0.2 and 0.3. Each 100 W of mechanical power production thus requires 333–500 W (287–430 kcal/h) of energy supply in food intake above that needed to maintain life. Thus where a normal meal of 500–1,000 kcal might be considered to supply eight hours of sedentary energy needs, an additional meal is needed for each one to two hours of cycling effort.

We acknowledge that our focus is on the limits and potentialities of top athletic performances, generally involving people who may have embraced sport because they are naturally constituted for it. Of course, most pedaled miles are actually traveled by average persons at a far easier pace than that pursued in athletic competition. Questions about how to make low-power riding more pleasant, or more energy efficient, or perhaps better at countering obesity are important ones and deserve to be addressed in the future.

The effects of pedaling position
Early tests showed that the maximum rate of oxygen uptake in the recumbent position was slightly less than in the conventional, upright position. However, this did not in itself prove that there was any difference in maximum steady-state power. And later, more exhaustive studies have failed to discern any significant difference. One of these was carried out by Bussolari and Nadel (1989) to choose the best pilot position for the four-hour flight of the Daedalus aircraft. Figure 2.11 shows typical results of power output versus oxygen uptake for the two positions, in which the energy cost (as revealed by oxygen uptake) is effectively identical.

Animals are fuel cells
Heat engines (such as steam-turbine plants and internal-combustion engines) produce work by heating and expanding a "working fluid" (steam,

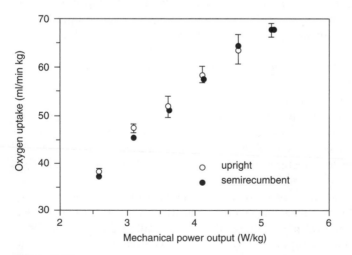

Figure 2.11
Typical tests comparing power output by conventional and semirecumbent pedaling. (From Bussolari and Nadel 1989.)

air, and/or gas). Thermodynamics teaches us that the work capacity of a given amount of heat is tied to its temperature relative to that of the environment: a small amount of very hot gas can theoretically produce more work than a much larger but cooler amount of gas with the same total thermal energy. The fraction of heat energy transformed to work, W, is given by the definition of thermal efficiency (W/Q_2) using the principle of the conservation of energy.

Because $W = Q_2 - Q_1$, where Q_2 is the heat input at a high temperature T_2 and Q_1 is the heat output at a lower temperature T_1, thermal efficiency $= (Q_2 - Q_1)/Q_2 = 1 - (Q_1/Q_2)$. In the case of pedaling, it is known that this heat-based quotient is approximately 0.25. According to thermodynamic reasoning applied to ideal heat engines (the second law of thermodynamics), this quotient must always be less than or equal to the last term in

$$\eta_{\text{th}} \equiv \frac{Q_2 - Q_1}{Q_2} \leq \frac{T_2 - T_1}{T_2},$$

when temperatures are measured on the absolute scale. Therefore a heat engine as efficient as the human body (25 percent) and working in a room-temperature environment (approximately 300 K) would have to have an upper working temperature given by $0.25 \leq 1 - 300/T_2$, therefore $300/T_2 \leq 3/4$, therefore $T_2 \geq 400$ K, at least as great as 400 K ($127°C$). (Automotive engines, with similar overall efficiency, have a far higher working temperature than this, because of severe power losses due to friction, gas turbulence, etc.)

Such a high temperature (well above the boiling point of water) is clearly impossible for the body to withstand: therefore, the human body does not function as a heat engine. It is instead a type of fuel cell in which chemical energy is converted directly to mechanical power. The energy not converted to power must appear, as for heat engines and fuel cells, as heat. (All animals, including man, also excrete wastes that have some calorific, heating value that should be included.)

Breathing
The amount of oxygen absorbed by the lungs of a person of average weight, at rest and not using any voluntary muscles, is about 5.5 ml/s (one-third of a liter per minute). This quantity is additional to any other absorption required by exercise. At the upper limits of steady-state aerobic athletic performance, over 80 ml/s may be used.

In ordinary air, a liter of oxygen is found in about 5 l of air. However, when air is breathed, about 24 l must be passed through the lungs for each liter of oxygen absorbed (Knipping and Moncrieff 1932). Thus, about 380

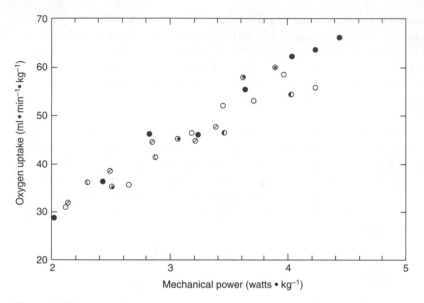

Figure 2.12
Measured oxygen uptake versus power delivered for five pilots. (From Bussolari 1987.)

percent more air than is needed to produce energy is used in the human engine. (Air taken in per minute is commonly termed the "minute ventilation.") Most other engines, such as internal-combustion and steam engines, require only 5–10 percent "excess" air to ensure complete combustion of the fuel. Gas turbines more nearly approach human lungs, taking in about 200 percent excess air.

The relationship between oxygen absorbed and mechanical power delivered to the pedals for five volunteers piloting the human-powered aircraft Daedalus is shown in figure 2.12 (Bussolari 1987). Both the oxygen uptake and the power are given per kilogram of body weight, because of the importance of the power-weight ratio. For this series of tests a woman (other data from whom are given in figure 2.10) produced the best power-weight ratio, where "power" was defined as 70 percent of the person's maximal aerobic power, an output sustainable for hours by most endurance athletes.[12] (The final choices for volunteers were made among bicycling champions who were taught piloting, which turned out to be easier than picking pilots and trying to make them into outstanding endurance athletes.) The variation in oxygen uptake among five individuals in good condition is not large. It would be interesting to gather more data, including some from the general population.

Although not a strict determinant of physical work capacity, maximal oxygen uptake is commonly used as a rough indicator of potential and a useful normalizing quantity. Everyone should be able to work easily at one-third of maximal uptake, but exceeding two-thirds of maximal uptake for a long duration may require considerable training. For a nonathletic person, the maximum oxygen-absorption rate (i.e., VO_2max) is assumed to be about 50 ml/s or 3 l/min (approximately 60 percent that of an elite competitor). Table 2.1 shows that when a rider is using about a third of his oxygen-breathing capacity, the power output is about 0.1 hp (75 W). It is thought that an average fit man or woman could work under these conditions (oxygen uptake of 50 ml/s, power output of about 0.1 hp) for several hours without suffering fatigue to an extent from which reasonably rapid recovery is not possible. Experience has also shown that 75 W propels a rider at approximately 12 mile/h (5.4 m/s) on a lightweight touring bicycle on level ground in moderate-wind conditions. As this speed can ordinarily be maintained by average touring-type riders, the numbers given in table 2.1 seem sound. Miscellaneous data on caloric expenditure of bicyclists given by Adams (1967), Harrison (1970) and others are collected in figure 2.13.

Breathing ability decreases with age. An athlete's peak breathing capacity is reached at about age twenty, and it is a rule of thumb that breathing capacity is halved by age eighty (Carpenter et al. 1965). Results of the 1971 U.K. 50-mile amateur time trials, in which the ages of the best "all-rounders" and of the "veterans" were given, are consistent with a breathing-based theory of performance. The average speed for each rider is plotted against the rider's age in figure 2.9. There is no recognizable falloff in performance up to age forty, after which there is a steady drop to that for the oldest competitor, aged seventy-seven. These performances have been converted to breathing capacity, estimated by the method of Whitt (1971). When the curve is extrapolated to eighty years, the estimated breathing capacity is indeed very close to half the peak value. However, such reductions in performance with age could have a different explanation altogether: in today's society, even an athlete may be sedentary 85 percent of the time. Maybe the falloff shown in figure 2.9 occurs particularly when a person takes a sedentary job.

Up to a breathing rate of about 0.67 l/s (40 l/min), people tend to breathe through the nose (Falls 1968) if they have healthy nasal passages. Nasal passages usually open during exercise, even during a heavy cold. Above this rate, the resistance to flow offered by even a healthy nose becomes penalizing, and mouth breathing is substituted. For a normally healthy individual riding on the level in still air on a lightweight bicycle, this limiting rate for nasal breathing is reached around 14 mile/h (6.3 m/s).

Tests by Pugh (1974) on bicyclists riding on an ergometer and on a flat concrete track at speeds up to 27 mile/h (12.1 m/s) confirmed the data

Table 2.1
Breathing rates for cycling and walking, based on: air l/min $= 7 + W * 0.29$, O_2 l/min = air l/min/24, air l/min = metabolic watts/14.63

Cycling Speed				Tractive power		Breathing rate (l/min)		Metabolic heat	
Racer[a]		Tourist[b]							
mile/h	m/s	mile/h	m/s	hp	W	Oxygen	Air	kcal/min	W
27	12.1	22.5	10.1	0.5	373	4.8	115	24	1,680
25	11.2	21	9.4	0.4	298	3.4	93	19.5	1,365
22	9.8	18.5	8.3	0.3	224	3	72	15	1,050
19	8.5	16	7.2	0.2	149	2.1	50	10.5	735
14.5	6.5	12	5.4	0.11	82	1.2	29	6	420
10.5	4.7	8.3	3.7	0.05	37	0.75	18	3.75	263
7.2	3.2	6	2.7	0.025	19	0.53	13	2.65	186
3.2	1.4	1.8	0.8	0.008	6	0.38	9	1.9	133
0	0	0	0	0	0	0.3	7	1.5	105
Walking Speed									
mile/h	m/s			hp	W	Oxygen	Air	kcal/min	W
4.46	2			0.141	105	1.83	44	9.1	637
3.33	1.5			0.076	57	1.1	26	5.5	385
2.23	1			0.0415	31	0.71	18	3.5	245
1.1	0.5			0.0226	17	0.52	12.5	2.5	175
0	0			0		0.28	6.8	1.4	98

Sources: "Velox" 1869; Bekker 1962; Dean 1965.
[a]Total mass 77 kg (170 lb), frontal area 0.34 m^2 (3.6 ft^2), tire pressure 100 lbf/m^2 (689 kPa).
[b]Total mass 85 kg (187 lb), frontal area 0.511 m^2 (5.5 ft^2), tire pressure 50 lbf/m^2 (345 kPa).

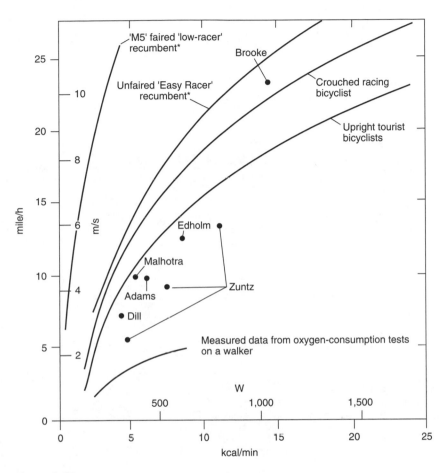

Figure 2.13
Gross caloric expenditure of bicyclists. (Curves from Whitt 1971.)

on pedaling speeds presented in table 2.2. Pugh's work also confirmed that the work produced accounted for about 25 percent of the extra fuel used. This figure was used by Whitt (1971) to calculate riders' metabolic heat production from the tractive forces at the driving wheel.

Pedaling forces

Average thrust

Table 2.2, compiled from data given in other parts of this book, compares the recorded pedaling rates of bicyclists of all types with estimates of the

Table 2.2
Pedaling speeds

	Distance (mile)	Time	mile/h	Gear (inches)	Crank (inches)	Crank speed (rpm)	Foot speed (ft/min)	Est'd power (hp)	Est'd thrust (lbf)
Ordinary, track	0.25	30 s	30	53	5	190	493	1.35	91
	0.5	72 s	25	56	5	150	392	1.05	88
		60 min	20.1	59	5.5	116	330	0.5	50
Safety, track	0.125	12.4 s	36.3	90	6.25	136	446	1.6	120
	0.125	12.2 s	37	68	6.5	182	619	1.6	85
	0.25	29 s	29.8	64	6.25	170	520	1.3	83
	0.125	11.5 s	39	90	6.5	145	473	1.65	115
Safety, track, motorcycle paced		60 min	40.1	106	6.75	126	445	0.5	37
		60 min	56	139	6.5	134	456	0.5	36
		60 min	61.5	144	6.5	143	488	0.5	35
		60 min	71	180	6.5	133	454	0.5	36
		60 min	76	191	6.5	134	454	0.5	36
Train-paced	1	57 s	62	104	6.5	198	670	1.2	59
Road safety bicycle	25	52 min	28.8	90	6.875	102	370	0.6	54
	100	4 h	25	85	6.875	99	368	0.5	45
	480	24 h	20	80	6.875	84	310	0.25	26
	100	4 h 28 min	22.4	81	6.5	93	316	0.5	52
Road, tourist			10	68	6.875	50	180	0.09	16
			12	68	6.875	61	220	0.11	16
			16	75	6.875	74	266	0.2	24
			18.5	75	6.875	85	305	0.3	32

Sources: Davison 1933, 55–56; England 1957, 326–327; "Vandy" 1964, 8; De Leener 1970, 26.

resulting power outputs. These estimates, in turn, have led to estimates of the average tangential forces applied to the pedals during cycling.

A steadily riding racing bicyclist tends to use very consistent but moderate pedal thrusts, amounting to mean applied tangential forces of only about one-third to one-sixth of the rider's weight. The rider's peak vertical thrusts are greater (approximately 1.5 times the mean) but still relatively small. No doubt this action enables the rider to maintain a steady seat position and steer steadily.

It is easy to calculate from a bicycle's crank length and a given pedaling speed in revolutions per minute what constant value of pedal thrust is required to achieve a given horsepower output on that bicycle. The peripheral pedal speed around the pedaling circle can be used in the equation

$$average\ propulsive\ force\ \text{(newtons)} = \frac{power\ \text{(W)}}{foot\ speed\ \text{(m/s)}}.$$

(The calculation presumes that only one foot is pushing at a time.) Foot speed is determined as revolutions per second times the circumference of the pedal circle (typically 1.07 m). To convert from newtons to pounds of force, divide by 4.45.

Detailed pedal-force data

Precise pedal-force measurements have been brought to a high art by Hull and his colleagues (Newmiller, Hull, and Zajac 1998; Boyd, Hull, and Wootten 1996; Rowe, Hull, and Wang 1998). To permit such measurements, specially designed pedals are instrumented with strain gages and calibrated to measure force components in up to three directions and possibly also twisting moments tending to bend the pedal spindle. Angle sensors are used to determine the orientation of each pedal relative to the crank and of the crank relative to the bicycle. All channels of data are logged by a computer, typically hundreds of times per second. Some pedal-force diagrams may be found in Coyle et al. 1991 and Hull and Jorge 1985, and one developed by Okijama (1983) is shown in figure 2.14. Because of pedal- and crank-orientation issues, this is not the simplest way to measure pedaling power (sprocket torque or chain tension is easier).

Some care is required for proper interpretation of the results (Papadopoulos 1987). In principle, if someone not exerting any muscle forces (apart from keeping the ankles from flopping) is strapped onto a bicycle, for each stationary orientation of the cranks, the feet will apply some force to the pedals, primarily because of the weight of the thighs and the elasticity of the uncontracted muscles.[13] The direction of that force is roughly along the line from knee to pedal. If the cranks are then driven by a motor (while the passive person is properly strapped to the saddle), additional

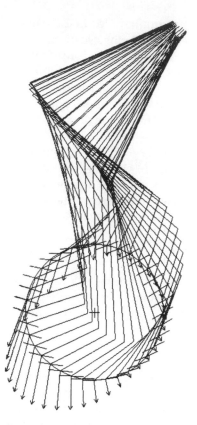

Figure 2.14
Typical pedal-force pattern. (From Okajima 1983.)

pedal forces will be observed, mostly relating to the acceleration and deceleration of the thighs. These forces act in roughly the same directions, unless the calves are heavy. (These dynamic foot forces can become very large as the pedaling cadence is increased, particularly for the leg when it is nearly straight, which approximates a force-multiplying toggle mechanism.)

If these purely mechanical, non-power-producing forces are subtracted from the actual measured forces of a pedaling person, what remains are the muscular forces, which alone create propulsive power. The mechanical forces almost totally obscure the muscular forces at the top and bottom dead centers of the pedaling motion and also on the upstroke. (In steady-state power production, a person's slight tendency to lift while pedaling doesn't usually overcome the weight of the leg.)

Once the partially obscured nature of the pedaling forces is appreciated, one may ask about optimal magnitudes and directions for pushing around the pedaling cycle. With the many different muscles in the leg, each with its own size, fiber makeup, and state of fatigue, such optimalities may never be specified in general. However, there is a very important observation to be made: it is widely supposed that muscular force in pedaling should ideally be oriented along the pedal path (i.e., perpendicular to the crank); otherwise some amount of force will be "wasted." (In fact, the most common suggestion is that the total measured force should be kept tangent to the pedal path.) This supposition is generally invalid: the example of a piston engine shows that there is nothing inefficient about exerting a force along the connecting rod. As a general rule, better performance (power, efficiency, endurance) can be expected if the muscular force applied to the pedal is permitted to deviate somewhat in direction from tangency to the pedal path. In fact, Papadopoulos (1987), assuming only certain sets of muscles to be active, showed that constraining the force direction leads to the performance of negative work by some muscles and the irreversible absorption of mechanical energy. Constrained motions (e.g., a fixed-length crank forcing the pedal to move in a circle) allow existing muscles to furnish their maximum power. Constrained forces (e.g., a crank that freely telescopes, requiring the pedaler to exert a total foot force exactly perpendicular to the crank) should severely reduce the rider's power, although as a training aid, they may encourage certain underused muscles to develop greater strength.

An upright-seated pedaler can turn the pedal cranks via any of a variety of distinct pedaling styles (Papadopoulos 1992–1994). Some styles involve strong tangential forces when the pedals are at their upper and lower extremes, or in contrast a "thrusting" style with brief high forces during the downstroke, or perhaps an unusual degree of lifting force (or leg-weight reduction) on the upstroke. Others involve additional phased pedal thrusts to counterbalance high-cadence bouncing at the saddle or control of foot-force direction to avoid slippage when there is nothing holding the foot to the pedal. Sideforce at the saddle and/or handlebars or a rotational couple of forces at the handlebar may result from high pedaling torque. Upper-body bobbing or fore-aft sliding are not unusual at high effort levels. Many other techniques or styles may be recognized, only some of which are for extreme (high-torque or high-cadence) situations. Standup pedaling offers even greater scope for a wide range of energy interchanges, as do brief rest periods related to stepping or leaping. We surmise that ergometry subjects may avail themselves of various styles based on either habit or fatigue and recommend that efforts be made to recognize and control technique in pedaling research.

Close examination of pedaling mechanics reveals some surprises. For example, take the case of high-cadence seated pedaling. The total potential energy of the two legs varies little throughout the pedaling cycle, so it may be ignored in comparison to the kinetic energy, which increases rapidly as the thighs swing into motion and decreases just as rapidly when they come to rest. (See Hull, Kautz, and Beard 1991 for specific numerical examples.) On a free-wheel-equipped bicycle (or a fixed-wheel bicycle on which the cyclist strives to keep the chain taut), the cyclist's muscles must supply the power required to accelerate the legs. The muscles could also absorb that energy to keep chain tension from varying, but it is far more efficient to allow the chain to slow the legs, so that their kinetic energy is transmuted into propulsive work. There is no obvious inefficiency in this pedaling style: the cyclic interchange of energy transmits a specific average power (proportional to the cube of pedaling rpm) to the rear wheel, on top of which one may perform additional work.

But now imagine the case of the rider's wishing to pedal more gently (after entering a paceline, for example). If the required pedaling power happens to be lower than the average dictated by the pedaling cadence and the rider's leg mass,[14] any excess of power will have to be absorbed by the brakes or by the rider's muscles through eccentric contraction. These are both very inefficient, so the cyclist is better advised to pedal only inter-mittently or to shift to a higher gear.

As another example, consider stand-up pedaling. If all the cyclist's body weight during stand-up pedaling is applied to each pedal in turn, then crank torque is a simple rectified (i.e., positive-only) sinusoidal func-tion with a fixed amplitude. Even if the rider's arms share in the work by tipping the bicycle (this is a good example of a nonmechanism way to add arm work), power is easily calculated from body weight times average speed of pedal descent (i.e., revs/s times twice the pedal-circle diameter). How can one change the stand-up pedaling torque so as to adjust pedaling speed? To increase pedaling speed, it is obvious that one can pull up on the rising pedal, adding more force to the other pedal and increasing crank torque to any level. But pedaling more slowly is a problem: if the rider fixes her center-of-mass (CoM) height (as if sitting on a seat) then any downforce applied by the rising leg is immediately translated into eccentric contrac-tion (negative work) that absorbs output from the other leg.

Actual stand-up pedaling involves a hearty range of vertical body motions. One option is for the cyclist to let the lower of her two legs remain straight while she straightens the upper. This motion exerts no net crank torque, as the body is lifted half the diameter of the pedal circle. Then the downward-moving leg, now straight, is held rigid while the body falls (no muscular work performed). By this means, the power output of the legs can be halved without any negative work. The contributing author has been unable to think of any stand-up pedaling scheme that could reduce energy expenditures still further while still supporting all the rider's weight on the pedals.

Effects of pedaling motion, body position, and rpm

Up to this point, the chapter has been concerned with the overall physiology of muscles and exercise and some general background on pedaling. Now we take up a variety of questions related to the specifics of pedaling. There is almost no theory to guide us in this area, so the main thrust will be to report on efforts to devise improved pedaling mechanisms.

Pedaling and rowing motions

Harrison's (1970) curve for short-duration pedaling or cycling (figure 2.15, curve 1) was developed based on measurements taken from a group of active men, not record athletes. The significance of his results lies, therefore, in measurements of the relative power produced by the same individuals using different motions and mechanisms. Harrison's findings seem to be particularly significant because his subjects produced, in some cases,

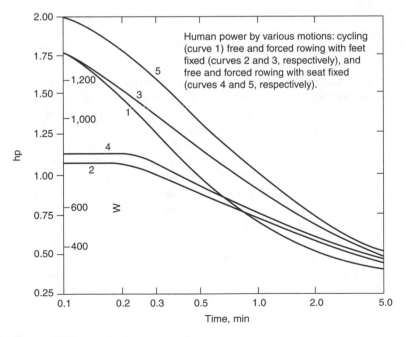

Figure 2.15
Peak human power output by various motions. (From Harrison 1970.)

more power through motions to which they were unaccustomed than through bicycle pedaling, with which they were all familiar. The curves for linear ("rowing") foot and hand motion (curves 2 and 4 in figure 2.15) lie considerably below the cycling curve for shorttime durations but rise above it after one minute.

Rowing data resulting from measurements taken on an ergometer have an additional reason for showing a diminished power output: if the subject's feet are fixed with respect to the ground, as they are normally fixed with respect to the boat when one is rowing, there are large energy variations from the rower's accelerating and decelerating his body from rest positions at the ends of the stroke, something that occurs to only a minor extent in actual rowing (where the light boat, rather than the heavy rower, does the accelerating and decelerating). It is actually possible for the rower to convert backward-moving kinetic energy to propulsive work, as long as the arms rather than the leg or trunk muscles are used to come to rest. However, to decelerate a boat's forward motion probably requires some negative work (in addition to elastic energy storage) in the leg and trunk muscles, particularly at high stroke rates. It is an interesting open question when such additional (but uncounted) power production, by a different set of muscles, is likely to reduce the desired power output. The simplest expectation is for fixed-power rowing endurance to be less when the feet are fixed to the stationary frame (curve 2) than when the seat is fixed and the feet are allowed to move (curve 4) (as Harrison found; see Harrison 1970).

Of great interest are Harrison's results for what he called "forced" rowing. (This has nothing to do with the slave galleys mentioned in chapter 1.) Harrison set up a mechanism that defined—or "forced"—the ends of the rower's stroke and conserved the kinetic energy of the moving masses, either with the feet fixed (curve 3) or the seat fixed (curve 5). The piston-crank-flywheel mechanism of a car engine is of this type. With forced rowing and the seat fixed, considerably longer durations of power than with normal pedaling were obtained at all power levels. We know of no case in which this apparently significant finding has been used to break any human-powered-vehicle record.[15]

Pedaling combined with hand cranking

The question frequently arises as to whether or not one can add hand cranking to pedaling and obtain a total power output equal to the sum of what one would produce using each mode independently. Kyle, Caizzo, and Palombo (1978) showed that, for periods of up to a minute, 11–18 percent more power could be obtained with hand and foot cranking than with foot cranking alone. The power produced was greater when the arms and legs were cranking out of phase than when each arm moved together with the leg on the same side. In later work, Powell and Robinson (1987) found that power production in a ramp test could be increased by more than 30 percent over pedaling alone when arm cranking was combined with pedaling. They tested thirty-two subjects, seventeen males and fifteen females, in power production using arms only, legs only, and combined arms and legs. Arm-power tests started at 25 W and were increased by 25 W every two minutes until rider exhaustion. The leg-power tests were similar, except that the increments were 33.3 W. The same increments and intervals as in the leg-power tests were used in the combined-arm-and-leg tests. VO_2max was higher for the combined-arm-and-leg power than for legs alone, supporting the statements made above about the use of this measure. We stated earlier that about half the advantage of the combined-arm-and-leg power over leg power alone was due to aerobic metabolism and half to anaerobic metabolism.

Upright and recumbent pedaling
Although early measurements showed an apparent small reduction in power when a bicyclist switches from a conventional pedaling position to a recumbent one, most modern research shows virtually no difference. Bussolari and Nadel (1989) tested twenty-four male and two female athletes in the two positions and found no significant difference in oxygen efficiency (figure 2.12). There are two pitfalls in particular to be avoided in such a comparison. One is in the definitions. The word "recumbent" is sometimes taken to mean "flat on one's back" but more often to mean sitting as one does driving a car, a style more accurately referred to as "semirecumbent." One would expect to produce a lower power when on one's back. The "upright" posture can be taken as that used on an all-terrain bicycle or "sit-up" bike. One might expect a reduction in maximum aerobic power for the crouched racing position because of the restriction in breathing the position imposes, as has been speculated elsewhere. The other pitfall involves the question of accustomization, which is always difficult when a "new" position is being tested. It might take months of practice before one's muscles are adapted to a new position, yet in tests one is usually allowed only minutes to accustom oneself to a shift in position.

Antonson (1987) studied the oxygen efficiency of recumbent and conventional bicycling positions at less than maximum workloads by testing thirty men: ten recumbent cyclists, ten cyclists used to conventional machines, and ten physically active noncyclists. Each pedaled for six minutes at 51.5 W, followed by six minutes at three times this power level (154.5 W), for each of the two positions, while being measured for oxygen consumption, ventilation, and heart rate. There were no significant differences in oxygen consumption or ventilation among the three groups. The noncyclists had a higher heart rate than those in the other two groups. Antonson found no indication that either group of bicyclists benefited from being accustomed to one position or the other.

Drela (1998), quoting results found by Nadel and Bussolari, confirmed that there was no significant difference in power output between recumbent and conventional bicycling. He also quoted a remarkable range in efficiency, defined as the ratio of the power output to the heat of glucose oxidation multiplied by the change in oxygen consumption. The so-defined efficiency for Olympic-class athletes ranged from 18.0 percent to 33.7 percent. This seems to validate the hypothesis that the fuel efficiency of high-power pedaling is virtually irrelevant to performance. Drela also reported a large range in the percentage of VO_2max (60–90 percent) that could be sustained by these athletes without buildup of lactate. These are truly remarkable differences for athletes capable of similar outputs. (The first, and so far the only, person to fly a human-powered airplane across the English Channel, Bryan Allen, discussed his preflight training regimen— bicycle rides of alternating 160 km and 300 km per day—with the MIT group planning the Daedalus long-distance flight. He said that the muscles that gave him the greatest problems were those for chewing, during the evening carbohydrate and protein reloading meals. One wonders, impolitely, if he was at the lower end of the efficiency range and had to eat more to produce his magnificent output.)

Too (1990) found that the configuration of semirecumbent pedaling for which subjects attained greatest power in a ramp test was one with a vertical seat back and the cranks on a line sloping down 15° from the seat. In 1991 Too showed that this configuration also maximized PP and AP in a Wingate test.

Backward pedaling

The concept of pedaling backward instead of forward seems unnatural. However, Spinnetti (1987) experimented with low-power backward pedaling, then carried out careful measurements that showed he could produce higher levels of short-duration maximum power pedaling backward (215 W) than forward (179 W) (figure 2.16). He explained this by demonstrating that he could produce more static torque pushing backward rather than forward (figure 2.17), and he included photos in his article to show that, in his opinion, more muscle groups were involved in the former than in the latter. One should not draw conclusions on the basis of one series of tests on one person, but the power differential Spinnetti found is intriguing.

The next topic is a similarly unusual pedaling system that seems to allow increased power to be produced through involvement of more muscle groups.

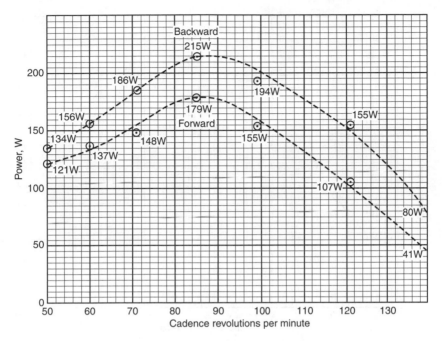

Figure 2.16
Maximum power produced versus rpm in forward and backward pedaling.
(From Spinnetti 1987.)

PowerCranks and the active involvement of the lifting muscles

One clever approach to involving additional major muscles in pedaling
is that given by Frank Day's PowerCranks. These are built with one-way
clutches so that each leg has to lift itself (helped neither by the counter-
balancing weight nor by the downpush of the other leg). Used only in
training, they force the development of some large muscles that most of us
are content to leave uninvolved, with the following results claimed: "Elite
mountain bikers have shown documented power improvements of about
20% in 7 months (440 watts to 520 watts using Conconi protocol) and
almost 40% improvement after about 9 months of use (440 watts to 580
watts)" (Day 2001). The testing protocol, originally developed to reveal LT,
reports the maximum power attained for the final full minute of a ramp
power test, in which the ramp rate is initially 40 W/min until a pulse
rate of 145 is reached, then 20 W/min until exhaustion. Although this is
a very promising result, with just a single ramp rate it is not possible to
determine whether it is anaerobic work capacity, critical power, or both
that improved.

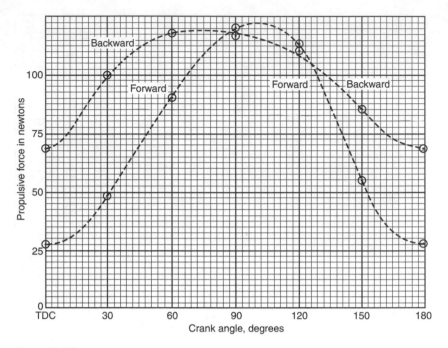

Figure 2.17
Static propulsive force versus crank angle for forward and backward pedaling.
(From Spinnetti 1987.)

Effect of saddle height

Using a single subject (a thirty-nine-year-old man, obviously not very athletic), Müller (1937) obtained the results shown in figure 2.18. For durations less than half an hour, he found that at least one and a half times greater endurance was achieved at each power level when the saddle was raised 40–50 mm above the "normal" height (that for which the heel can just reach the pedal with the leg stretched and the posture upright). Equivalently, the power that could be tolerated for each duration was increased by about 7 percent. No less important, perhaps, is the dramatic 15–30 percent reduction in power, or 80 percent reduction in endurance, when the saddle was set 100 mm lower than normal.

Thomas (1967a, 1967b; Hamley and Thomas 1967) tested one hundred subjects on a Müller ergometer (figure 2.1) and also found that maximum sprinting power output (averaged over the duration of the short task, and thus similar to the AP determination of a Wingate test) was obtained with the saddle set at a height about 10 percent greater than leg length. He defined saddle height as the distance from the pedal spindle at its lowest point to the top of the saddle, so that about half of the thickness of the

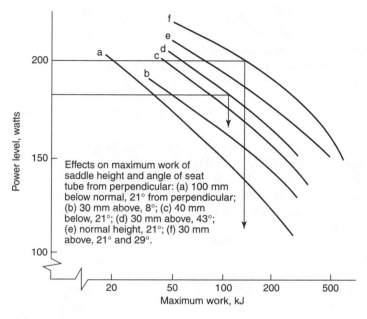

Figure 2.18
Effects on maximum work of saddle height and angle of seat tube. (From Müller 1937.)

pedal would reduce the effective height. Unfortunately, cadence was not optimized in these tests, and to achieve higher power requires sustaining a higher cadence than used.

Effect of crank length

The length of the cranks in safety bicycles is fixed within narrow limits. With the saddle at the normal height above the pedals, as defined by Müller (1937), and with the pedals at a distance above the ground such that in moderate turns (when the bicycle will be inclined toward the center of the turn) the pedals do not contact the ground, the saddle is at a height at which the rider can just put the ball of one foot on the ground when stopped while still sitting on the saddle. The crank length is then chosen so that almost all riders will feel comfortable. This length is normally, for adult riders, taken as 165 mm (6.5 inches) or 170 mm (6.7 inches). Thus, the height above the ground of the bottom-bracket axle is fixed. An attempt to fit longer cranks will lead to a reduction of pedal clearance when cornering.

Few riders, then, have an opportunity to try long cranks, because each crank length strictly requires a frame designed specially for that length. In this respect, bicycles designed for off-road use, when used on-road, and even more so, recumbent bicycles, have an advantage. Most data on the effects of crank length are based on tests that have been taken on ergometers. But ergometer data can be regarded with suspicion, as we have implied, and this has certainly been true with regard to data on long cranks. So few people have been able to experiment with significantly longer cranks on actual bicycles that their impressions must also be treated with reserve.

Two people writing for a bicycling magazine in 1897 advocated shorter cranks (DeLong 1978). One, Perrache, experimented with 160-, 190-, and 220-mm cranks on a bicycle over a 5-km course and found that, in maximum-speed runs, he could get about 9 percent more power output with the 160-mm cranks than with the 220-mm cranks. It is not known whether the gear ratios were changed for different crank lengths. It would obviously penalize longer cranks if the gear ratio were not increased to give approximately similar ratios of pedal speed to road speed. It would also be a disadvantage if the rider were accustomed to using short cranks.

Grosse-Lordemann and Müller (1936) tested the effect of crank lengths on an ergometer, employing only one subject. Their approach was to set the power output the subject had to produce and to measure the maximum duration for which this output could be sustained. They also used three crank lengths: 140, 180, and 220 mm. In this case, and for all power levels, the subject was able to produce the most total work (that is, work for the longest periods) when using the longest cranks. At the highest powers, the body efficiency (work output divided by energy input in food) was also highest when the longest cranks were used.

Harrison (1970) gave his five subjects an initial choice of crank length and found that they preferred the longer cranks (177 and 203 mm; 7 and 8 inches). The subjects were not particularly tall. Harrison intended to perform all of his tests at two different crank lengths; however, he found from initial tests that "crank length played a relatively unimportant role in determining maximum power output" and used just one (unspecified) length for most of his tests.

The world champion Eddy Merckx used 175-mm cranks for the world's one-hour record and has used 180-mm cranks for time trials and hill climbs in the Tour de France (DeLong 1978, 192).

More recent data confirm these earlier findings. Too (1999b) measured the anaerobic power outputs of six male subjects, aged twenty-four to thirty-five, employing the Wingate protocol, in conventional and recumbent positions, using cranks of lengths from 110 to 265 mm. The highest APs were given for 180-mm cranks for both positions. This crank length also gave the highest PP for the conventional position, whereas the shortest cranks, 110 mm, allowed the recumbent bicyclist to produce the highest

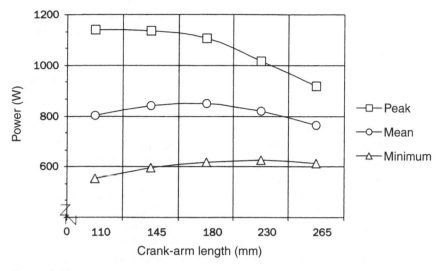

Figure 2.19
Peak, mean, and minimum power in recumbent pedaling as functions of crank-arm length. (From Too and Williams 2000.)

PP. For all crank lengths, PP and AP were higher in the recumbent position. This result seems at variance with earlier data quoted above. The recumbent data are summarized in figure 2.19 (Too and Williams 2000).

The same author (Too 1999a) discusses crank-arm length in an earlier note, pointing out that the optimum crank-arm lengths depend on individual proportions and probably are functions of hip- and knee-joint angles that optimize muscle lengths. Also, shorter cranks (145 mm) gave higher powers at the start of a thirty-second test, for instance, whereas longer cranks (230 mm) allowed the riders to produce the highest power levels of all crank lengths at the end of the period, when they were tiring.

In summary, crank-arm length is not of major importance in the quest for producing maximum power. No manufacturer is currently making cranks whose lengths can be varied during use. However, for racers, even factors of seemingly minor importance can produce a win. To choose the optimum among all the factors involved is too detailed a topic for this book: we recommend a study of the references quoted and of others, existing and, undoubtedly, coming.

Nonround chainwheels
Elliptical chainwheels can be fitted to normal cranks, in which case the pedal motion remains circular, but of varying speed or "gear ratio." The usual purpose of elliptical chainwheels is to reduce the supposedly useless

time during which the pedals are near the top and bottom "dead centers." This topic has some similarity to that of long cranks, in that there are fierce proponents and antagonists and few reliable data. Four of Harrison's (1970) five subjects produced virtually identical output curves (power versus duration) using circular and elliptical chainwheels. One, apparently Harrison himself, gave about 12.5 percent more power with the elliptical chainwheel. All preferred the elliptical chainwheel for low-speed, high-torque pedaling. The degree of ovality was not specified, but Harrison stated that the foot accelerations required were high. An illustration in Harrison's paper shows a chainwheel of a very high degree of ovality (about 1.45; see discussion of ovality below).

The degree of ovality of an elliptical chainwheel can be specified by the ratio of the major to the minor diameter of the ellipse. In the 1890s, racing riders using elliptical chainwheels with ovalities of about 1.3 became disillusioned with their performances, and these chainwheels fell out of favor. In the 1930s the Thétic chainwheel, with an ovality ratio of 1.1, became quite popular. Experiments with chainwheels having ovalities up to 1.6 confirm that high ovality (perhaps 1.2 or greater) decreases performance (Whitt 1973). With a Thétic-type chainwheel, no deterioration of performance compared with that on a round chainwheel was recorded, and a small proportion of riders improved their performances by a few percent.

In the 1980s Shimano introduced a nonround chainwheel, called "Biopace," that was not elliptical. The scientific background was given by Okajima (1983), who showed a typical pedal-force pattern (figure 2.14), which, with the dynamic model of figure 2.20, enabled his group to determine the leg-joint torques for normal circular-chainwheel pedaling, shown in figure 2.21. Okajima pointed out that the knee has a period of strongly negative torque: "We saw two specific restrictions to be solved:

1.   the difficulty of spinning, both in the motion and in the direction the force must be applied, restricts the speed of muscle contraction during pedaling to a rather slow rate, and requires the force to be on the high side, and
2.   the knee joint is overused, while the hip joint is underused (the ankle is rather passive)."

Okajima writes further, "We decided that an appropriately uneven angular velocity pattern would reduce the loss of kinetic energy, and also make it easier for the rider to switch between the firing of different muscle groups at appropriate times (to be specific, at the reversal of knee torque)."

The shapes of three chainwheels resulting from the Shimano study (used together in a triple chainwheel) are shown in figure 2.22. Unfortunately, we know of no evidence that these made any significant difference to performance, and they are no longer widely available.

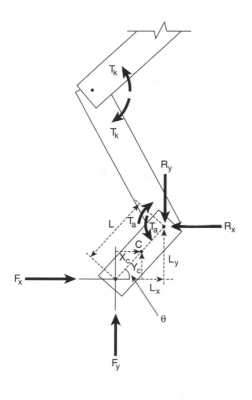

Dynamic model — calculation of Joint torques:

Equations:
Balance of forces:
$$F_x - R_x = m\ddot{x} \qquad (1)$$
$$F_y - R_y = m\ddot{y} \qquad (2)$$
Balance of torques (around center of gravity C):
$$T_a + F_x y_c + R_x (L_y - y_c) \\ - F_y x_c - R_y (L_x - x_c) = I\ddot{\theta} \; (3)$$
where
$\ddot{x} = d^2x/dx^2$, that is, the *x-component of acceleration; and similarly* $\ddot{y} = d^2y/dy^2$ *and* $\ddot{\theta} = d^2\theta/d\theta^2$
and m is mass, I Is moment of inertia, and other terms are as shown in the diagram.

Procedure:
F_x, F_y, x, y, and $\ddot{\theta}$ are measured data.
m, I, x_c, y_c, and L are deduced from anthropometric statistics.
R_x and R_y are calculated from equations (1) and (2).
Given all these values, equation (3) can be solved to give joint torque T_a.
The procedure can then be repeated with values appropriate for the knee-to-ankle segment (including equal and opposite reactions to the T_a, R_x, and R_y just found, acting at the ankle) to find knee torque, and again to find hip torque.

Figure 2.20
Dynamic model of pedaling leg. (From Okajima 1983.)

A more versatile mechanism giving the same effect as a nonround sprocket was the Brown SelectoCam, also sold as the Stronglight Power-Cam. A bell-crank riding around a fixed central cam advanced and retarded the round chainring relative to the crank, twice each revolution, without the manufacturing and chain-shifting disadvantages of a variable sprocket radius (see U.S. patent 4,281,845).

Lever or linear drives
Many people have invented and reinvented forms of the linear drive, in which each foot pushes on (for instance) a swinging lever, with a strap or

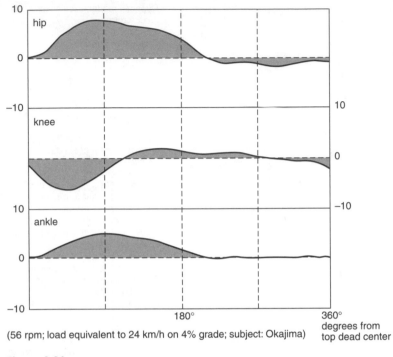

(56 rpm; load equivalent to 24 km/h on 4% grade; subject: Okajima)   degrees from
top dead center

Figure 2.21
Leg-joint torques during one crank revolution. (From Okajima 1983.)

cable attached to the lever at a point along it that can be varied to give different gearing ratios. The cable is then attached, perhaps through a length of chain, to a free-wheel on the back wheel and to a return spring (figure 2.23). The drive of the American Star high-wheeler (figure 1.19) was of this type, although the gear of the American Star was not variable. Pryor Dodge has been gracious enough to allow us to reproduce the jacket photograph of his 1996 book *The Bicycle* (figure 2.24) showing a superb example of swinging-lever drive with a gear-changing mechanism, apparently on a bicycle from the late 1890s.

The overwhelming disadvantage of swinging-lever drives is that the feet and legs must typically be accelerated and subsequently decelerated by the muscles in the same way as in shadowboxing (Wilson 1973). Harrison (1970) found rather low outputs for motions of this type (figure 2.15). However, some believe that this disadvantage holds only for the most primitive embodiments: careful design should make it possible to oscillate the feet at high cadence with no energy cost. With geometrical slowing

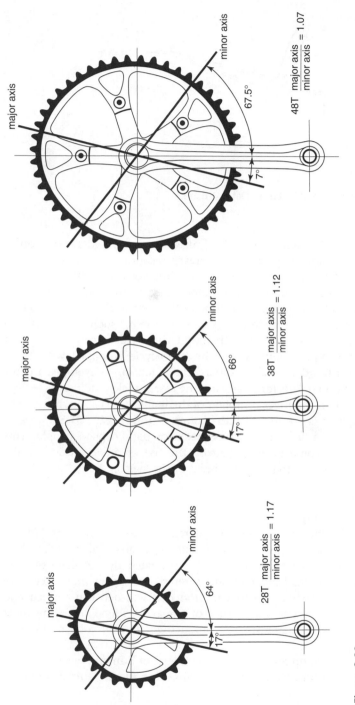

Figure 2.22
Biopace chainwheels of twenty-eight, thirty-eight, and forty-eight teeth. (From Okajima 1983.)

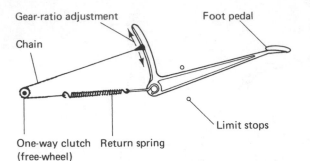

Figure 2.23
Swinging-lever drive. (Sketch by Dave Wilson.)

(a reducing sprocket radius, or a drive linkage approaching its condition of zero mechanical advantage) kinetic energy is recaptured at the stroke end. (See figure 9.24 for the mechanism of the Thijs Rowcycle.) With coupling between the left and right pedals, one foot may lift the other in the same way as with a rotating crank. Even though the feet would then be brought to rest simultaneously, simple energy-storing systems such as springs or a flywheel can serve to reaccelerate them.[16] Other defects in swinging-lever drives are the impossibility of wheeling a bicycle equipped with these drives backward and the extraordinary chain tensions developed in low gear, if there is no means for increasing the rear sprocket radius.

Pedalers of lever drives complain of the inability to use ankle motions for propulsion, as is possible with the common rotary drives.[17] Some years ago in Germany a "foot cycle" was made for handicapped people. This machine, which could be propelled by the use of ankle motions only, demonstrated the help that the lower part of the legs can be to the ordinary bicyclist.

Noncircular cranking
Harrison (1970) showed that a rowing (straight-line) motion, with kinetic-energy conservation at the ends of the stroke, enabled riders to produce greater short-term power than could be generated by circular pedaling. There has been a small but constant interest over more than a century in the question of whether a foot motion that was between circular and straight would be better than either of the two individually. The most common form of mechanism for producing elliptical foot motions is shown in figure 2.25.

We have seen no results of ergometer tests of human power produced on such mechanisms. However, Miles Kingsbury in the United Kingdom

Figure 2.24
Sophisticated example of swinging-lever drive on an early bicycle. (With the kind permission of Pryor Dodge, who provided a transparency.) (From Dodge 1996.)

Figure 2.25
Mechanism to produce elliptical pedal paths. (From an 1890 German text.)

has manufactured a modern form of such mechanisms, under the name "K-drive," and it has been used to win several races (Larrington 1999). Perhaps the primary advantage of the K-drive lies in reducing the area swept out by the moving foot, so that a smaller, streamlined fairing may be used. In its present embodiment it adds weight and friction (because of several additional moving links), so the winning performances achieved with it must be regarded as significant.

Some other forms of power input
Mechanisms such as rotating hand cranks or rocking handlebars have been developed to allow the rider to employ muscles other than the legs for propulsion. But perhaps surprisingly, this facility is already present to some degree in conventional upright bicycles:

•   When the rider is standing, substantial arm work is easily performed by tilting the bicycle away from the descending pedal. The amount of arm power exerted can be calculated from the diminution of pedal displacement at the given crank torque. Example: the bicycle is tipped from 15° right to 15° left, a total of 30°. The pedal's offset from the frame plane is about 5 inches (125 mm). So foot motion is reduced by 65 mm from its original 170×2 or 340 mm; that is, the arms are doing 19 percent of the work. Presumably the legs can then push harder or move faster, for an overall increase in power output.[18]
•   The arms can also be used powerfully at very low cadences and high torques by pulling the torso forward (i.e., sliding along the saddle toward the handlebars) during each downstroke.
•   Especially when standing, the cyclist can leap upward with the assistance of the torso's uncoiling and push or pull somewhat vertically with the arms. (Not only can this technique add work produced by other muscles, but it makes it possible to convert low foot speeds to high, after which the rider descends on a straight, nonworking leg.)

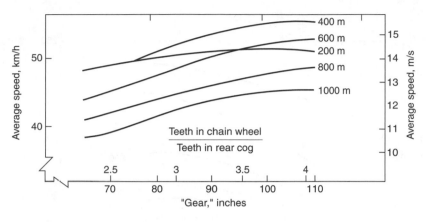

Figure 2.26
Effect of gear ratio on the average speed of a racing bicyclist at various distances in a standing-start 1-km time trial.

· The torso can be used, to a lesser extent, by bouncing when seated, which pulls on the upper ends of the biceps femoris and gluteus muscles of the descending thigh.

Recumbents may be disadvantaged because they do not permit the rider to use additional muscles in this way. A spring-preloaded, rearward-slidable seat on a recumbent might provide a useful analog to the stand-up pedaling permitted by an upright bicycle.

Measurements made during actual bicycling

Thorough and accurate data relating oxygen consumption, heart rate, pedal torque, pedaling rate, bicycle speed, gear ratio, and crank length have been gathered by the Japanese Bicycle Research Association by equipping several riders with instruments and telemetrically recording their behavior during actual riding (Bicycle Production and Technical Institute 1968). Some of these data are shown in figures 2.26, 2.27, and 4.17. Methodical investigation of a series of gear ratios used in a standing-start 1-km time trial tended to show a distinct performance optimum for rear sprockets of thirteen to fifteen teeth (the front sprocket had forty-nine). However, the gear ratio for minimum oxygen usage at steady speeds of about 75 percent of racing speed was not clearly correlated to that for greatest speed. Indeed, one rider exhibited a maximum of oxygen usage at the best gear ratio. Surprisingly, gear ratio for minimum oxygen usage at various speeds was not clearly tied to pedaling cadence. (It is unknown whether the oxygen-measuring system contributed to the bicycle drag, or how carefully the test

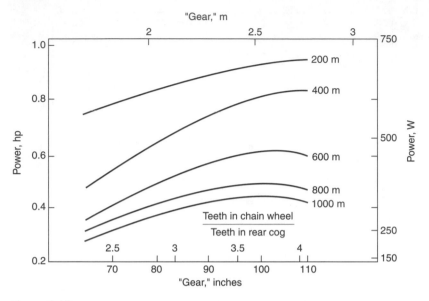

Figure 2.27
Power required for various gear ratios at various distances in a standing-start 1-km time trial.

subjects maintained a fixed aerodynamic drag in their trials.) Each given speed in the figures represents a fixed power level. At each power level, the lowest fuel usage occurred at the greatest gear ratio.

Conclusions

A rider's ability to produce power cannot be described with a single number. In practice endurance is measured at a variety of power levels. The highest levels of power call upon anaerobic work capacity, namely, the finite amount of energy that can be produced without fuel oxidation. Lower levels of power are tied to the virtually steady-state ability of the muscles to oxidize fuel.

For each way that human muscles can convert fuel to mechanical work, there are limits to performance, for example, the maximum rate of fuel utilization, the store of available fuel, and the rate of lactate production. Understanding these limits, and when they apply, will go a long way to explaining actual performances in competition. Recent thinking has returned to the old speculation that involving more major muscles could enhance both short-term and long-term performances.

Once a person has achieved basic fitness, it appears that further training cannot increase her or his power more than about 30 percent. The main gains in speed must be achieved through aerodynamics.

Pedaling as it is done on conventional bicycles enables riders to approach their maximum power output. However, mechanisms that give noncircular foot motions or nonconstant foot velocities, or that allow hands and feet to be used together, seem to be required if the absolute maximum power output is to be obtained. Such mechanisms may also increase the power or comfort of low-effort utility bicycling.

Set against the inevitable difficulties of describing human performance, it is encouraging to realize that pedaling is one of the simplest and most studied exercise tasks. (In running, for example, there is no way to know the actual power produced by the muscles.) So the potentially relevant literature is vast.

Notes

1.   Actually, it is better to average over crank rotations rather than wheel rotations, to smooth out the cyclic power variations occurring in each crank revolution.

2.   The so-called immediate fuels are adenosine triphosphate and phosphocreatine (also called creatine phosphate).

3.   This number presumes that the gear ratio and wheel diameter produce 5.9 m "development" (i.e., slippage past the brake in one pedal revolution).

4.   The PowerTap actually reports average power from the previous 3.78 seconds, updated each 1.26 seconds. Near the end of the trial, quickly switching the display from actual power (i.e., MP), to AP, and then less hurriedly to maximum power (i.e., PP), permits all three Wingate parameters to be determined.

5.   The best equivalent to the foot force and apparent inertia of this ergometer would be riding up a 19 percent slope in a 33.5-inch gear (2.67 m development).

6.   The fact that isometric work has a maximum tolerable duration is believed to arise from the blood vessels' being squeezed down, thus restricting the muscle's blood supply.

7.   It might be appropriate to term pyruvate a carbohydrate fuel also, but since it apparently is not stored or transported, we present it here as a mere temporary intermediate compound.

8.   To relate this to a power level in watts, the conversion is (600 kcal/h)/3600 = 167 cal/s = 700 W energy supplied. With an efficiency of 25 percent, mechanical power is 175 W.

9.   Sometimes other factors such as dehydration or cramping also play a role.

10.   Perhaps a fairer statement is that VO_2max doesn't show much increase from activities, such as cycling, that don't elicit it.

11.   As an example, consider the vertical lift of a barbell behind the neck of a standing person, with her forearms slightly above horizontal. The upper arms articulate up and back, a motion that would lower the barbell if the elbows didn't also extend. Therefore, the shoulder joint appears to be absorbing energy.

12.   The 70 percent figure should not be considered a definite endurance limit, since some athletes can actually work at well above 80 percent. A more appropriate physiological determination would be the power output at OBLA, but it should also suffice simply to determine the power level sustainable for the requisite duration.

13.   The effect of muscle elasticity can be demonstrated by sitting relaxedly on a bicycle with no chain. The at-rest crank orientation is affected by trunk inclination.

14.   For example, Hull's measurements suggest that leg kinetic-energy exchanges at 120 rpm require putting at least 87 W into the cranks—more for a heavier person, and proportional to the cube of the rpm.

15.   Rowing a lightweight land vehicle, as in sculling, approximates a seat-fixed activity: it is the vehicle rather than the rower's body that alters velocity with each extension of the legs. "Forced" motion is something of a common-place in bicycle pedaling: it corresponds to a "fixed-gear" transmission, that is, one without a free-wheel.

16.   Unfortunately this approach creates dead centers of no propulsion, making it preferable that both feet not come to rest simultaneously.

17.   This observation, offered in the first edition by Frank Whitt, calls for clarification. A possible interpretation is that restricting foot motions to a straight-line path prevents riders from occasionally using muscles that alter length only when leaving that path.

18.   A possible mechanical aid to arm work that we have not encountered is a laterally slidable saddle. When unlocked, such a saddle would bear the rider's weight but still permit the bicycle to be tipped forcefully by the arms.

References

Adams, W. C. (1967). "Influence of age, sex and body weight on the energy expenditure of bicycle riding." *Journal of Applied Physiology* 22:539–545.

Ahlquist, L. E., D. R. Bassett, R. Sufit, F. J. Nagle, and D. P. Thomas. (1992). "The effect of pedaling frequency on glycogen depletion rates in type I and type II quadriceps muscle fibers during submaximal cycling exercise." *European Journal of Applied Physiology* 65:360–364.

Andersen, Jesper L., Peter Schjerling, and Bengt Saltin. (2000). "Muscle, genes, and athletic performance." *Scientific American* 283, no. 3 (September):48–55.

Antonson, Ingrid. (1987). "Oxygen cost of submaximal exercise in recumbent and conventional cycling positions." *Human Power* 6, no. 3:7, 17–18.

Astrand, Per-Olof, and Kaare Rodahl. (1977). *Textbook of Work Physiology*, 2d ed. New York: McGraw-Hill.

Beelen, A., and A. J. Sargeant. (1991). "Effect of fatigue on maximal power output at different contraction velocities in humans." *Journal of Applied Physiology* 71:2332–2337.

Bekker, M. G. (1962). *Theory of Land Locomotion*. Ann Arbor: University of Michigan Press.

Bergstrom, J., L. Hermansen, E. Hultman, and B. Saltin. (1967). "Diet, muscle glycogen, and physical performance." *Acta Physiologica Scandinavica* 71:140–150.

Bergstrom, J., and E. Hultman. (1966). "Muscle glycogen synthesis after exercise: an enhancing factor localized to the muscle cells in man." *Nature* 210:309–310.

Bicycle Production and Technical Institute. (1968). Report. Japan.

Boyd, T., M. L. Hull, and D. Wootten. (1996). "An improved accuracy six-load component pedal dynamometer for cycling." *Journal of Biomechanics* 29, no. 8 (August):1105–1110.

Brooks, G. A., T. D. Fahey, and T. P. White. (1996). "Energetics and athletic performance." Chapter 3 in *Exercise Physiology*, 2d ed. Mountain View, Calif.: Mayfield.

Bussolari, Steven R. (1987). "Human factors of long-distance HPA flights." *Human Power* 5, no. 4:8–12.

Bussolari, Steven R., and Ethan R. Nadel. (1989). "The physiological limits of long-duration human power production: Lessons from the Daedalus project." *Human Power* 7, no. 4:1–10.

Carpenter, R. C., et al. (1965). "The relationship between ventilating capacity and simple pneumonosis in coal workers." *British Journal of Industrial Medicine* 13:166–176.

Coyle, E. F. (1995). "Integration of physiological factors determining endurance performance ability." In *Exercise and Sport Sciences Reviews*, vol. 23, ed. J. O. Holloszy. Baltimore: Wilkins & Wilkins.

Coyle, E. F., M. E. Feltner, S. A. Kautz, M. T. Hamilton, S. J. Montain, A. M. Baylor, L. D. Abraham, and G. W. Petrek. (1991). "Physiological and biomechanical factors associated with elite endurance cycling performance." *Medicine and Science in Sports and Exercise* 23, no. 1:93–107.

Coyle, E. F., W. H. Martin, A. A. Ehsani, J. M. Hagberg, S. A. Bloomfield, D. R. Sinacore, and J. O. Holloszy. (1983). "Blood lactate threshold in some well trained ischemic heart disease patients." *Journal of Applied Physiology* 54:18–23.

Davies, C. N. (1961). *Design and Use of Respirators*. Oxford: Pergamon.

Davison, A. C. (1933). "Pedaling speeds." *Cycling* (January 20):55–56.

Day, Frank. (2001). Quotation from a now unavailable Web site, but confirmed in a personal communication with Jim Papadopoulos, July 29, 2001.

Dean, G. A. (1965). "An analysis of the energy expenditure in level and grade walking." *Ergonomics* 8, no. 1:31–47.

De Leener, M. (1970). "Theo's hour record." *Cycling* (March 7):28.

DeLong, F. (1978). *DeLong's Guide to Bicycles and Bicycling*. Radnor, Penn.: Chilton.

Dodge, Pryor. (1996). *The bicycle*. Paris: Flammarion.

Drela, Mark. (1998). "Oxygen uptake, recumbent vs. upright" (technical note). *Human Power* 13, no. 2:17.

England, H. H. (1957). "I call on America's largest cycle maker." *Cycling* (April 25):326–327.

Falls, H. B. (1968). *Exercise Physiology*. New York: Academic Press.

Forester, John. (1984). "The physiology and technique of hard riding." Chapter 25 in *Effective Cycling*, 6th ed., 219–243. Cambridge: MIT Press.

Gaesser, Glenn A., Tony J. Carnevale, Alan Garfinkel, Donald O. Walter, and Christopher J. Womack. (1995). "Estimation of critical power with non-linear and linear models." *Medicine and Science in Sports and Exercise* 27:1430–1438.

Gehan, E. A., and S. L. George. (1970). "Estimation of human body surface area from height and weight." *Cancer Chemotherapy Reports* 54:225–235.

Gollnick, P. D., R. B. Armstrong, W. L. Sembrowich, R. E. Shepherd, and B. Saltin. (1973). "Glycogen depletion pattern in human skeletal muscle fibers after heavy exercise." *Journal of Applied Physiology* 34, no. 5:615–618.

Gollnick, P., D. Karin Piehl, and B. Saltin. (1974). "Selective glycogen depletion pattern in human muscle fibres after exercise of varying intensity and at varying pedaling rates." *Journal of Physiology* 241:45–57.

Grosse-Lordemann, H., and E. A. Müller. (1936). *Der Einfluß der Leistung und der Arbeitsgeschwindigkeit auf das Arbeitsmaximum und den Wirkungsgrad beim Radfahren* (The influence of power and working speed on maximum power and efficiency for bicycles). Dortmund-Münster, Germany: Kaiser Wilhelm Institut für Arbeitsphysiologie.

Hamley, E. J., and V. Thomas. (1967). "The physiological and postural factors in the calibration of the bicycle ergometer." *Journal of Physiology* 191:55–57.

Harrison, J. Y. (1970). "Maximizing human power output by suitable selection of motion cycle and load." *Human Factors* 12, no. 3:315–329.

Henneman, E., G. Somjen, and D. Carpenter. (1965). "Excitability and inhibitability of motoneurons of different sizes." *Journal of Neurobiology* 28:599–620.

Hermansen, L., E. Hultman, and B. Saltin. (1967). "Muscle glycogen during prolonged severe exercise," *Acta Physiologica Scandinavica* 71:129–139.

Hermina, W. (1999). "The effects of different resistance on peak power during the Wingate anaerobic test." M.S. thesis, College of Health and Human Performance, Oregon State University, Corvallis, Ore.

Hill, David W. (1993). "The critical power concept." *Sports Medicine* 16, no. 4:237–254.

Holloszy, J. O., and E. F. Coyle. (1984). "Adaptations of skeletal muscle to endurance exercise and their metabolic consequences." *Journal of Applied Physiology* 56, no. 4:831–838.

Hull, M. L., and M. Jorge. (1985). "A method for biomechanical analysis of bicycle pedaling." *Journal of Biomechanics* 18:631–644.

Hull, M. L., S. Kautz, and A. Beard. (1991). "An angular velocity profile in cycling derived from mechanical energy analysis." *Journal of Biomechanics* 24, no. 7:577–586.

Inbar, O. Bar-Or, and J. S. Skinner. (1996). "The Wingate anaerobic test." *Human Kinetics*. Champaign, Ill.

Jenkins, David G., and Brian M. Quigley. (1990). "Blood lactate in trained cyclists during cycle ergometry at critical power." *European Journal of Applied Physiology* 61:278–283.

Jenkins, David G., and Brian M. Quigley. (1992). "Endurance training enhances critical power." *Medicine and Science in Sports and Exercise* 24, no. 11:1283–1289.

Knipping, H. W., and A. Moncrieff. (1932). "The ventilation equivalent of oxygen," *Queensland Journal of Medicine* 25:17–30.

Kyle, C. R., V. J. Caizzo, and M. Palombo. (1978). "Predicting human powered vehicle performance using ergometry and aerodynamic drag measurements."

Paper presented at conference Human Power for Health, Productivity, Recreation and Transportation, Technology University of Cologne, September.

Lanooy, C., and F. H. Bonjer. (1956). "A hyperbolic ergometer for cycling and cranking." *Journal of Applied Physiology* 9:499–500.

Larrington, Dave. (1999). "Different strokes?" (editorial). *Human Power*, no. 48:25–27.

Martin, J. C., B. M. Wagner, and E. F. Coyle. (1997). "Inertial-load method determines maximal cycling power in a single exercise bout." *Medicine and Science in Sports and Exercise* 29, no. 11 (November):1505–1512.

McArdle, W. D., F. I. Katch, and V. L. Katch. (1996). *Exercise Physiology: Energy, Nutrition, and Human Performance*. Baltimore: Williams and Wilkins.

McMahon, Thomas A. (1984). *Muscles, Reflexes, and Locomotion*. Princeton, N.J.: Princeton University Press.

Morton, R. Hugh. (1994). "Critical power test for ramp exercise." *European Journal of Applied Physiology* 69:435–438.

Morton, R. Hugh, S. Green, D. Bishop, and D. Jenkins. (1997). "Ramp and constant power trials produce equivalent critical power estimates." *Medicine and Science in Sports and Exercise* 29, no. 6:833–836.

Morton, R. Hugh, and David J. Hodgson. (1996). "The relationship between power and endurance: A brief review." *European Journal of Applied Physiology* 73:491–502.

Müller, E. A. (1937). *Der Einfluß der Sattelstellung auf das Arbeitsmaximum und den Wirkungsgrad beim Radfahren* (The influence of saddle height on maximum power and efficiency of bicycling). Dortmund-Munster, Germany: Kaiser Wilhelm Institut für Arbeitsphysiologie.

Newmiller, J., M. L. Hull, and F. E. Zajac. (1988). "A mechanically decoupled 2 force bicycle pedal dynamometer." *Journal of Biomechanics* 21, no. 5:375–386.

Okajima, Shinpei. (1983). "Designing chainwheels to optimize the human engine." *Bike Tech* 2, no. 4:1–7.

Papadopoulos, Jim. (1987). "Forces in bicycle pedaling." In *"Biomechanics in Sport: A 1987 Update,"* ed. R. Rekow, V. G. Thacker, and A. G. Erdman. New York: American Society of Mechanical Engineers.

Papadopoulos, Jim. (1992–1994). Column in *Cycling Science* on many different pedaling styles.

Passfield, L., and J. H. Doust. (1998). "Effect of endurance exercise on 30 s Wingate sprint in cyclists." *Journal of Physiology*, 506P, 100P.

Powell, Richard, and Tracey Robinson. (1987). "The bioenergetics of power production in combined arm-leg crank systems." *Human Power* 6, no. 3:8, 9, 18.

Pugh, L. G. C. E. (1974). "The relation of oxygen intake and speed in competition cycling and comparative observations on the bicycle ergometer." *Journal of Physiology* (London) 241:795–808.

Reiser, Raoul F., Jeffrey P. Broker, and M. L. Peterson. (2000). "Inertial effects on mechanically braked Wingate power calculations." *Medicine and Science in Sports and Exercise* 32, no. 9:1660–1664.

Rowe, T., M. L. Hull, and E. L. Wang. (1998). "A pedal dynamometer for off-road bicycling," *Journal of Biomechanical Engineering, Transactions of the ASME* 120, no. 1 (February):160–164.

Spinnetti, Ramondo. (1987). "Backward versus forward pedaling: Comparison tests." *Human Power* 6, no. 3:1, 10–12.

Thomas, V. (1967a). "Saddle height." *Cycling* (January 7):24.

Thomas, V. (1967b). "Saddle height—Conflicting views." *Cycling* (February 4):17.

Too, D. (1990). "The effect of body configuration on cycling performance." In *Biomechanics in Sports*, vol. 6, ed. E. Kreighbaum and A.-R. McNeill. Stutgart Germany: The International Society of Biomechanics in Sports, University of Stuttgart.

Too, D. (1991). "The effect of hip position/configuration on anaerobic power and capacity in cycling." *International Journal of Sports Biomechanics* 7, no. 4:359–370.

Too, Danny. (1999a). "Crank-arm length" (technical note). *Human Power*, no. 48:17–19.

Too, Danny. (1999b). "Summaries of papers" (technical note). *Human Power*, no. 46:14–20.

Too, Danny, and Chris Williams. (2000). "Determination of the crank-arm length to maximize power production in recumbent-cycle ergometry." *Human Power*, no. 51:3–6.

Van Bolhuis, B. M., W. P. Medendorp, and C. C. Gielen. (1997). "Motor unit firing behavior in human arm flexor muscles during sinusoidal isometric contractions and movements." *Experimental Brain Research*, 117, no. 1 (October):120–130.

"Vandy, the unbeaten king." (1964). *Cycling* (March 11):8.

"Velox." (1869). *Velocipedes, Bicycles, and Tricycles: How to Make and Use Them.* London: Routledge.

Von Döbeln, W. (1954). "A simple bicycle ergometer." *Journal of Applied Physiology* 7:222–224.

Whitt, F. R. (1971). "A note on the estimation of the energy expenditure of sporting cyclists." *Ergonomics* 14, no. 3:419–424.

Whitt, F. R. (1973). Personal communication with D. G. Wilson.

Wilson, S. S. (1973). "Bicycling technology." *Scientific American* 228(3) (March):81–91.

Wirhed, Rolf. (1984). *Athletic Ability and the Anatomy of Motion*. London: Wolfe Medical.

3 Thermal effects on power production (how bicyclists keep cool)

Introduction

Bicycling can be hard work. For each unit of work put into the pedals, a bicyclist must get rid of about three units of heat. It is as important for the body as for any engine that it not become overheated when producing power. We pointed out in chapter 2 that the measurement of the power output of bicyclists on ergometers is open to criticism because the conditions for heat dissipation on ergometers are critically different from those occurring on bicycles. The performances of bicyclists riding in time trials and other long-distance races are, however, very amenable to analysis. Such time trials are of far longer duration than the few hours usually assumed (see, e.g., Willkie 1960) as the maximum period over which data on human power output are available. Time trials (unpaced) are regularly held for twenty-four-hour periods; distances of 775 km (480 miles) are typical. The "Race across America" that has been held at various intervals under different names lasts from five to ten days, and it is usual for aggressive bicyclists to try to gain a forbidding lead over their rivals by riding continuously for over thirty hours at the start.

The relative air flow generated by bicycling is of such magnitude that it bears little resemblance to the drafts produced by the small electric fans often used for cooling people pedaling ergometers. As a consequence it can be said that under most conditions of level cycling, the bicyclist works under cooler conditions than does an ergometer pedaler. At high speeds, most of the rider's power is expended in overcoming air resistance. At 9 m/s (20 mile/h) about 150 W (0.2 hp) are dissipated in the air. The cooling that occurs is a direct function of this lost power. Even if cooling fans of this power level were used for ergometer experiments, the cooling effect would be much less than that for the moving bicyclist, because most of the fan power is dissipated as air friction in areas other than around the subject's body.

The science of "convective" (nonevaporative) heat transfer between a surface and a gas in relative motion is based largely on the analogy between fluid friction and conduction heat transfer derived by Osborne Reynolds in 1874 (see, e.g., Eckert 1963, 134–137). Reynolds' analogy states that the heat transferred between the surface of a body and the "attached" air flowing past is proportional to the air friction at the surface multiplied by the difference of temperature between the surface and the air. Therefore, we can think of surface air friction as partly useful in terms of cooling, at least in warm weather. However, much of the air friction that slows a bicyclist occurs as eddies in the wake behind the body. These are therefore in "separated" or "unattached" flow, and they do not contribute to heat transfer in any way. (See chapter 5 for a fuller description of types of air flow.)

Reynolds' contributions to aerodynamics and heat transfer are epitomized by his finding that the flow of fluids around bodies and in ducts is correlated with a nondimensional number that has become known as the Reynolds number. It is composed of the relative velocity of the fluid multiplied by its density and a length (e.g., the duct diameter) and divided by the fluid viscosity, all in a consistent set of units. The viscosity of air is very small in any system of units, so that Reynolds numbers of all but "creeping" flows are usually in the millions. For our bicycling purposes, we can regard the Reynolds number as a measure of speed. Very low Reynolds numbers produce "sticky" or "viscous" flow, as would be obtained if you moved a spoon through syrup. This type of flow is referred to as "laminar." Higher Reynolds numbers produce "turbulent" flow, in which the flow immediately against solid surfaces breaks spontaneously into small intense vortices that greatly increase the air friction and the heating or cooling of the surfaces. In most weather conditions we live in the turbulent "boundary layer" of the wind against the ground, trees, buildings, and so forth, and we become used to the continual vortices of the wind.

The effect of adequate cooling may be inferred from Wilkie's finding, from experiments with ergometer pedalers, that if it is necessary to exceed about half an hour's pedaling, the subject must keep her or his power output down to about 150 W (0.2 hp). However, peak performances in twenty-four-hour time trials can be analyzed using wind-resistance and rolling-resistance data from chapters 5 and 6 to show that about 225 W (0.3 hp) are being expended over that period (see data in chapter 2). It seems that the exposure of the pedaler to moving air is principally responsible for the improvement in cooling. It is also known that an ergometer pedaler who attempts a power output of about 375 W (0.5 hp) in normal laboratory ambient temperatures can expect to give up after perhaps ten minutes and will be perspiring profusely. That is the same power output required to propel a racing bicyclist doing a "fast" 40-km (25-mile) distance trial of nearly one hour. Again the effect of moving air upon a pedaler's performance is very apparent.

Local and mean heat transfer

At moderate temperatures it seems to matter little how and where the heat is removed from the human body to prevent overheating. The intensity of heat transfer, measured by the heat-transfer coefficient, varies a great deal from place to place on the body, however. The most intense heat transfer affecting a bicyclist is normally that at the front of the body and at the limbs, the head and so on. Each such point is called a "stagnation point": it is where the relative air flow comes momentarily to rest before accelerating in one direction or the other around the body. (For transverse flow around cylinders and other two-dimensional bodies, there will be a "stagnation line" rather than a stagnation point.) Winter skiers are apt to acquire frost-bite first on the fronts of their noses, where there is a stagnation point. If they travel at higher than 10–15 m/s (22–33 mile/h), however, the most intense local heat transfer will be where the smooth laminar flow under-goes a transition to turbulent flow and a sharp increase in both friction and heat transfer, as will be seen below. (Also see chapter 5.)

The flow around human bodies on utility bicycles is frequently mod-eled as the flow around a circular cylinder. The variations of the pressure and the heat-transfer coefficient around a cylinder in cross-flow are shown in figure 3.1. The local pressure is at a maximum at the stagnation point and falls to a minimum at between 70° and 85° around from that maxi-mum. The location of this minimum depends on the relative speed of the airflow to the body, represented by the Reynolds number. At this point the pressure begins increasing, and the thin flow near the surface, the boundary-layer flow, becomes sharply turbulent. The scrubbing action of the vortices generated by this turbulence greatly increases the intensity of local heat transfer, and at the higher speeds the starting value of the tur-bulent heat transfer will exceed that at the stagnation point.

Heat-transfer data and deductions

Let us examine the literature for suitable correlations with established heat-transfer data in order to find quantitative reasons for the above observations. There is no published information concerning heat-transfer experiments with riding bicyclists, and it is therefore necessary to make calculations with suitable approximations of a cyclist's shape. The approx-imate forms used are a flat plate and a circular cylinder. In addition, data from experiments with actual human forms (Nonweiler 1956; Colin and Houdas 1967; Clifford, McKerslake, and Weddell 1959) can be examined, although the flat and upright postures employed in these experiments were not those of bicycling.

The results of many calculations using established correlations for both convective and evaporative heat transfer are given in figure 3.2. Also shown is the heat evolution of a rider at various speeds and power outputs on the level. The figure indicates that the effect of shape on the heat flux for a given temperature difference is not great in the case of convective heat transfer. In the case of evaporative transfer, the difference between results with models and with an actual human body is 20 percent. It appears that a midway value can be obtained from data on cross-flow over a wetted cyl-inder of 150-mm diameter or over a plate. For the same driving potential,

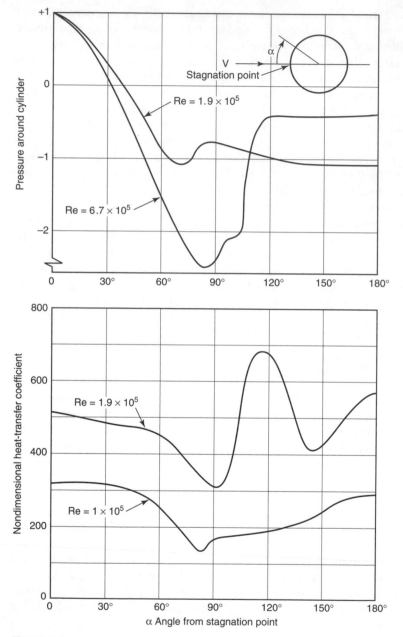

Figure 3.1
Cross-flow around a circular cylinder. (From various sources, including Eckert 1963.)

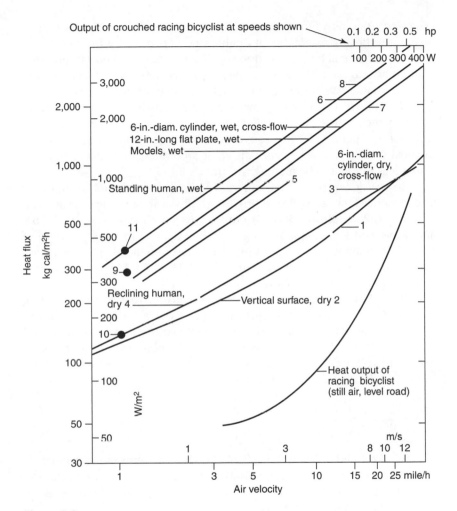

Figure 3.2
Convective and evaporative heat flows.

expressed as water-vapor pressure or temperature difference, the rate of evaporative heat transfer is about double that of convective transfer.

Under normal free-convection conditions (that is, when the airflow is induced by buoyancy alone), data given by Rohsenow and Choi (1961) and Cox and Clarke (1969) lead to the conclusion that air must move at about 0.5 m/s (1.5 ft/s) or more to provide cooling. This is supported by figure 3.2, in which line 6, for forced convection over a cylinder at 0.5 m/s (1.5 ft/s), and point 9, for free convection, both predict about 325 W/m² as the heat flux for that air speed. This value would be greater for a bicyclist, whose legs would also be moving the air.

In the design of heating and ventilating plants, the maximum heat load produced by a worker doing hard physical labor has long been accepted as 2,000 Btu/h (586 W) (*Kempe's Engineer's Year Book* 1962, 761, 780; Faber and Kell 1943). This figure, when applied to a body surface of 1.8 m², also amounts to 325 W/m². It is recommended that such hard work be done at a room temperature of 55°F (12.8°C). Most of the heat is lost through evaporation of sweat.

The above evidence leads to the conclusion that a rider pedaling in such a manner that her or his body gives out a total of 2,000 Btu/h (586 W), in average air conditions where free convection holds, does not suffer a noticeable rise in body temperature no matter how long she or he works. If the pedaler's efficiency is 25 percent, the power output W for a heat loss Q is calculated as follows:

$$0.25 = \frac{W}{W + Q} = \frac{1}{1 + Q/W};$$

therefore $W = Q/3 = 195$ watts $= 0.26$ hp. Thus, it seems that a pedaler on an ergometer working for long periods and not "driven" by strong inducements produces only about 150 W (0.2 hp) because of an unwillingness to tolerate a noticeable rise in body temperature.

In chapter 2 it was shown that many cyclists can exert 373 W (0.5 hp) for periods of up to an hour. According to Japanese data (Bicycle Production and Technical Institute 1968), that corresponds to a speed of about 27 mile/h (12.2 m/s). At that speed the heat flow from the moving bicyclist is about 707 W/m² (figure 3.2). If the cyclist exerts 0.5 hp (373 W) pedaling on an ergometer, all the heat lost by convection and evaporation in moving air—all of the heat in excess of 325 W/m²—must be absorbed by the pedaler's body. Thus, the ergometer pedaler with a body area of 1.8 m² absorbs $(707 - 325) \times 1.8 = 688$ W if the small heat losses through breathing are neglected. If the pedaler weighs 70 kg and has a specific heat of 3.52 J/g/°C, and if a rise in body temperature of 2°C is acceptable before physical collapse, the tolerable time limit for pedaling at 0.5 hp is

$$\frac{70{,}000 \times 3.52 \times 2}{688 \times 60} \cong 12 \text{ minutes}$$

For highly trained racing bicyclists attempting to pedal ergometers at a power output of 0.5 hp (373 W), a common range of endurance is five to fifteen minutes (Whitt 1973). Hence, the above estimates have some validity. The fact that all the racers observed were capable of outputs of nearly 0.5 hp (373 W) in one-hour time trials demonstrates vividly the value of flowing air in prolonging the tolerable period of hard work.

Experimental findings supporting the vital importance of cooling in human-power experiments are given in a paper by Williams et al. (1962) concerning the effect of heat on the performances of ergometer pedalers.

Minimum air speed

From figure 3.2 it can be seen that a racing bicyclist producing 0.6 hp (450 W) emits heat at about 850 W/m². According to curve 5 in the figure, heat generated at such a rate could be absorbed by air moving at about 3 m/s (7 mile/h). Verification of the value of this prediction is found in an ascent (by Bill Bradley) of the Gross Glockner hill climb. He rode at about 5.4 m/s (12 mile/h) at a power output of 450 W (0.6 hp), demonstrating that it is not necessary to have a road speed above 12 m/s (27 mile/h) for riding at 450 W, when nonevaporative heat transfer can cool if the air is at a lower temperature than the body. Bradley's ride was completed in high-air-temperature conditions, but these were well compensated for by a low atmospheric humidity of about 40 percent; despite the high temperatures, he could sweat freely and achieve efficient evaporative heat loss.

Chester Kyle and coworkers at California State University carried out extensive trials with streamlined fairings for riders of various machines (Kyle 1974). An interesting outcome was that, even on short rides, a fairing that was skirted almost to ground level caused the rider to overheat grossly, almost certainly because of a lack of air flow over the rider. This problem seems to have been appreciated even with the earliest fairings for bicyclists, developed in the early 1900s. Bryan Allen also suffered from overheating in the pedaled airplane Gossamer Albatross because of insufficient through-ventilation and insufficient water during his nearly three-hour flight across the English Channel on June 12, 1979.

Bicycling in cold and hot conditions

A problem faced by advocates of bicycling as a means for daily commuting to and from work is that even temperate regions have days, and sometimes weeks, of extreme weather conditions during which bicycling may be unpleasant for many and impossible for some. There is no one set of temperature boundaries at which bicycling becomes impossible. Many "fair-weather" cyclists put their machines away for the winter when the morning temperatures drop to 10°C (50°F) and will not ride in business clothes at temperatures above 25°C (77°F). However, many hardier folk find bicycling to be still enjoyable at −15°C (5°F). The main problem at temperatures below this seems to be the feet. The size of insulated footwear is limited to that which can fit on bicycle pedals, and it is fairly common experience that, at −18°C (0°F), even when the trunk of the body is becoming overheated through exertion, the feet can become numb with cold.

The effects of cold air are intensified by wind. Weather forecasters often express these effects in terms of the "windchill factor": the air temperature that would have to exist, without wind, to give the same cooling to a human body as a particular combination of actual temperature and actual relative wind. The windchill factors tabulated by the U.S. National Weather Service, National Oceanic and Atmospheric Administration are plotted in figure 3.3. Using this chart, one can find the effect on a rider's

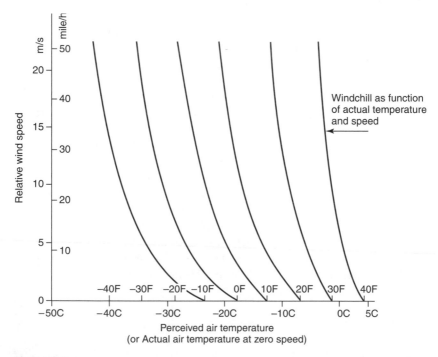

Figure 3.3
Windchill factors. (From data from the U.S. National Weather Service, plotted by Dave Wilson.)

perceived temperature of bicycling into a relative wind. For instance, if the air temperature is $-18°C$ ($0°F$) and one is bicycling into a relative wind of 5 m/s (11 mile/h), one is subjected to the same amount of cooling as if one were in calm conditions at a temperature of $-30°C$ ($-22°F$). The feet are periodically traveling at a higher relative velocity (as they come over top dead center) and then at a lower velocity relative to the wind. Because the cooling relationship to relative wind is nonlinear, the average cooling effect seems to be more severe. (The formula for calculating windchill was modified in 2001, but the relative effects of wind speed on temperature perception remain unchanged.)

At higher temperatures, humidity becomes very important. The bicycle is highly prized for personal transportation and for local commerce throughout Africa and Asia. In northern Nigeria, for example (where the author lived for two years), the air is so dry throughout most of the year that one's range on a bicycle is limited more by the availability of water than by the temperature. The long-distance bicyclist Ian Hibell rode through the Sahara (principally at night), limited again by his water supplies. He could not carry sufficient water for the longer stages between oases and relied on gifts of water from passing motor travelers.

During the record heat wave of July 1980, Houston, Texas, had over four weeks of temperatures over $100°F$ ($38°C$), coupled with very high humidity, with $111°F$ ($44°C$) reached on several occasions. Yet some bicyclists continued to ride to work. What makes this especially remarkable is that on American roads, crowded with cars, trucks, and buses with air conditioners going at their maximum, the ambient temperature that bicyclists experience must be far above the local off-highway values.

There are three lessons to be learned from the experience of the hardier riders who brace themselves for cycling against what seem to be extreme conditions. First, the promotion of good circulation through exertion helps the body cope with high temperature and high humidity as well as with cold weather. Second, the relative airflow that bicycling produces is a major factor in making riding in hot weather tolerable and usually enjoyable. Third, the fact that so many riders choose to bicycle in extreme conditions (rather than being forced to do so by economic necessity) shows that many other healthy but more timid cyclists could push their limits with regard to conditions conducive to or comfortable for cycling without fear of harm.

Physiology of body-temperature regulation

Falls (1968) includes a survey of recent experimental work on the complex processes involved in body-temperature regulation and a large bibliography.

Heat-transfer comparison of swimming, running, and bicycling

Swimmers are believed to maintain 65 percent of top velocity for one hour, runners only 55 percent, as deduced from comparisons of record speeds of athlete swimmers and runners. Such figures show that bicyclists maintain even higher degrees of efficiency than swimmers.

Water is a far better heat-removal fluid than air. Thus, with appropriate water temperatures, a swimmer can keep cool more easily than a runner. It can therefore be concluded that the swimmer uses a smaller proportion of his cardiac output to dissipate heat, and a larger proportion to transport oxygen to the muscles, than a runner. This statement appears to be just as appropriate to a bicyclist as to a swimmer, in comparison with a runner.

Conclusions

The heat-removal capacity of the air surrounding a working human is a key factor in the duration of his effort. Static air conditions are apparently such that at low air speeds with free convection, air is capable of removing 585 W (2,000 Btu/h) from the average body surface. Hence, if more heat than that is given out from working at rates higher than about 150 W (0.2 hp), the body temperature rises. (An ambient temperature of 55°F, or 12.8°C, is assumed.)

The fast-moving air around a bicyclist traveling on the level can be estimated to have a heat-removal capacity much greater than that of the stationary air surrounding an ergometer pedaler. Quantitative estimates can be made using established heat-transfer correlations based on air flow over wet cylinders of about 150-mm diameter (in cross-flow) (Sherwood and Pigford 1952, 70, 87–89) or from data given by Clifford, McKerslake, and Weddell (1959) on air flow over a standing person who is perspiring.

The heat-removal capacity of the air around a moving cyclist, at most speeds on the level, is such that much more heat can be lost than the amount produced by the cyclist's effort. Hence, a rider can wear more clothing than the amount that would be tolerable to a static worker giving out the same mechanical power.

Some speculations

At least two ergometers used for testing the power capacities of racing bicyclists have incorporated air brakes in the form of fans. However, few manufacturers appear to have thought of directing the air from such air brakes onto the body of the pedaler and measuring the effect of the fast-moving air on performance. It is improbable that an air flow from such an arrangement could give anything very much above, say, half the flow rates surrounding an actual riding bicyclist giving out the same power, but the results of tests conducted under such conditions would be interesting.

Pedaling on an ergometer out of doors should result in higher power output than pedaling on the same ergometer at the same speed indoors. Even in calm conditions, air is likely to be moving faster than the 0.5 m/s (1.5 ft/s) quoted above for free-convection conditions around a heated body.

In view of the fact that, at 150 W (0.2 hp) output, to maintain tolerable body temperatures, the body must get rid of its heat through an evaporative process, indoor exercise seems rather unhealthful compared with riding a bicycle in the open air. Maybe the designers of home exercisers should put less emphasis on instrumentation and more on self-propelled cooling equipment.

References

Bicycle Production and Technical Institute. (1968). Report. Japan.

Clifford, D., D. McKerslake, and J. L. Weddell. (1959). "The effect of wind speed on the maximum evaporative capacity in man." *Journal of Physiology* 147:253–259.

Colin, J., and Y. Houdas. (1967). "Experimental determination of coefficient of heat exchanges by convection of the human body." *Journal of Applied Physiology* 22, no. 1:31–38.

Cox, R. N., and R. P. Clarke. (1969). "The natural convection flow around the human body." *Quest*:9–13.

Eckert, E. R. G. (1963). *Introduction to Heat and Mass Transfer.* New York: McGraw-Hill.

Faber, O., and J. R. Kell. (1943). *Heating and Air Conditioning of Buildings.* Cheam, U.K.: Architectural Press.

Falls, H. B. (1968). *Exercise Physiology.* New York: Academic Press.

Kempe's Engineer's Year Book. (1962). Vol. 11. London: Morgan.

Kyle, C. R. (1974). "The aerodynamics of man-powered land vehicles." Paper presented at Third National Seminar on Planning, Design, and Implementation of Bicycle and Pedestrian Facilities, San Diego, Calif.

Nonweiler, T. (1956). "Air resistance of racing cyclists." Report no. 106, College of Aeronautics, Cranfield, U.K.

Rohsenow, Warren M., and Harry Y. Choi. (1961). *Heat, Mass and Momentum Transfer.* Englewood Cliffs, N.J.: Prentice-Hall.

Sherwood, T. K., and R. L. Pigford. (1952). *Absorption and Extraction*. New York: McGraw-Hill.

Whitt, Frank R. (1973). Personal communication with D. G. Wilson.

Williams, C. G., et al. (1962). "Circulatory and metabolic reactions to work in heat." *Journal of Applied Physiology* 17:625–638.

Willkie, D. R. (1960). "Man as an aero-engine." *Journal of the Royal Aeronautical Society* 64:477–481.

Recommended Reading

Craig, A. G. (1963). In *Journal of Sports Medicine and Physical Fitness* 3:14.

Edholm, O. G. (1967). *The Biology of Work*. New York: McGraw-Hill.

Nadel, E. R. (1977). *Problems with Temperature Regulation during Exercise*. New York: Academic Press.

Rees, W. H. (1969). "Clothing and comfort." *Shirley Link* (Summer):6–9.

"World champion Hugh Porter drops only 3.2 seconds over 5 kilometres wearing Trevira jersey, and clothes." (1973). *Daily Mail*, May 9. See also issues of May 8, 11, and 12.

II SOME BICYCLE PHYSICS

4 Power and speed

Introduction

One of the first lessons we learn from bicycling is that more effort is required to ride fast, or uphill, or against the wind (than to ride at a more moderate speed on the level in calm conditions or with the wind at one's back). The power available from the pedaler for various durations was characterized in chapter 2. In this chapter we discuss what speed will be achieved at a given power level and under what circumstances cycling will be perceived as difficult. The chapter introduces the various kinds of drag, some of which are treated more fully in later chapters. It also explores the potential for a rider to increase his speed.

The object of pedaling, in scientific terms, is to exert a propulsive force (F_P) against the ground. To maintain a constant speed, the average magnitude of that force must equal the total force resisting forward motion. The force is composed of

- air resistance (F_A), from the motion of the bicycle relative to the air (the relative air speed is bicycle speed (V) relative to the ground, plus headwind speed (V_W) relative to the ground);[1]
- slope resistance (F_S): what one would measure in terms of resistance if stationary on a hill, restrained from rolling by a spring scale parallel to the road surface;
- rolling resistance (F_R) from deformation and friction of the rolling rubber tire (also from deformation of the ground, if it is soft); and
- average bump resistance (F_B) on rough surfaces: mounting a bump immediately reduces forward velocity, and descending the other side of the bump restores only part of it. Most of the lost energy is dissipated in the rider's body.

Any propulsive force in excess of (or less than) the sum of these resistances will accelerate (or decelerate) the bicycle plus rider. When a rider briefly exerts a force more than (say, two to four times) that needed for propulsion, there results a brisk acceleration (a) of the system mass (m):

$$F_P - (F_A + F_S + F_R + F_B) = F_{Acc} = ma.$$

(The mass m is so large that even a "brisk" acceleration is never very great.) This excess force is termed the "acceleration force" (F_{Acc}).

A bicycle's gearing system is a way of altering the "leverage" or mechanical advantage between the rider's foot and the ground. The ratio of ground velocity to pedal velocity[2] (if friction is disregarded, this is the inverse of the ratio of ground propulsive force to foot force) can be calculated using the following formula:

$$\frac{ground\ velocity}{pedal\ velocity} = \frac{R_W}{crank\ length} \times \frac{front\ sprocket\ teeth}{rear\ sprocket\ teeth}.$$

The velocity ratio typically ranges between 1.5 and 7.5, corresponding to a force ratio of 0.67 to 0.13.

In principle, by choosing a low enough velocity ratio (equivalently, a great enough leverage or force ratio), with a given foot force a rider could exert any desired magnitude of propulsive force F_P. But a high force magnitude does not necessarily equate to a high riding speed! In addition to exerting force, the pedaler is providing power, usually just a few tens of watts.

The power level exerted by a cyclist equals the product of (wheel) force and (forward) velocity. Therefore, in steady riding at a given power level, a high force is possible only when the velocity is low. Ultralow gear ratios are useful for extremely steep slopes, thick mud, or strong headwinds, situations in which high resistive force is present at low speed. But in level riding of a typical bicycle, the force required for propulsion is quite low at slow speeds and would become very large (above 180 N) only at speeds above 30 m/s (see "Air Resistance"). Achievement of such a great propulsive force, multiplied by the high speed needed to achieve it, would require exertion of a tremendous power (5400 W), far exceeding any rider's capabilities.

Determining a cyclist's achievable speed with given levels of power requires an understanding of what propulsive power is required for every possible speed the rider can achieve. The power available to the rider dictates what speed is possible. (It is assumed that the rider will adjust his gear ratio appropriately to permit pedaling at his preferred cadence while traveling at the given speed.) The speed-determining process can be demonstrated both graphically and in equation form.

In this discussion, power is defined as power delivered by the driving wheel, $\dot{W}_W$. This is slightly less than the power $\dot{W}_R$ the rider produces by pedaling (which is equal to foot force times foot velocity), because of losses through transmission inefficiency (chain-joint friction, rubbing due to misalignment, bearing losses, etc.).[3] Such losses will be discussed in more detail in chapter 9. We are making an approximation here in taking propulsive power to be the same as rider power. When greater precision is required, the equation to use is $\dot{W}_W = \dot{W}_R \times \eta_m$, where the transmission efficiency η_m is usually between 0.85 and 0.97 (Kyle and Berto 2001; Spicer et al. 2000).

Air resistance

In riding on smooth, level pavement in still air, at or below jogging speeds (3 m/s), the main resistance acting on a bicycle is the rolling friction of the tires (see below). But as the cycle's forward speed is increased, aerodynamic drag grows quickly and becomes far more important as a source of resistance.

The kind of air resistance most relevant to bicycling can be envisioned as consisting of two main components. One is that of pushing into and accelerating the air directly ahead of the rider, or bluff-body pressure drag. The other is that of sliding past the air, or skin-friction drag.

Air has a density of roughly one kilogram per cubic meter. (See chapter 5 for a more precise measurement.) Each cubic meter of air "swept" by the cyclist is roughly comparable to colliding with a liter (or quart) container of milk: a parcel of air of that magnitude is brought roughly to the vehicle's speed, then is pushed aside.

The force the cyclist must exert on the air is the mass encountered per unit time, multiplied by its change in velocity. In elementary physics texts, this kind of calculation is referred to as "momentum conservation in steady flow." The mass being struck by the rider per unit time is proportional to the product of air density ρ, velocity V, and frontal area A, as indicated in figure 4.1. The speed with which this mass is struck is also proportional to velocity. The resulting expression approximates air drag as $\rho V^2 A$, and the force exerted by the drag is seen to increase as the *square* of the velocity.

A more careful analysis (to be found in any text on fluid dynamics) includes a factor 0.5 in recognition that streamlining allows the cyclist to "penetrate" the air without striking so much of it: the force expression includes a drag coefficient C_D that is usually less than one. This coefficient will be discussed in much more detail in chapter 5. For now, we'll simply express aerodynamic drag force as aerodynamic-drag factor K_A times relative air velocity squared. K_A is defined as $C_D A\rho/2$.

In standard international units (SI), the drag factor K_A is in units of kg/m, or equivalently $N/(m^2/s^2)$. For a rider on a bicycle, the drag factor is typically between 0.1 (small person, recumbent position, snug clothing, in hot, low-pressure, humid air) and 0.3 (large person sitting upright, with bulky loose clothes, in cold, high-pressure, dry air). For a streamlined vehicle with substantial laminar (low-drag) flow enclosing a prone rider, K_A might be as low as 0.01.

As an example, consider a rider with drag factor $K_A = 0.2$, traveling at a speed of 10 m/s (22.4 mile/h). The aerodynamic drag force at that speed is 0.2×10^2 or 20 N (4.5 lbf). The power level is 20 N $\times$ 10 m/s = 200 W, which can be sustained for hours by a fit cyclist (see figure 2.4).

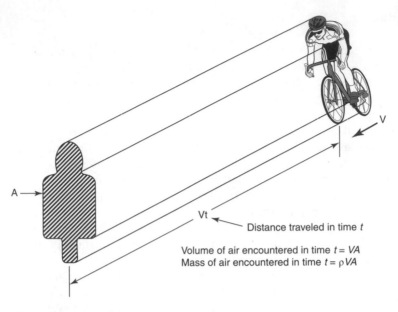

A →

V

Vt

Distance traveled in time *t*

Volume of air encountered in time *t* = *VA*
Mass of air encountered in time *t* = ρ*VA*

Figure 4.1
Mass of air encountered per second in cycling.

Only when the bluff-body drag is virtually eliminated by a stream-lined fairing (smooth, with a rounded nose and tapered afterbody) is the effect of skin-friction drag (intrinsically much lower in magnitude) worth considering. Skin friction can be reduced by minimizing the fairing's surface area, improving its surface smoothness, and trying to forestall turbulence in the thin layer of fluid flowing along the surface.

Conclusions on air resistance

The main conclusion of this elementary discussion is that aerodynamic drag force is proportional to the square of the velocity V relative to the air. (If a headwind V_W is present, the force involves the square of $[V + V_W]$.) Thus, if the drag force at 10 m/s is 20 N, then at 5 m/s it is only about 5 N (1 lbf), whereas at 50 m/s it is close to 500 N (112 lbf). The aerodynamic forces developed at high wind speeds are therefore so great that they can knock over a person, lift an airplane, or destroy a building.

Speed achieved at a given power
Power equals drag force times speed. The simplest way to see what speed can be maintained for a given rider power is to plot the power required at

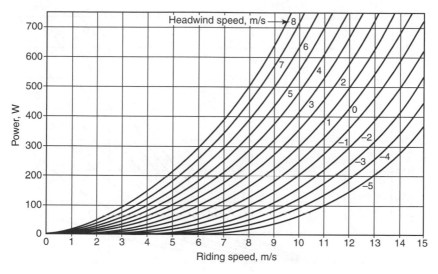

Figure 4.2
Air-drag power for a rider of $K_A = 0.25$ at various headwind speeds. (Plotted by Jim Papadopoulos.)

each riding speed. Initially, we make the approximation that of the forces of resistance that the cyclist encounters, only air drag is important. This is fairly accurate above speeds of about 7 m/s, if streamlining is poor and the road is level. Then the expression for drag power is $\dot{W}_W = K_A V(V + V_W)^2$, where V_W is headwind speed.

Figure 4.2 shows air-drag power as a function of speed for a fairly large rider in a nonaerodynamic position ($K_A = 0.25$) encountering various headwind speeds. Figure 4.3 shows air-drag power as a function of speed for various aerodynamic drag factors. It is obvious that to achieve high speeds on a level road at any given power requires a low value of K_A. Unfortunately, after the modest improvements afforded by tight clothing (which lessens C_D), a "racing crouch" (which lessens A), and high altitude (which lessens ρ), the only significant ways to diminish drag further are a recumbent position (i.e., minimum A) and a streamlined fairing (minimum C_D).

Slope and rolling resistance

Slope resistance
In typical level-road riding, aerodynamic drag is the most important source of resistance that the rider encounters. However, on noticeable hills, resistance from slope becomes the main factor in the resistance acting against

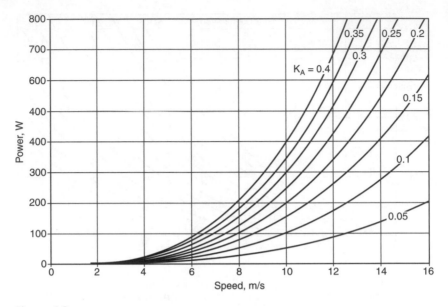

Figure 4.3
Air-drag power versus speed for various aerodynamic drag factors K_A. (Plotted by Jim Papadopoulos.)

the rider, in part because the rider slows on hills, dramatically reducing the magnitude of the air drag. Figure 4.4 shows at what point slope becomes the dominant part of the drag, at various power levels, for one particular rider.

In contrast, tire rolling resistance is never very great, so the only time it provides most of the drag acting on a rider is at low speeds on level surfaces.[4] This is often the situation for very casual low-power (50 W) pedalers, who also tend to have inexpensive, high-resistance tires. As will be shown below, rolling resistance and slope play similar roles to one another. Therefore figure 4.4 can be used to show the effect of rolling resistance, if the vertical axis is read as rolling resistance rather than slope (consider values around 0.006).

Slope resistance (F_S) is based on the weight (mass × gravitational acceleration) (mg) of a bicycle and rider and the slope of the hill up which they are traveling. Steepness can be defined either as an angle or as a ratio of altitude increase per unit distance traveled. This ratio is referred to as the slope (s). (Often slope is expressed as a "percent grade" [$s_\%$].)[5] Whether distance traveled is measured horizontally, or along the slope, makes little difference for ordinary hills. (The old British designation of a slope as being, for example, "one in four" referred to the distance traveled along

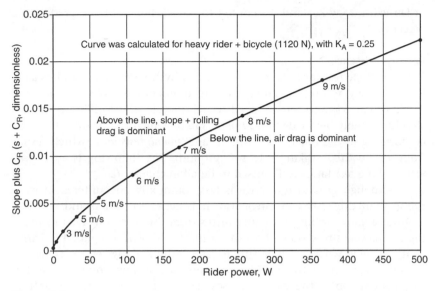

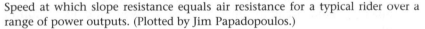

Figure 4.4
Speed at which slope resistance equals air resistance for a typical rider over a range of power outputs. (Plotted by Jim Papadopoulos.)

the road per unit of height.) Slope resistance is measured based on the vertical weight vector of the bicycle and rider not being perpendicular to the road—what is relevant is its component parallel to the road surface: $mg \sin(inclination\ angle)$, which is equal to mgs under the above definition. Slope resistance F_S is constant; that is, it does not vary with speed. Hill-climbing power, or $F_S V$, is therefore proportional to speed.

A slope of 0.001 (i.e., a grade of 0.1 percent) can barely be detected by human senses. Typical modest hills have slopes ranging up to 0.03 (3 percent). A hill with a slope of 0.06 (6 percent) is considered significant, one with a slope of 0.12 is hard to ascend, and some roads have brief stretches on which slopes reach 0.18 or even 0.24. The slope of rough terrain can exceed this, but tire-to-track friction must be good in order to permit climbing or braking on such terrain.

Two places in the United States that are infamous for the steepness of their slopes are Mt. Washington in New Hampshire, with an average slope of 0.115 (grade of 11.5 percent) over 12.2 km, and one block of Filbert Street in San Francisco, with a slope of 0.315 (grade of 31.5 percent.) The author remembers riding up a hill with a maximum slope of 1 in 3.5 (grade of 30 percent), possibly Porlock Hill in Devon, United Kingdom, on a three-speed heavy bike (i.e., one having a low gear of around 36″). Baldwin Street in Dunedin, New Zealand, is reputed to have a slope of 0.33 (grade of 33 percent).

Rolling resistance

Whereas slope resistance is rooted in basic physical laws and can be calculated precisely, rolling resistance is founded on empirical observations that it takes some force to roll a loaded wheel. Bicycle-wheel rolling resistance should probably be divided into *tire resistance* (which results from the tire's conforming to the much harder road) and *ground resistance* (which results from a hard tire's sinking into soft ground). The following is an overall view: rolling resistance is discussed in detail in chapter 6.

Although ground resistance is less commonly encountered in most riding than tire resistance, in a way it is easier to understand. On soft ground or snow, rolling resistance arises from the work involved in pressing the bicycle's tires down into the surface as shown in figure 4.5. Large-diameter wide wheels reduce this resistance by sinking less to achieve a footprint for which the load-bearing pressure is low enough that further penetration of the ground ceases. The drag force exerted by ground resistance at low speeds is approximately the ground-contact pressure (essentially, ground strength) times the area of a vertical cross section of the rut.

Figure 4.5
Appearance of some grades.

Tire rolling resistance is more mystifying. It is most conspicuous when one is riding on training rollers. Exacerbated by the small-diameter rollers, tire rolling resistance accounts for virtually all of the drag in such a situation. Even when one is riding on the road, it is far greater in magnitude than the bearing drag and the aerodynamic drag of the rotating spokes. It evidently arises from two factors.

• Energy loss within the materials of construction. When a rubber inner tube, tread, or sidewall is bent or stretched through the application of a force, it doesn't spring back with the same force: some energy has been transformed into heat. This loss goes by names such as hysteresis, viscoelasticity, or relaxation. It usually depends strongly on the rate or frequency of deformation and on the tire temperature.

• Energy loss due to rubbing of two materials (inner tube against tire at very low pressure, tire tread against road, or perhaps one textile fiber against another in the tire cords).

Both ground resistance and tire resistance are found to increase when additional load is carried. As a rough empirical description of all rolling-resistance factors, it is usual to define the rolling force as weight carried times a coefficient of rolling resistance: $F_R = C_R mg$. If m is given in kg, then g (at sea level) is 9.807 m/s^2, and F_R is in newtons. Loosely, F_R can be described as a fraction of a percent of system weight.

There is no particular reason to think that the force of rolling resistance should be exactly proportional to weight, nor that it should be independent of velocity, as this expression implies. Unfortunately this is an area in which too few careful measurements have been made, so there is no good alternative to using this expression. Tire resistance (i.e., rolling resistance on hard roads) is described by C_R as low as 0.002 for high-quality racing tires at high pressure and as great as 0.008 for utility tires at low pressure. Even greater values may be experienced with a smaller-than-usual wheel radius.

The expression for F_R, namely, a fixed number times system weight, is similar to that for F_S. In other words, the rolling-resistance coefficient should "feel like" a simple augmentation of the slope. For example, a slope of 0.03 and a rolling-resistance coefficient of 0.004 can be combined into a corresponding apparent slope of 0.034. The resistive force is $0.034mg$, and the power is $0.034mgV$. In most equations and graphs, slope s and rolling resistance coefficient C_R can be combined in this way, into an apparent or "effective slope."

Figure 4.6 provides the power required for a given rider (aerodynamic drag factor 0.2) at various speeds, to cycle up effective slopes of 0.02 (2

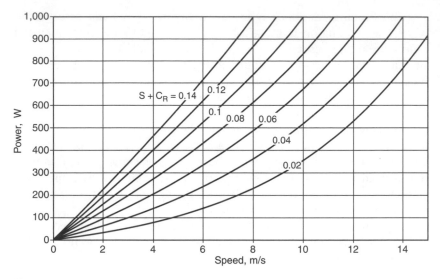

Figure 4.6
Power requirement versus speed and slope. (Plotted by Jim Papadopoulos.)

percent) through 0.10 (10 percent). It can be seen that these curves depart from straight lines only when air resistance becomes significant, that is, at higher powers and lower grades.

Rolling resistance from bump losses
It is unquestioned that bumps, when encountered, retard forward progress. However, we are unaware of any good measurements clearly delineating the magnitude of the resistance bumps present for bicycling.

Bumps cause energy loss in various ways.

▪   If a bump is gentle and causes neither loss of wheel contact with the ground nor shock due to sudden onset (whether either of these occurs depends strongly on speed), then it need cause no energy loss. However, the energy required can be affected by rider body motions: a floppy upper body will absorb energy when a bicycle on which it is riding encounters a bump, at least in comparison to a firmly braced torso. On the other hand, a rider can actually add to the energy of forward motion by "pumping" appropriately over the bump. The phasing is critical in determining whether this can occur. The rider's mass must be at its lowest (relative to the bicycle) and accelerating upward while the bicycle is on a downslope.
▪   If a bump is gentle but causes either wheel to leave the ground, a sudden shock is likely when the wheel regains contact, unless the ground

slope is coincidentally parallel to the landing direction, like a ski jump, allowing the system to recapture the kinetic energy transferred into vertical motion. The shock causes energy loss partly through the vibration of the bicycle on its tires that the shock induces, but mostly through the vibration of the rider's body that results from the shock.

·   If a bump has a sudden onset, whether or not contact is lost, it will cause the rider's body (and also the tires) to undergo vibrational deformation, leading to substantial energy loss.

A series of papers on human vibration written by Pradko and collaborators (Pradko and Lee 1966; Pradko, Lee, and Kaluza 1966) at the U.S. Army Tank-Automotive Center correlated the rate of energy absorption in the human body (i.e., lost power) with a hard seat's vibration amplitude and frequency, as shown in figure 4.7. The authors also established that energy absorption correlated well with perceived discomfort on the part of the rider. *This is a very important result, because it implies that improving vibration comfort will also reduce losses of energy that result from encountering bumps.*

Based on data in Pradko and Lee 1966, vibrational power absorbed by the test subjects (in W) may be represented very approximately by a three-part function (frequency in Hz, and displacement amplitude in mm):[6]

·   For vibration frequencies between 1 Hz and 5.5 Hz, $W = (Hz^6/1000)(mm^2)$.
·   For vibration frequencies between 5.5 Hz and 9 Hz, $W = 28(mm^2)$.
·   For vibrational frequencies in the range of interest to bicyclists, between 9 Hz and 50 Hz, $W = (Hz^{2.6}/10.75)(mm^2)$.

The highest power absorption charted by the researchers is 2000 W, and the lowest is 2.7 W. Evidently, intense bumpiness can dissipate thousands of watts and potentially slow a speeding bicycle in seconds. More widely relevant is the case of a road on which bump losses approach or exceed rolling resistance (say, 30 W).

Of course, riding over most bumpy surfaces will not generate a single frequency of vertical seat movement. A spectrum of vibration frequencies is expected, perhaps with fairly distinct peaks at a few different frequencies. To calculate each frequency's own displacement amplitude, it should be treated alone, and the results then added together. Pradko's work implies that the energy loss of a rider on either a rigid or suspended bicycle can be predicted by computer (either by incorporating measured human response to vibration or by building in a multipart mechanical model of the rider, as in Wilczynski and Hull 1994).

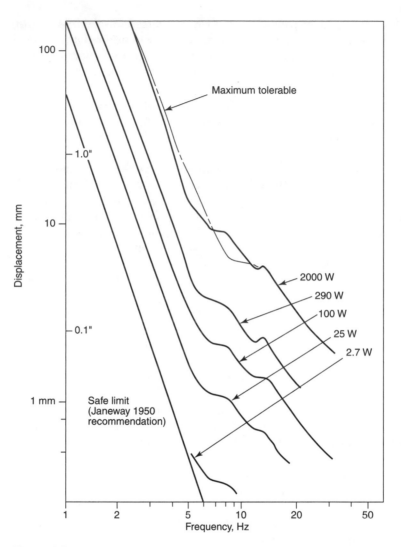

Figure 4.7
Power lost in vibration as function of amplitude and frequency. (From Pradko,
Lee, and Kaluza 1966.)

Minimizing bump losses   The speed loss caused by a bump is minimized by reducing the force of the bump on the bicycle via tires, suspension when present, and most commonly intentional rider control of upper-body motion or stiffness.

The functions performed by tires and suspensions in respect to minimizing the speed loss from bumps are not exactly the same. The most valuable attribute of tires is that they "swallow" small road irregularities such as pebbles; a rigid steel wheel with very low rolling resistance encountering the same small irregularities would be launched into the air and possibly be dented as well. Enveloping a pebble with a tire produces very little extra force and so does not lift or jolt the rider. In addition, the slight retarding force produced when absorbing the pebble becomes a nearly equal propulsive force when leaving it (because of the low mass of the small portion of the tire and tube affected). When the obstacle encountered has much bigger surface area, like a step change in pavement height, tires are far less "soft": unless large-cross-section, low-pressure tires are used, tires are inferior to bicycle suspensions in this regard.

By far the greatest suspension capability is that inherent in the human body. The body's range of travel and ability to absorb energy far outstrips the hardware of any ordinary bicycle suspension. In addition, human adaptability or even active compensatory motion makes a huge difference. Jim Papadopoulos reports having had the experience of striking a bump in the dark unprepared and being thrown from the bicycle, whereas the same bump, traversed in the day, was barely felt. Part of the difference is simply a matter of "softening up" the arms and torso or even standing slightly with the intention of allowing the bicycle to move up independently. A further strategy is to "lift" the bicycle over the obstruction, which involves far less of an energy change than having one's entire body suddenly accelerated upward. (This is one respect in which riders of unsuspended recumbent bicycles are at a disadvantage: they cannot use much body language to reduce bump losses.)

Determining bump losses   We have no sensible, simple formulas to express bump resistance as a function of road condition, tire construction, inflation pressure, suspension details, and speed, and anyway, most roads are reasonably smooth, so bump losses have been left out of this chapter's graphs and expressions for power. The omission is a shortcoming, particularly for those whose riding is primarily on irregular surfaces.

Measurements of bump losses are likely to show a dramatic rider influence, as different individuals will display a greater or lesser tendency to adjust elbow angle and muscle stiffness to minimize bump disturbances. Standardizing "rider qualities" in the test is likely to be a major problem.

It appears that there could be two "limits" for measuring losses attributable to bumpy conditions. The simple one would involve high bump frequencies, mostly from slight irregularities traversed at high speed. Once rider stiffness has been set by muscle tensions and arm angles, then the energy loss could be calculated from passive dissipative rider properties, as described in von Gierke 1964, and suspension properties. Rider "skill" and forethought would not have to be controlled. Indeed, modern computer software for measuring multibody dynamics could conceivably calculate energy losses credibly, if the body's elastic and energy-absorbing properties are known.

The other limit would involve a single bump, or no more than one bump every few seconds (low frequency). Here it seems self-evident that rider skill and planning could really minimize energy loss, by permitting (or even helping) the bicycle to surmount the bump. One might try to measure losses for both unskilled and skilled riders, with the awareness that significant "bump experience" during the test might alter their responses. The computational route seems utterly impossible, unless one is simulating a cadaver with no active participation. The single bump is discussed briefly by Sharp (1896), who cited Scott (1889). In making such measurements it is necessary to wait for upper-body swaying to settle before determining residual speed.

If "averaged" bump resistance is to be measured in towing or coasting experiments, it is crucially important to have an energy-absorbing "rider," or at least a standard damped, suspended load, otherwise the experiments will be irrelevant. The results of such experiments are likely to be speed-dependent in surprising ways.

The best way to measure bump losses may be with an on-road power-measuring device as described below. Riding the same path with and without standardized bumps should result in a finding of significant differences in average power for a given speed. Indoor testing is preferred in this instance to eliminate the effect of wind variations.

Steady-speed power equation

The aerodynamic resistance (including headwind), slope resistance, and rolling resistance can be combined into the following equation for calculating bicycling power:

$$\dot{W}_R \eta_m = \dot{W}_W = [K_A(V + V_W)^2 + mg(s + C_R)]V = [K_A(V + V_W)^2 + F]V,$$

where F refers to the fixed force arising from both slope and rolling resistance. This equation was known a century ago. A somewhat more comprehensive version was given in the second edition of this book (at the top of page 157) with an incorrectly restrictive introductory description. A recent validation of this equation was published by Martin et al. (1998).

The bicycling power equation can be used to make power or speed estimations in cases not covered by the graphs in this chapter. Here is an example, adapted from a calculation by Sandiway Fong.[7] Marco Pantani has a mass of 55.9 kg, and his bicycle has a mass of 7.3 kg. In the Tour de France, he climbed the notorious Alpe d'Huez (road distance 13.84 km, average slope 0.079, or 7.9 percent grade) with an average road speed of 6.12 m/s.

- System weight is 620 N, for a slope force $F_S = 49$ N.
- Assume no wind, and $K_A = 0.3$ (Pantani is small but climbs hills standing up). Then $F_A = 11$ N.
- Assume $C_R = 0.003$, so $F_R = 2$ N.

The total resisting force is 62 N (of which 79 percent is due to the slope), and the total required wheel power is 379 W. If the transmission efficiency was 0.95, the rider would have put out an average of almost 400 W for almost thirty-eight minutes.

The typical way to use this formula is simply to plot power as a function of speed with your own choice of parameters. This is easily achieved with a computer spreadsheet. One useful way to lay out the calculation is to let each row of the spreadsheet correspond to a different speed. Then within one row, the first column could be F_A, the second F_S, and the third F_R. (This would make it easy to see the relative contribution of each term.) The fourth column would be $\dot{W}_W$ (see table 4.1). When power is plotted as a function of speed, it becomes simple to read off the speed to be expected for any given power level.

This tabular approach has been used to compare five categories of bicycles: roadsters (i.e., heavy utility bicycles with high-loss tires and upright rider position), racing bicycles (with the best tires and crouched rider position), sports bicycles (intermediate between those two), and both a utilitarian "commuting" human-powered vehicle (HPV) and the "ultimate" HPV conceivable with today's technology. The parameters for each category are given in table 4.2 and the power curves in figure 4.8.

The power equation can also be used by selecting a given power level and solving for a parameter such as K_A. For any given speed V, the K_A needed to achieve that speed can be thus determined.

Yet another approach to relating speed to power is to sidestep graphs and to solve the power equation for V when power is given. But the equation is a cubic polynomial, the solution of which involves many subsidiary calculations and is not so easy for the math-rusty. A simple and versatile alternative is to create a simple iterating spreadsheet that can solve for V

Table 4.1
Power versus speed for 1.5 percent downslope

V (m/s)	F_A	F_S	F_R	$\dot{W}_W$
2	1	−12.0626	2.975444	−16.1743
4	4	−12.0626	2.975444	−20.3487
6	9	−12.0626	2.975444	−0.523
8	16	−12.0626	2.975444	55.30267
10	25	−12.0626	2.975444	159.1283
12	36	−12.0626	2.975444	322.954
14	49	−12.0626	2.975444	558.7797
K_A, kg/m	0.25			
mass, kg	82			
s, m/m	−0.015			
C_R, N/N	0.0037			

Note: Negative power due to downslope means brakes must be applied to hold speed!

numerically. The theory here is that given a poor guess for V, the equation returns an improved guess. Repeating this process ten or twenty times (i.e., using the "goal-seeking" tool in a spreadsheet application) usually produces an unchanging number eventually, the desired solution.

The iterative calculation is initiated by rewriting the power equation to put one of the factors of V alone. One simple way to do this is

$$V = \dot{W}_W/[K_A(V + V_W)^2 + mg(s + C_R)].$$

A guess for V is inserted into the right-hand side, and value of V generated on the left-hand side is considered to be a new guess for V, which is then inserted anew into the right-hand side. The only flaw in this approach is that sometimes the answer doesn't converge quickly enough—or at all. It helps to add a "convergence parameter," K_C, as follows:

$$V = \{\dot{W}_W/[K_A(V + V_W)^2 + mg(s + C_R)] + K_C V\}/(K_C + 1).$$

Adjusting K_C somewhere between 0.5 and 2 generally yields a converged answer within twenty iterations (See table 4.3).

In one of the marvelously unexpected ways that the Internet has encouraged and facilitated the distribution of information, a Web page ⟨http://www.analyticcycling.com/⟩ has been developed by Tom Compton to perform a wide variety of bicycling speed computations in response to

Table 4.2
Characteristics of five types of bicycle and rider

	Roadster (Utility) bicycle	Sports bicycle	Road racing bicycle	Commuting HPV	Ultimate HPV
Frontal area, A (m^2)	0.5	0.4	0.33	0.5	0.4
Drag coefficient, C_D	1.2	1	0.9	0.2	0.12
Bicycle mass (kg)	15	11	9	20	15
Rider mass (kg)	77	75	75	77	75
Rolling resistance coefficient, C_R	0.008	0.004	0.003	0.003	0.002
Force of rolling resistance, F_R (N)	7.218	3.374	2.471	2.854	1.765
Aerodynamic drag factor, K_A (kg/m)	0.368	0.245	0.182	0.061	0.029

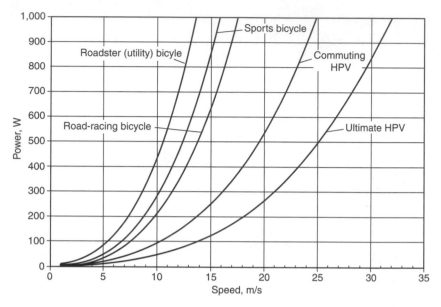

Figure 4.8
Power required to propel various bicycles in still air on the level.

information supplied by the user. Although we have not verified the results obtained using the Web page's calculators in detail, it appears to be an outstanding resource.

Rules of thumb

One use of the power equation is to understand quantitatively how various changes in equipment, fitness, or conditions will affect cycling speed (see Bassett et al. 1999) to improve performance, perhaps for a competition, or for the simple pleasure of understanding one's own riding. Although exact solutions to the equation usually require a computer, Papadopoulos 1999 provided a series of approximations, based on the power equation, but potentially simple enough to remember and apply mentally.

- The slight level-ground speed lost because of rolling resistance can be calculated as $F_R/3K_AV$. As an example, at a speed of 10 m/s, a typical K_A of 0.25, and a typical rolling resistance force of 3 N, the expected speed loss is 0.4 m/s.
- If a rider can reduce system weight, the percentage increase in speed climbing a hill is

Table 4.3
Example of successive calculations and the use of the convergence parameter

Parameters to find velocity			Iterative calculation of velocity	m/s	mile/h
Rider power (W)	$\dot{W}$	683.25	Initial velocity guess, V_a	10	22.36936292
Drag factor (kg/m)	K_A	0.19	$V_b = \{\{\dot{W}/[(V_a + V_w)^2 + (mg)(s + C_R)]\} + K_C V_a\}/(1 + K_C)$	18.95256	42.39567614
Headwind velocity (m/s)	V_W	0	Repeat calculation going downward: V_c from V_b	15.25651	34.12783975
Rider + bicycle mass (kg)	m	71.38	V_d from V_c	15.05357	33.6738751
Acceleration of gravity (m/s^2)	g	9.807	V_e, etc.	15.0845	33.74306248
Slope of hill (sine of angle)	s	0		15.07941	33.73168105
Tire rolling resistance coefficient	C_R	0.003		15.08024	33.73353233
				15.0801	33.73323064
Convergence parameter (usually 0.5–2) (select for fast convergence)	K_C	1.5		15.08013	33.73327979
				15.08012	33.73327178
Given the velocity (m/s)	V	15.08		15.08012	33.73327309
Find the power (W)	$\dot{W}$	683.25		15.08012	33.73327288
				15.08012	33.73327291
				15.08012	33.73327291
				15.08012	33.73327291
				15.08012	33.73327291
				15.08012	33.73327291
				15.08012	33.73327291
				15.08012	33.73327291

$$(\textit{percent weight reduction})\left(\frac{1-u^3}{1+2u^3}\right),$$

where

$$u \equiv \left(\frac{\textit{speed on hill}}{\textit{speed on level}}\right)$$

at the rider's normal power level. In other words, on a hill that halves the rider's speed (at a fixed power output), u is 0.5, and a 1 percent weight reduction will allow a 0.7 percent speed increase.

- A constant wind always reduces the average speed of a level out-and-back ride at constant power on an unfaired bicycle. If the wind is not too great compared to riding speed, the fractional decrease is $(1/3)(V_W/V)^2$. For a steady crosswind, just half as much reduction occurs. (For other wind angles, see Isvan 1984, 1, figure 2.) (This applies only to unfaired bicycles and riders. Faired (streamlined) vehicles can generate thrust from crosswinds, i.e., they can sail.)

- If a rider can reduce the aerodynamic drag factor K_A (by crouching, or wearing tighter clothing), then speed will increase both in downhill coasting and in level pedaling. The percentage increase in level-pedaling speed will be approximately two-thirds of the percent increase found in downhill coasting.[8]

- The effect of a 1 percent increase in mass on a long sprint begun at low speed, not quite so easy to describe generally, typically involves dropping back 0.25–0.5 m.

Acceleration

One confusing aspect about pedaling a bicycle is that whereas the power required to travel at any given speed can be defined in a power curve, we also know intuitively that it is possible to apply almost any power level at a given speed (from zero to our maximum effort). How can these ideas be reconciled? The answer is acceleration: *any* amount of power applied in excess of the requirement for maintaining a steady speed is absorbed in changing the speed.

When rider power exceeds that required at a given speed, propulsive force is increased, and acceleration $[a = (dV/dt)]$ occurs. The acceleration force ma enters into the power equation, which now can be written:

$$\dot{W}_W = \left[K_A(V+V_W)^2 + mg(s+C_R) + m_{\text{eff}}\left(\frac{dV}{dt}\right)\right]V.$$

In essence, we have added the rate of increase of kinetic energy to the power equation. The value of m_{eff}, the "effective" mass, is slightly greater than that of m because it includes the kinetic energy of the rotation of the bicycle's wheels. (Approximately, total system mass m is increased by the weight of the tires, the rims, and one third of the spokes. For most purposes, this slight difference is unimportant.)

This unsteady version of the power equation is a differential equation; that is to say, it includes not only the speed but (dV/dt), the rate of change of speed with time. To study acceleration behavior involves solving the differential equation.

For a given fixed power output, in constant conditions of slope and headwind, speed will first increase and then level out at the value defined by the steady power equation. However, in maximum-power sprinting, the rider becomes exhausted long before steady speed is reached (see Reiser, Broker, and Peterson 1999 on power drop-off measured by the Wingate anaerobic test), so the acceleration and speed achieved over a brief interval are what determine sprinting performance.

We briefly mention two analytical solutions to the differential equation. But for most readers, the numerical solution offered will be more usable, as well as more versatile in handling time effects (the pedaler's power may decrease, or a wind may spring up) and distance effects (the slope may change as he descends a hill).

1.   Coasting equation: the bicycle accelerates or decelerates under the effect of aerodynamic drag (including a direct head- or tailwind) and slope plus rolling resistance. The power equation is rewritten to describe the derivative of $(V \mid V_W)^2$ with respect to distance. The coasting velocity approaches terminal velocity based on a "settling distance," defined as $m_{eff}/2K_A$, which typically exceeds 100 m. It is necessary to coast for several times the settling distance to approach the theoretical terminal velocity. (In coasting tests it is best to pedal to the expected speed and then carefully note any changes.)

2.   Constant-power pedaling equation, no headwind or rolling resistance: there is a propulsive force inversely proportional to speed. The bicycle accelerates at first as if there is no air resistance. Then the terminal velocity is approached as if propulsive force were constant. The complicated solution for velocity as a function of time can be found analytically.

3.   Numerical integration: the easiest numerical scheme is to solve the power equation for dV/dt (acceleration) in terms of the current velocity. Then for a tiny increment of time such as 0.01 s, that value of acceleration permits the calculation of a new velocity, which in turn updates the power equation to yield a new value for dV/dt. Repeating this simple calculation thousands of times permits tracking the evolution of V as time passes,

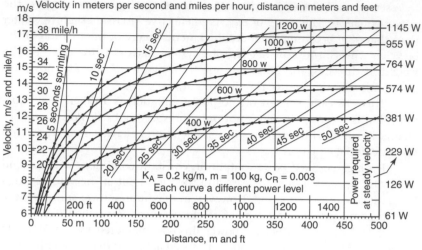

Figure 4.9
Sprinting distance versus velocity and time. (From Papadopoulos 1999.)

for the given specified power. (If *position* is also important, as for a sprint against a competitor, for each time step it is necessary to use the current value of *V* to calculate how far the bicycle has moved during the time step.) This approach, known as *Euler integration*, is extremely simple; however, it is necessary to perform an important check: to repeat the calculation with a smaller time step to make sure the results do not change. (If they do, the time step originally chosen was too large.)

Figure 4.9, from Papadopoulos 1999, can be used in real-world sprints to estimate roughly the rider's peak power.

Measurement of on-road power

The problem of determining the parameters of bicycle resistance is not a simple one. Part of this is because of the conditions in which cycling typically occurs: the air is rarely still, or even constant in velocity (wind can stop or start within a few seconds). The road is rarely level, or even constant in slope (it is common to vary one percent up or down). These considerations mean that casual outdoor coast-downs or terminal-velocity trials with an electronic speedometer yield only the crudest estimates of power. One is forced either to measure the rapidly changing road conditions or to experiment in a large building. The very best data regarding the parameters that affect bicycle resistance are derived from wind tunnels and tire drum testers.[9]

In previous decades, some careful on-road drag tests have been performed. Glenn Brown (Miller 1982) installed a precise electronic pendulum, damped to reduce the effect of vibrations, and on a bicycle coasted down a long hill. Unavoidable dips in the road cause no spurious reading (apart from the relatively tiny rotational inertia of the wheels, downhill acceleration causes no pendulum displacement), and the result of the test is a pendulum angle related to air drag plus rolling drag. Wind must be absent or at least steady. Chester Kyle (1974) performed coast-down tests on corridors inside a large building between timing traps and Kyle and Burke (1984) developed a heavy tricycle of low frontal area for coast-down studies of C_R. Doug Milliken (1991) made aerodynamic comparisons by simultaneously coasting two bicycles (presumably with equal weights and tires) down a long hill. Such an approach subjects both vehicles to the same wind gusts.

In the 1990s practical on-bicycle power-measuring equipment became commercially available. If used carefully, it can allow bicycle resistance parameters to be determined with reasonable accuracy. An example is the PowerTap system from the Tune Corporation ⟨www.etune.com⟩, shown in figure 4.10. A torque-sensing transducer and a speedometer are built into the rear hub of the system, and the data are wirelessly transmitted to a bicycle computer. A competing system is the SRM ⟨www.srm.de⟩ which is built into the crankset. Similar products have been offered before, and no doubt more will be developed. Such instruments will make it possible to "rewrite the book" when it comes to measuring rider power, rolling resistance and other variables.

Pedaling power is highly variable, so all instruments for measuring pedaling power require the ability to average data. For example, during each revolution of the crank, the instantaneous power of a seated rider may vary over a range of five to one. Also, a rider will occasionally ease off for a few pedal strokes even while riding on the level. On top of this, virtually unnoticeable little rises or wind gusts easily lead to a doubling or halving of power. For testing purposes, it is best to seek out nominally constant conditions, and then to determine the average power under those conditions for several minutes at least.

A power-measuring (and -averaging) instrument can be used for two different purposes. One is to measure the power a rider can produce for various durations (i.e., the power-duration curves of chapter 2). The other is to evaluate the power required to ride in a certain fashion (e.g., at a certain speed, or accelerating in a sprint). To achieve good results with such an instrument, careful protocols are needed. As an example, see Broker, Kyle,

Figure 4.10
PowerTap measuring instrument. (Copyright Graber Products, 2001.)

and Burke 1999, which determines the reduction in power requirements that result from drafting behind another rider.

On-road determination of aerodynamic and rolling drag

Slope drag on any particular road can be calculated from surveyed slopes and basic physics equations. Around a circuit, in particular, the average slope drag is zero. However, aerodynamic drag and rolling drag must actually be measured. Lacking a wind tunnel and a tire test stand, a reliable on-bicycle power meter is the next-best method for measuring aerodynamic and rolling drag.

The approach for obtaining such measurements is to find a riding circuit on which it is safe to travel without any use of the bicycle's brakes.[10] The circuit should be as level as possible, and preferably out in an open area, so that wind is neither blocked nor funneled at any location around the circuit.[11] A running track can be quite good.

The scheme is to ride several laps at constant speed, with a flying start. Speed is best tracked with two digital speedometers, one set to show average speed since the start. The instantaneous speed reading can show whether one is a little fast (hence needing to ease off) or slow (needing to pedal slightly harder). The average reading can show if one is successfully spending equal times above and below the target speed: if average speed wanders from the target, then the rider may mentally choose a slightly different "nominal" target to bring it back.

There are several reasons for seeking to fix speed rather than power during a trial. Some are mathematical: for example, the effects of varied slope or headwind on power can be calculated easily for a fixed speed, but determining their effects on speed at fixed power requires the power equation to be solved. Furthermore, the instrument's response to power variations is slow, and control of power seems considerably harder than controlling speed.

Both average power and average speed should settle down to relatively constant numbers after a few laps. Still, there may be visible variation within a particular lap due to wind or slope, so it is important to end the measurement by crossing the starting line with the same speed as when starting out. The data to record are average speed and average power.

The essence of generating a drag curve is to repeat this test for a number of different speeds. It is essential that throughout the trial for a particular speed, and also for every different speed evaluated, body position and clothing should be identical. If wind cannot be avoided, it is preferable that its average velocity be the same for each trial.

However many different speeds are attempted (five is a good number, plus perhaps a repetition at one or two of the speeds chosen), they should be plotted according to the value of V^2. Therefore it is desirable to choose them with roughly equal intervals after squaring. (For example, speeds of 3, 6, 8, 9.6, and 11 m/s have squares of 9, 36, 64, 92, and 121). The order in which the trials involving the various speeds should be randomized so that a progressive change in temperature, wind, etc. will not add a change that might otherwise appear correlated with speed.

Data for all speeds should be plotted on a graph, but not initially on speed versus power axes. It is far more useful to plot propulsive force as a function of speed squared, because theoretically, propulsive force on the level follows the expression $F_P = K_A V^2 + F_R$, where $F_R = mgC_R$. In other words, if the theory holds and the data are of good quality, then the (V^2, F_P) points will fall on a straight line. The slope of the line is K_A, and the intercept is F_R, quantities that can easily be determined by linear-regression commands in most spreadsheets.

Early explorations of this technique for finding aerodynamic and rolling drag have suggested the following.

- Data points generally give a straight line, so any deviations point to a change in conditions or a recording error.
- Comparing trials with different temperatures reveals that temperature seems to have a large effect on rolling resistance. In particular the rolling

resistance results in warm conditions are impressively low. (This suggests it may be important to control, or at least measure, *road surface temperature* when comparing two tires. This hypothesis is bolstered by measurements on home-exercisers, in which rolling has been found to get easier at the same time as the tires become warm to the touch.)

•   Comparing trials on the same day shows a greater reduction of K_A from shedding winter clothing than from crouching.

Once the linear regression is performed on the data obtained to determine the rolling resistance and aerodynamic drag factor, it is appropriate to plot the entire power curve based on those parameters. Of course, the actual power readings could have been plotted, but the technique of fitting the force curve best minimizes uncertainty and allows the speed expected for any power level to be determined.

Comparing with trials in other environmental conditions

The apparent effect of temperature on rolling resistance was alluded to above. To state that one tire is better or worse than another obviously requires that the two tires be compared at the same temperature. If that is for some reason not possible, some means for extrapolating to a reference temperature must be developed.

Similar concerns arise for measuring aerodynamic performance. If the drag of two setups is compared based on trials from different days, what has happened to the air density between the two days? One needs to know the air temperature (using a thermometer shielded from radiation) and the barometric pressure (either from weather service reports plus a compensation for altitude, or from one's own barometer) to make this determination.

By far the greatest problem is wind. In many areas of the world it will not reliably vanish at a convenient time. Even if the average wind during a trial can be determined, how can the data obtained be corrected to reflect windless conditions?

A theoretically based scheme (as yet untested) involves a calculation of the change in average drag when a steady wind blows across a circular course. The main premise of the scheme, actually, is not that the wind is steady, but that it spends on average an equal time approaching the rider from each direction, which should be true if several circular loops are completed in an open area and speed is held constant.

A second premise is that the rider's shape acts roughly as if it were a circular cylinder (in an aerodynamic sense), so that wind approaching from any angle creates about the same force. Actually, as long as one is riding faster than the wind velocity, the wind always approaches one somewhat from the front. A cosine approximation to the retarding force offered by the wind has been shown to be reasonable for an unfaired bicycle[12] (see Milliken and Milliken 1980).

The resulting change in the average required propulsive force (over a lap) at every value of V^2, surprisingly, is not a departure from a straight line, or even a modification to the aerodynamic drag factor. Instead, it takes the form of a simple shift of the drag line: effectively, an increase in V^2 of $(2/3)V_W^2$. This effectively masquerades as an increase in the line's y-intercept (i.e., of the rolling drag F_R) by $(2/3)K_A V_W^2$. It is hoped that a wind-speed correction of this form can be used to make drag determinations more precise.

Real-world determination of drag parameters is complex because of the many irregular disruptions, primarily wind, but also slope. For a taste of the effort that can be required to achieve even modest accuracy in such determinations, see Lucas and Emtage 1987 (91). On-bicycle power instrumentation such as PowerTap offers a relatively simple way to get good (but approximate) numbers for drag parameters, although reliable resolution of small differences in drag may not be feasible.

On-road determination of rider power curve

Each point on a rider's power curve represents the maximum duration for which the rider can produce a given power level. The spectrum runs all the way from maximum acceleration at high speed (duration of about four seconds) to an all-day ride. The "long-term average" feature of any instrument is essential on rides of all but perhaps the shortest duration (of course, short-term averaging over a couple of pedal strokes is taken for granted). The average power produced during a ride is a useful measure of training intensity, and the occasional attempt to maintain a higher intensity may be a good training tool.

When the duration for which cycling at higher intensity (four to eight times the power of the long-term tests) can be maintained is to be determined, problems occur. For example, at very high power levels, a bicycle accelerates so quickly that the gears must be shifted; however, shifting interrupts pedaling. Also, for a maximum one-minute trial, although a steady-power riding condition may improve performance, even to attain the high trial velocity is already somewhat exhausting.

The most useful technique seems to be to use hills effectively. For a one-minute trial, the best option found so far has been to approach a steep hill easily on the flat, then hold the achieved speed up the hill. For brief maximum sprint power, the highest numbers have come from swooping down into a steep dip, then applying maximal effort to increase speed up the other side.[13]

Discussion of insights regarding power and drag

This chapter has been about the various forms of drag that act on a bicycle in motion, and how in combination with available power they determine speed. Future chapters will go into the specifics involving each type of drag, but at this juncture it is already possible to outline some general conclusions and recommendations.

Some prescriptions for increasing speed at medium- or high-power levels (150 W and above)

The greatest potential for improvement in cycling speed is aerodynamic. Tight clothes, a good body position, and an aerodynamically clean bicycle can, in combination, cut K_A by 50 percent or more. On a conventional bicycle, it appears that the body position that generates the lowest drag involves the center of the rider's back's being the highest part of his body, with the knees almost brushing the chest. Beyond that, finding a way to bring the arms inward seems to pay good dividends (according to figure 5 and table 5 of Bassett et al. 1999, a decrease of 15–20 percent is possible using aero bars). For already low-drag racers, Broker, Kyle, and Burke (1999) showed that close drafting in a pursuit paceline can reduce the power required to achieve or sustain a given speed by 30 percent, possibly more. Truly astonishing reductions in aerodynamic drag, up to a factor of ten or more, are possible if the recumbent body position and a streamlined fairing are used.

The second-greatest potential for increasing cycling speed arises from training. Although we have no specific data to support this, it seems likely that a basically fit rider could eventually achieve aerobic power increases of 30 percent through training; however, an increase as large as 50 percent seems unlikely.[14] For much shorter efforts, in line with the considerable improvements found possible by weight lifters, we speculate that the peak power for five seconds might be as much as doubled by extensive practice (leading to improved muscle-fiber recruitment) and strength training.[15]

The third-greatest potential for speed improvement (on a smooth road at least) is properly pressurized tires of low-loss construction. The reduction of (say) 4–8 percent of the total drag that can result from installation and proper pressurization of such tires comes at perhaps the least degree of pain of any of the recommendations.

There is virtually no speed improvement to be had from ordinary weight reductions (say, 15 percent of bicycle weight, or 1–2 percent of system weight). On level, smooth roads, such reductions might reduce drag by 0.1–0.3 percent. On long uphills, the drag reduction approaches 1–2 percent if the hill is steep. However, in this case the resulting speed increase equals the drag decrease (unlike in flat riding, in which the percentage speed increase is just one-third the percentage drag decrease). So a 1 percent weight reduction on a steep mountain is as valuable as a 3 percent aerodynamic drag reduction on the level.

Championship racing performances

It seems that top riders achieve success not primarily by exerting superhigh power levels (since power has only modest scope for improvement, and a power change has only a small effect on level-road speed),[16] but rather by achieving superlow drag levels. What remains unknown is the potential for an individual's evolution on each of these characteristics. A group of elite racers has much lower drag and somewhat higher power than a group of average sport cyclists. But how much of this difference is due to changes in individuals, and how much could be due purely to selection of those who intrinsically have low drag or high power? Leaving aside the interesting question of whether some individuals have more potential than others to improve in power or drag characteristics, it is of vital interest to track the racing power and speed of some beginning competitors to see what changes actually occur in those who go on to succeed competitively.

The relationship between power and speed

Having reviewed the power-output capabilities of humans and the various power losses associated with bicycles and similar vehicles, we can now combine these characteristics to arrive at the power requirements for traveling at various speeds on different types of bicycles. We can also place bicycling along the entire range of muscle-powered movement and compare it with other modes of wheeled transportation such as roller skating and walking. We can also give a scientific answer to a question that repeatedly raises itself to the touring cyclist in hilly country: when is it better to dismount and walk up a hill than to continue straining on the pedals?

It is easy to show that the bicycle is very energy-efficient. However, it is unscientific to claim that it is even more efficient than the dolphin (a frequently heard extravagance). The resistance to motion, and therefore the overall energy efficiency, is a strong function of speed for all modes of transportation. The way in which resistances vary with speed is peculiar to each vehicle, animal, or mode. Therefore, comparisons among vehicles, animals, or modes are valid only if they are made at the same speed. Even with this proviso, the bicycle still comes out well.

Figure 4.11 shows the world-record speeds for different durations for the principal forms of human-powered propulsion. Presumably the contestants who achieved these speeds were putting out about the same power in each mode for the same durations. The standard lightweight track bicycle is 2–4 m/s (4–8 mile/h) faster than the best speed skater. The

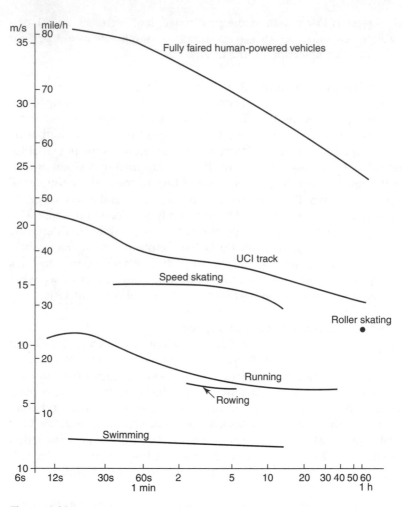

Figure 4.11
World-record speeds under human power in various modes. (From various sources; plotted by Dave Wilson.)

astonishing jump in record speeds from standard racing bicycles to machines using streamlined fairings in the IHPVA races adds another potential advantage to bicycling. World-record speeds may be derived reasonably accurately from the maximum power outputs of athletes for various durations (figure 2.4), the air-drag and rolling-friction-drag values of chapters 5 and 6, and an estimate of the other frictional resistances in the transmission and the wheel bearings (chapter 6 and 9).

Energy consumption as a function of distance

We can use the specifications in table 4.2 to find the energy consumed in bicycling various distances on level ground. In the physical sciences, energy is measured in joules (1 J/s = 1 W), but in nutrition, kilocalories are used to measure the energy content of food. A kilocalorie is the heat or work energy required to raise the temperature of a kilogram of water 1°C, and is equal to 4,186.8 joules. (Unfortunately, in nutrition it is usually abbreviated to "Calorie," which confuses physicists.)

We have used a reasonable mean value of the body's energy efficiency for fit people of 0.2388, or 23.88 percent, because when multiplied by 4,186.8 J/kcal, it gives 1,000 in the calculation of figure 4.12. For this value of net efficiency, a consumption of one kilocalorie of food energy produces one kilojoule of work.

We can see from figure 4.12 that a bicyclist racing at 9 m/s (20 mile/h) could travel more than 574 km/l (1,350 mile/U.S. gallon) if there were a liquid food with the energy content of gasoline. (Milk is mostly water but has enough energy to take a racing bicyclist about 40 km/l (95 mile/gal), so bicyclists could help to solve America's supposed energy shortage and milk surpluses simultaneously.)

Power needed for land locomotion

In order to survive, living species like animals and humans had to develop, early in their evolution, controllable movement, independent of gravitational and fluid forces that are the usual basis for movement of inanimate objects. The animal world developed systems involving levers that pushed against the ground in various ways from crawling, as do snakes, through bounding, like rabbits, to walking, as practiced by man, which in some ways is like the rolling of a spoked but rimless wheel. With the adoption of the wheel, yet another lever mechanism for movement, came the chance for using a separate, inanimate source of power besides that of the muscles of the moving creature. Vehicles powered by steam, internal-combustion engines, and electricity rapidly appeared once lightweight engines and motors of adequate power had been produced.

The bicycle is only one of the many human-developed lever systems for land transport, but it and roller or in-line skates are the sole remaining

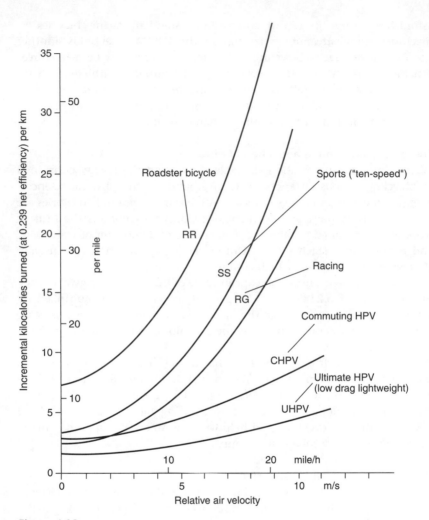

Figure 4.12
Energy consumption in bicycling over distances, if net metabolic efficiency is 23.9 percent. (Plotted by Dave Wilson.)

types that have a limited propulsive power. All other wheeled vehicles have, in general, been fitted with driving units of progressively increased power. In ancient times teams of horses or cattle succeeded single draft animals. The urge for more power and speed seems ever present in human activities.

Animals or wheels   The relative power needed to move a vehicle or animal against ground resistance by various means is shown in figure 4.13. At speeds of a few miles per hour, the sliding, crawling, leaping, or rolling motions by which these vehicles and animals move absorb almost all the power exerted by the subject, so that wind resistance can be neglected for purposes of approximate comparison. At higher speeds, the resistance to motion due to air friction assumes a dominant role and obscures the more fundamental difference between wheel motion and other systems of movement based on leverage.

Lever systems are intrinsically efficient, and figure 4.13 (which includes data from Bekker 1952) shows that nature, in developing walking for human progression, has provided a system more economical in energy use than that employed by many other animals. Nature has also arranged for the lever systems used by various animals to be adjusted automatically according to the resistance encountered. The stride of the walker changes, for instance, according to the gradient of the surface on which he is walking.

Bicycles versus other vehicles

The bicycle and rider have in common with most other wheeled vehicles and their passengers that they can move over hard smooth surfaces at speeds at which air resistance is significant, that is, at speeds greater than the 5 mile/h (2.2 m/s) upper limit of figure 4.13. The sum total of wind resistance, ground movement resistance, and resistance from machinery friction decides the rate of progress for a given power input to a vehicle. These resistances have been studied carefully over a long period for commonly used machines, such as those using pneumatic tires on pavement and steel wheels on steel rails.

Graphs showing how each of these resistances contribute to the total for a small automobile (figure 4.14) and, in watts per kilogram, the relative power requirements versus speed for walkers, bicycles, railway trains, and automobiles are given in figure 4.15. In each case, typical examples of vehicles without special streamlining have been chosen in order to bring out reasonable comparisons. Tricycles require an incremental effort for propulsion (up to 10 percent above that for the bicycle, as can be deduced from the times achieved in races).

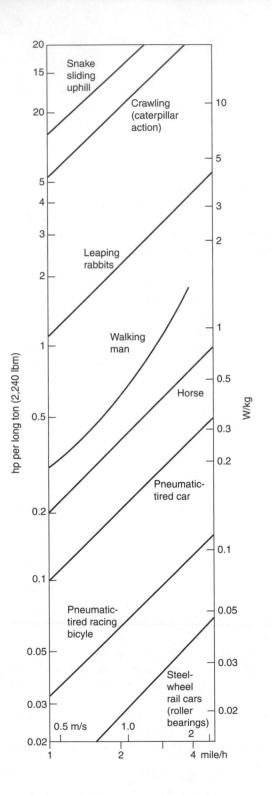

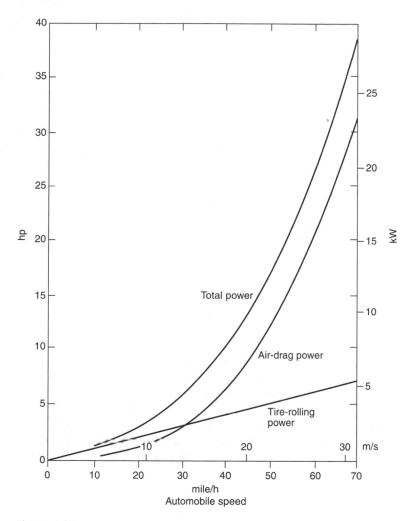

Figure 4.14
Propulsion power needed at the wheels of an automobile of about 1,000-kg weight and 1.9 m² frontal area.

◀ **Figure 4.13**
Power requirements of human walking and propulsion of various animals and vehicles. (Some data from Bekker 1952.)

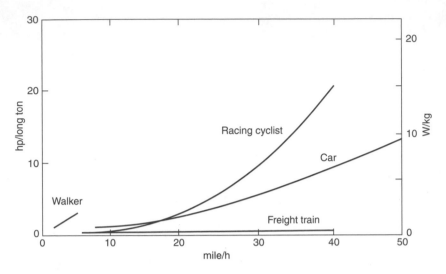

Figure 4.15
Power requirements of human walking and propulsion of a racing cyclist, an automobile, and a freight train over a range of speeds.

Our present purpose in comparing these various means of locomotion is to relate the bicycle to other common road vehicles. Some relative power requirements are shown in figures 4.13–4.15, and table 4.4 shows that, of all the vehicles compared, bicycles are impeded the most by wind if drag per unit weight is considered. A feature of modem automobiles is the relatively high power absorbed by the tires. In contrast, railway trains are hardly affected by wind resistance below 18 m/s (40 mile/h). With regard to the propulsion power required per unit weight, the bicyclist can be seen to need far less than the walker at low speeds.

Human versus animal muscle power

The power available for propelling a bicycle is limited to that of the rider. Let us study how human muscle power compares with that of other living things with similar muscle equipment.

For thousands of years—and even today in the less-developed parts of the world—horses, cattle, dogs, and humans have been harnessed to machines to turn mills, lift water buckets, and do other tasks. When the steam engine was invented, it was necessary to have handy a comparison between its power and that of a familiar source. Experiments showed that a big horse could maintain for long periods a power equal to that required to raise 33,000 lb (14,698 kg) one foot (0.3048 m) in one minute (745.6 W).

Table 4.4
Estimated forces opposing the motion of various vehicles on smooth surfaces in still air (typical cases)

Vehicle and weight	Origin of force	Resisting force, lbf (N)			
		5 mile/h (2.24 m/s)	10 mile/h (4.47 m/s)	20 mile/h (8.94 m/s)	40 mile/h (17.9 m/s)
Man walking, 150 lb (68 kg)	Wind	0.2 (0.89)			
	Rolling	13.0 (57.9)			
	Total	13.2 (58.7)			
Cyclist, 170 lb (77 kg) (racing type)	Wind	0.2 (0.89)	0.8 (3.6)	3.2 (14)	12.8 (57)
	Rolling	0.9 (4.0)	0.9 (4.0)	0.9 (4.0)	0.9 (4.0)
	Total	1.1 (4.9)	1.7 (7.6)	4.1 (18)	13.7 (61)
Automobile, 2,240 lb (1,016 kg)	Wind	0.9 (4.0)	3.5 (15.6)	14.0 (62)	56.0 (249)
	Rolling	37.0 (167)	37.0 (165)	37.0 (165)	37.0 (165)
	Total	37.9 (169)	40.5 (180)	51.0 (227)	93.0 (414)
Freight train, 1,500 tons	Wind	35 (156)	140 (620)	560 (2,490)	2,250 (10,010)
	Rolling	7,500 (33,370)	7,500 (33,370)	7,500 (33,370)	7,500 (33,370)
	Total	7,535 (33,530)	7,640 (33,990)	8,060 (35,860)	9,750 (43,380)

Table 4.5
Power outputs of horse and man

	Period	hp	kW
Horse			
Galloping at 27 mph (12 m/sec)[a]	2 min	2	1.5
Towing barge at 2.5 mph (1.1 m/sec)[b]	10 h	0.67	0.5
Man			
Towing barge at 1.5–3 mph (0.67–1.34 m/sec)[b]	10 h	0.11	0.08
Turning winch[b]	10 h	0.058	0.043
Working treadmill[b]	10 h	0.081	0.06
Climbing staircase[c]	8 h	0.12	0.09
Turning winch[c]	2 min	0.51	0.38

[a] *Source:* Burstall 1963.
[b] *Source:* D'Acres 1659.
[c] *Source:* Sharp 1896.

This value came to be universally accepted as "horsepower." Average horses could in fact work at a greater rate, but only for briefer periods.

Other information relating power output to duration of effort is given in table 4.5 and figure 2.4. An average human seems to adjust his power output to rather less than 75 W (0.1 hp) if he intends to work for other than very short periods and is not engaged in competition. This power level can be shown by experiments and by calculation to move a bicyclist and machine on the level at 4–7 m/s (9–15 mile/h), depending on wind resistance, type and weight of bicycle, and condition of road surface. This range of speeds has been associated with average cycling since the standardization of good rear-driven pneumatic-tired bicycles. Information on the energy cost of locomotion of animals other than man can be found in Kerkhoven 1963, Schmidt-Nielson 1972, Wilson 1973, and Rice 1972.

In a review of the energy used per ton-mile (or tonne-km) and passenger mile (km) for such varied means of transportation as the *S. S. Queen Mary*, the supersonic transport, a rapid-transit system, and oil pipelines, Rice (1972) points out that a bicycle and rider are by far the most efficient. He calculates that a modest effort by a bicyclist that results in 72 miles (116 km) being covered in six hours could require an expenditure of about 1,800 kcal (7.54 MJ), which is in agreement with figure 4.8 for something between a roadster and a sports bicycle. Assuming a mass of 200 lbm (90.6 kg) for rider and machine, Rice states that this figure is equivalent to 100 ton-miles (146 tonne-km) (or over 1,000 passenger-miles) per gallon (3.785 liters) of equivalent fuel. The *Queen Mary* managed, by contrast, 3–4

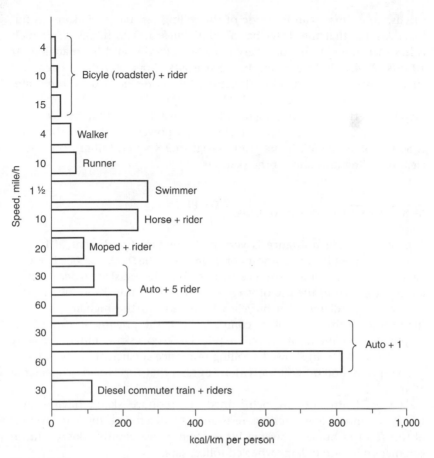

Figure 4.16
Energy cost of human movement and of the propulsion of various vehicles.

passenger-miles per gallon (1.27–1.70 passenger-km per liter). The energy consumption of other modes of transportation in comparison with that of a bicyclist is shown in figure 4.16.

Bicycling versus other human-powered locomotion

Roller skating

From figure 4.11 it can be seen that for one hour of maximum power output the record speed credited to a roller skater (10 m/s or 22.4 mile/h) is less than that of a track bicyclist (13.4 m/s or 30.7 mile/h). If it is assumed that such record makers exert equal powers at their respective relative speeds, an estimate can be made of the rolling resistance of skates as follows. Assume that the skater has a frontal area of 0.28 m² (3 ft²), which is less than the 0.34 m² of a very crouched bicyclist and his machine. At 10 m/s (22.4 mile/h) a racing bicyclist exerts about 200 W to overcome air resistance (see figure 4.8). Therefore, the power needed by the skater to overcome air resistance is $(0.28/0.34) \times 200 = 165$ W. At 13.7 m/s, the bicyclist exerts 500 W (see figure 4.8), and we assume that the skater at 10 m/s is exerting the same power. Hence, the power absorbed by the skates is $500 - 165 = 335$ W. If the skater weighs 69.8 kg (154 lbm), the coefficient of rolling resistance of the skates is

$$\frac{335 \text{ W}}{69.85 \text{ kg} \times 9.81 \text{ m/sec}^2 \times 10 \text{ m/sec}} = 0.049$$

The above rolling resistance is very high compared with that of bicycle wheels, assumed for the purposes of figure 4.8 to be 0.003–0.008. The very large increase in rolling resistance between bicycle and skate wheels can be partly attributed to the use of very small wheels in the skates (about one-thirteenth the diameter of bicycle wheels) and to the high resistance at high speed of the hard rollers compared with the pneumatic tires of the bicycle. Measurements of the pull required to keep a skater moving steadily made by Frank Whitt showed a rolling-resistance coefficient of about 0.060 at low speeds, and other information suggests that this would be greater at 10 m/s.

Several attempts are being made to produce skates having large wheels of much lower rolling resistance to determine the effectiveness of this form of human-powered locomotion. Cross-country skiers train in summer on a form of large-wheeled roller skate.

Walking

Dean (1965) gives data indicating that the maximum tractive resistance of a walker is about one-thirteenth of his weight. This figure was used as early as 1869 ("Velox" 1869). A higher resistance of two-fifteenths is, however, estimated from a simple geometrical model (Bekker 1952; "An Experienced Velocipedist" 1869, 5–6). The data show that for the same breathing rate, the bicyclist's speed is about four times that of the walker.

The metabolic heat figures for energy expenditures were obtained by multiplying the oxygen consumption, in liters per minute, by a calorific-value constant of 5 kcal per liter of oxygen, given by Falls (1968) as a reasonable value for the circumstances. This represents the total "burn-up" of human tissue that must ultimately be replaced by food. If each kilocalorie could be converted in one minute at 100 percent efficiency to mechanical energy (via muscle action), about 70 W of power should result. Dean (1965) shows that walking up a hill is slightly more efficient (in terms of energy consumption) than level walking, so the difference between cycling and walking is reduced in that case.

Running
The recorded times for sprint runners and racing bicyclists on level tracks in still air show that a cyclist can reach 18 m/s (40 mile/h) for 200 m and about 13.5 m/s (30 mile/h) for a mile (1.6 km), whereas a runner reaches only half these speeds. Assuming that the wind resistance acting on a bicycle and rider and that acting on a runner are similar at similar speeds, we can estimate that the power needed for cycling is only about a fifth of that needed for running at the same speed, in the range of 7–9 m/s (15–20 mile/h).

Effect of gradients and headwinds
Gradients and headwinds impede both the bicyclist and the walker, but to different degrees, compared with movement on the level in still air. It can be calculated that a gradient of 4 percent (1/25) or a headwind of 4.5 m/s (10 mile/h) slows a bicyclist exerting a constant 37 W (about one-twentieth of 1 hp) to about 1.1 m/s (2.5 mile/h). A walker exerting the same power would be slowed from about 2 mile/h to about 1.25 mile/h. The rider is slowed to 25 percent speed and the walker to about 62 percent. As a consequence, the rider notices difficult conditions more than the walker. On the other hand, with a tailwind or when going downhill, the bicyclist is aided to a far greater extent than the walker, and it is probably this virtue of the bicycle that will ensure its use even in country with hills so steep that the bicycle must be pushed up them.

When a bicyclist or a walker climbs a hill, his weight has to be lifted through a vertical distance, and as a consequence extra power is required above that needed for progress along the level. The additional power required for a bicycle and rider with a total weight of about 750 N (170 lbf) to climb a hill of 5 percent (1/20) at 11.2 m/s (25 mile/h) is

$$\frac{750 \text{ N} \times 11.2 \text{ m/s}}{20} = 420 \text{ W}.$$

Hence, it is seen from figure 4.6 that a racing bicyclist climbing a 5 percent hill must exert a power of about 725 W. He would be sorely stressed and could do this for only about two minutes, according to figure 2.4.

Bradley (1957, 90) gives interesting information about his climbing a one-in-twelve (8.5 percent) pass on the Gross Glockner more than 20 km (12.5 mile) long in about 57 minutes. The gear used was 47 inches (3.75 m "development"), and it can be deduced that he exerted at least 448 W

(0.6 hp), pedaling at a rate of about 90 rpm. This performance is remarkably close to fast 25-mile (40-km) time-trial performances and provides convincing proof that there is sound evidence for all the power-requirement estimates based on wind-resistance calculations (as distinct from the more easily accepted simple weight-raising calculations associated with hill-climbing bicyclists).

Should one walk or pedal up hills?

Noncompetitive bicyclists have the option of walking up steep hills. Some prefer doing so to cycling up such hills, alleging that a change of muscle action is agreeable to them. Some bicyclists, however, prefer to fit low gears to their bicycles and to ride as much as possible. Whether it is easier to ride or to walk up steep gradients is often debated among bicyclists. We will use data developed previously to show that it should be more efficient to ride up to an approximately limiting gradient (determined below).

If we confine attention to the everyday bicyclist, we can assume that he is unlikely to wish to use more than about 0.1 hp (74.6 W) in cycling. A commonly encountered steep hill is one with a gradient of 1 in 6.7, or 15 percent. It is assumed that the road speed, which is thereby fixed as 0.67 m/s (1.5 mile/h), gives no difficulties in balancing.

There have been many experiments on the oxygen consumption of pedalers (Bicycle Production and Technical Institute 1968; Rice 1972; Dickenson 1929). The data given in chapter 2 appear typical in that, for a power output of 75 W (0.1 hp) at the bicycle's wheel, a metabolic gross efficiency of 21 percent is reasonable. The cyclist will be "lifting" a machine weighing, say, 130 N (30 lbf) in addition to his body (150 lbf, or 667 N), so a factor is necessary to account for the efficiency when compared with body weight alone. This factor can be calculated as

$$21 \times 150 \text{ lbf}/(150 + 30) \text{ lbf} = 17.5 \text{ percent},$$

if one assumes that there is negligible rolling or wind resistance at 0.7 m/s and if one neglects power losses in the low gear.

McDonald (1961) gives a summary of experimental work concerning the oxygen consumption of walkers going up various gradients at various speeds. For a walking rate of 0.67 m/s (1.5 mile/h) up a grade of 15 percent, it appears that a metabolic gross efficiency of 15 percent is accepted as typical. This efficiency assumes as a basis the body weight being lifted against gravity. The bicyclist pushing his machine will be in a semi-crouched position, so an adjustment to the efficiency must be made. Data from Dean (1965) and McDonald (1961) concerning the effects of walking in stooped positions and when carrying small weights show that pushing a 30 lbf (14 kg) bicycle absorbs 30 percent extra effort, so that the walker's

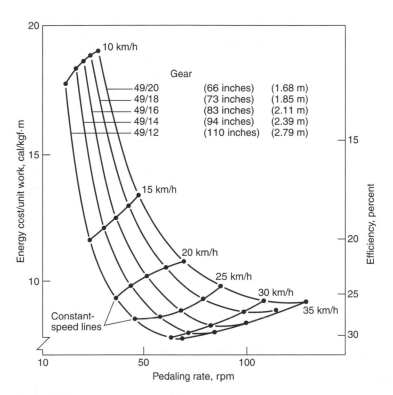

Figure 4.17
Effect of gearing on energy efficiency. (Bicycle and Production and Technical
Institute 1968.)

muscle efficiency based on his body weight alone is decreased to $17.5 \times$
$(100 - 30)/100 = 12.3$ percent. From the estimations above, it appears that
it is about 30 percent (12.3/17.5) easier to ride up a 15 percent gradient
than to walk up the gradient at the same speed of 0.67 m/s (1.5 mile/h),
pushing the bicycle.

However, in practice, the lowest gear available to the rider may be
20 inches (1.6 m), which gives a pedaling rate of 26 rpm—not optimal,
according to figure 4.17. A decrease in the previously assumed overall pedaling efficiency of 21 percent is bound to occur. Let us estimate this
decrease at about 18 percent. As a consequence, the 30 percent difference
quoted above between the energy required to pedal up a given gradient and
that required to walk the bicycle up the same gradient should be taken as
about 18 percent. This estimate of the difference may, however, need to
be increased, because recent data suggest that a very low derailleur gear is
more efficient than a higher gear. Calculations along the lines of the above

Table 4.6
Energy cost of movement by various means

	Speed		Energy consumption[a] per person		
	mile/h	m/s	kcal/km	mile/gal[b]	km/l
Bicycle (roadster) + rider	4	1.79	8.4	2440	1037
	10	4.47	15.6	1310	557
	15	6.70	24.4	840	357
Walker	4	1.79	55.3	370	157
Runner	10	4.47	68.3	300	127
Swimmer	1.5	0.67	269.6	76.0	32.3
Horse + rider	10	4.47	245.4	83.5	35.5
Moped + rider	20	8.94	88.3	232	98.6
Auto + 5 riders	30	13.4	120.5	170	72.3
	60	26.8	183.0	112	47.6
Auto + 1 rider	30	13.4	539	38	16.2
	60	26.8	820	25	10.6
Diesel commuter train + riders	30	13.4	112	183	77.8

[a] For the metabolic energies, these figures give the incremental consumption above the resting level.
[b] Equivalent miles per U.S. gallon of 33,000-kcal/gal fuel (gasoline per person, calculated as follows: mile/gal = 33,000 × 0.621/kcal-km.

show that the 15-percent gradient may be a critical one, and that at gradients of 20 percent there is no really appreciable advantage in riding the bicycle, even in a low gear.

A matter not given prominence in this type of discussion is the lack of wind cooling for the cyclist's relatively high heat output. At a power output of 82 W (0.11 hp) (that is, 0.1 hp plus an allowance for gear friction), a rider on the level would be traveling at some 6.2 m/s (14 mile/h) and would receive considerable cooling. When climbing a hill at 1.5 mile/h for, say, fifteen minutes, it is certain that an averagely clothed bicyclist would feel himself getting hot. Unpublished data suggest a body temperature rise of appreciable magnitude would result: 1°F (0.55°C). It is probable that such considerations influence bicyclists to get off and walk at very low cycling speeds (say, less than 1 mile/h) when the smaller heat loss from the lowered power output is more tolerable. Proponents of very low gears for hill climbing can claim not only a higher metabolic efficiency but also a much needed heat-removal effect from the more rapid movement of the legs at low forward speeds.

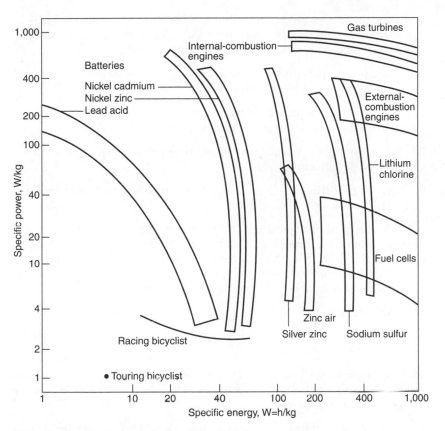

Figure 4.18
Specific energy versus specific power of human and mechanical systems. (From Gouse 1970.)

Specific energy versus specific power of various power sources
Data on various heat engines and human performances are given in table 4.6 and figure 4.18.

Notes

1.   It is rare for the wind to blow exactly parallel to the road. Some effects of wind direction are discussed below.

2.   Both velocities are defined relative to the frame of the bicycle.

3.   This inefficiency is typically 2–4 percent but is sometimes as great as 10 percent.

4.   Actually, a more precise statement is "when $V + V_W$ is small," which can also be true when one is riding with the wind. (Tire rolling resistance also predominates when one is riding stationary training rollers.)

5.   As a rough approximation, each percentage point in grade $s_\%$ is about half a degree in angle of inclination from horizontal.

6.   Regrettably, it is not possible to determine from the reference whether "displacement amplitude" refers to "peak to peak" or half that.

7.   Fong has a power calculator at ⟨http://www.neci.nj.nec.com/homepages/sandiway/bike/climb.html⟩

8.   As will be discussed below, it can take a very long time to achieve maximum ("terminal") velocity on a hill.

9.   Of course, such approaches still have some deficiencies. Wind tunnels generally don't have a moving ground plane, and wheels of bicycles used in wind-tunnel tests often don't rotate (and rotation torque may not be measured when they do). Drum testers are not flat like the road, rarely are rough like the road, and may not be able to furnish temperature extremes.

10.   Any application of the brakes adds to the apparent drag.

11.   In fact, the ideal is no wind whatever; however, a possible correction for wind will be described below.

12.   For a streamlined vehicle, the graph of the retarding force for off-axis wind looks like a *modified* cosine. Such measured behavior can also be incorporated into the above scheme.

13.   For such a short time the "average" function of the available instruments is not even usable. However, the PowerTap has a "maximum" function that seems to average over several pedal strokes.

14.   As a basis for this supposition Papadopoulos has noted that his own maximum hour-long power (essentially untrained) of 300 W is roughly 70 percent of championship levels. Consistent with this is the very limited increase possible in an individual's maximum oxygen uptake.

15.   What is not clear is the extent to which a good position aerodynamically compromises power, with the possibility that good riders are actually intrinsically more powerful than racing measurements reveal.

16.   Of course this is less true for mountain riding, in which speed is based on the ratio of power to weight.

References

"An Experienced Velocipedist." (1869). In *The Velocipede*. London: J. Bruton Crane Court.

Bassett, David R., Jr., Chester R. Kyle, Louis Passfield, Jeffrey P. Broker, and Edmund R. Burke. (1999). "Comparing cycling world hour records, 1967–1996: Modeling with empirical data." In *Medicine and Science in Sports and Exercise*. Indianapolis, Ind.: American College of Sports Medicine.

Bekker, M. G. (1952). *Theory of Land Locomotion*. Ann Arbor: University of Michigan Press.

Bicycle Production and Technical Institute. (1968). Report. Japan.

Bradley, B. (1957). "My Gross Glockner ride." *Cycling* (July 25):90.

Broker, Jeffrey P., Chester R. Kyle, and Edmund R. Burke. (1999). "Racing cyclist power requirements in the 4000-m individual and team pursuits." In *Medicine and Science in Sports and Exercise*. Indianapolis, Ind.: American College of Sports Medicine.

Burstall, A. F. (1963). *A History of Mechanical Engineering*. London: Faber and Faber.

D'Acres, R. (1659). *The Art of Water Drawing*. London: Henry Brome; reprint, Cambridge, U.K.: Heffer, 1930.

Dean, G. A. (1965). "An analysis of the energy expenditure in level and grade walking." *Ergonomics* 8, no. 1:31–47.

Dickenson, S. (1929). "The efficiency of bicycle pedaling as affected by speed and load." *Journal of Physiology* 67 (1929):242–245.

Falls, H. B. (1968). *Exercise Physiology*. New York: Academic Press.

Gouse, S. W. (1970). [Article title unavailable.] *Science* 6, no. 1:50–56.

Isvan, Osman. (1984). "The effect of winds on a bicyclist's speed." *Bike Tech* 3, no. 3 (June):1.

Janeway, R. N. (1950). "Vertical vibration limits for passenger comfort." In *Ride and Vibration Data*, a set of reference charts, Special Publication SP-6. Warrendale, Penn.: Society of Automotive Engineers.

Kerkhoven, C. L. M. (1963). "Kenelly's law." *Work Study and Industrial Engineering* 16 (February):48–66.

Kyle, Chester R. (1974). "Factors affecting the speed of a bicycle." *Bicycling* (July):22–24.

Kyle, Chester R., and Frank Berto. (2001). "The mechanical efficiency of bicycle derailleur and hub-gear transmissions." *Human Power*, no. 52:3–11.

Kyle, Chester R., and E. M. Burke. (1984). "Improving the racing bicycle." *Mechanical Engineering* 109, no. 6:35–45.

Lucas, G. G., and A. L. Emtage. (1987). "A new look at the analysis of coast-down test results." *Proc. I. Mech. E.* 201, no. 2:91.

Martin, J. M., D. L. Milliken, J. E. Cobb, K. L. McFadden, and A. R. Coggan. (1998). "Validation of a mathematical model for road-cycling power." *Journal of Applied Biomechanics* 14, no. 3:276–291.

McDonald, I. (1961). "Statistical studies of recorded energy expenditures of man. II. Expenditures on walking related to age, weight, sex, height, speed and gradient." *Nutrition Abstracts and Reviews* 31 (July):739–762.

Miller, Crispin Mount. (1982). "Testing for aerodynamic drag: A new method." *Bike Tech* 1, no. 4 (December):1–2.

Milliken, D. (1991). Personal communication.

Milliken, D. L., and W. F. Milliken. (1983). "Moulton bicycle aerodynamic research program." In *Proceedings of the Second International Human-Powered-Vehicle Scientific Symposium*, ed. Allan V. Abbott. San Luis Obispo, Calif.: IHPVA.

Papadopoulos, J. M. (1999). "Simple approximations for the effects of tire resistance, wind, weight and slope." *Human Power*, no. 48 (Summer):10–12.

Pradko, F., and R. A. Lee. (1966). "Vibration comfort criteria." Society of Automotive Engineers paper no. 660139, Automotive Engineering Congress, Detroit, Mich.

Pradko, F., R. A. Lee, and V. Kaluza. (1966). "Theory of human vibration response." American Society of Mechanical Engineers (ASME) paper no. 66-WA/BHF-15. 1966 Winter Annual Meeting.

Reiser, Raoul F., Jeffrey P. Broker, and M. L. Peterson. (1999). "Inertial effects on mechanically braked Wingate power calculations." In *Medicine and Science in Sports and Exercise*. Indianapolis, Ind.: American College of Sports Medicine.

Rice, R. A. (1972). "System energy and future transportation." *Technology Review* 74 (January):31–48.

Schmidt-Nielson, K. (1972). "Locomotion: Energy cost of swimming, flying and running." *Science* 17 (July 21):222–228.

Scott, R. P. (1889). *Cycling Art, Energy and Locomotion: A Series of Remarks on the Development of Bicycles, Tricycles, and Man-Motor Carriages*. Philadelphia: J. B. Lippincott.

Sharp, A. (1896). *Bicycles and Tricycles*. London: Longmans, Green; reprint, Cambridge: MIT Press, 1977.

Spicer, James B., C. J. K. Richardson, M. J. Ehrlich, and J. R. Bernstein. (2000). "On the efficiency of bicycle chain drives." *Human Power*, no. 50:3–9.

"Velox"(pseudonym). (1869). *Velocipedes, Bicycles and Tricycles: How to Make and Use Them*. London: Routledge.

Von Gierke, H. E. (1964). "Biodynamic response of the human body." *Applied Mechanics Reviews* 17:951–958.

Wilczynski, H., and M. L. Hull. (1994). "A dynamic system model for estimating surface-induced frame loads during off-road cycling." *J. Mech. Design* 116, no. 3 (September):816–822.

Wilson, S. S. (1973). "Bicycle technology." *Scientific American* 228 (March):81–91.

5 Bicycle aerodynamics

Introduction

This chapter is about aerodynamic drag and other aerodynamic phenomena such as the flow effects when people ride side by side and one behind the other, wind buffeting from vehicles, and the effects of side winds. It is a large and complex subject: I hope to explain some (but a long way from all) of the complexities.

"Wind resistance" is an everyday experience, particularly to bicyclists: at normal biking speed it is the largest component of drag apart from that due to hills. It is caused by two main types of forces: one normal to the surface of the resisted body (that could be the human body, or the body of a vehicle) (felt as the pressure of the wind) and the other tangential to the surface (which is the true "skin friction" and is dissipated in immediate slight heating of the air) (figure 5.1a). For a nonstreamlined body, such as a bicycle and rider, the pressure effect is much the larger of the two, and the dissipated pressure energy appears initially as kinetic energy in the wake that dissipates also into heating of the air. Figure 5.1b shows this kinetic energy appearing as eddies at the rear of a cylinder. As can be seen in figure 5.1c, a streamlined shape produces lower kinetic energy in the wake, because there is "diffusion" or pressure recovery along the aft (downstream) surfaces. Most of the drag that affects bicyclists is caused by actual friction, again called for some reason "skin friction," against the surface of the body.

Vehicles intended for high speeds in air are almost always constructed to minimize pressure drag. Streamlined shapes incorporate gradual tapering from a rounded leading edge. The exact geometry of shapes that maximize the possibility of the flow's remaining attached (rather than separating in local jets and eddies) and that minimize the skin friction can be approximated using rather complex mathematics. Alternatively, it is usual in aeronautics either to refer to one of a family of published "low-drag" shapes (one is given in figure 5.2) or to test models in a wind tunnel (Abbott and Doenhoff 1959; Simons 1999).

Although wind-tunnel experiments can yield good data for motor vehicles, the interaction of the airflow surrounding the moving bicyclist with the stationary ground and with the usually whirling legs is relatively more important for bicyclists than for those traveling in motor vehicles. This reduces the validity of wind-tunnel data on bicyclists. More accurate information can be obtained with actual riders on a road or track.

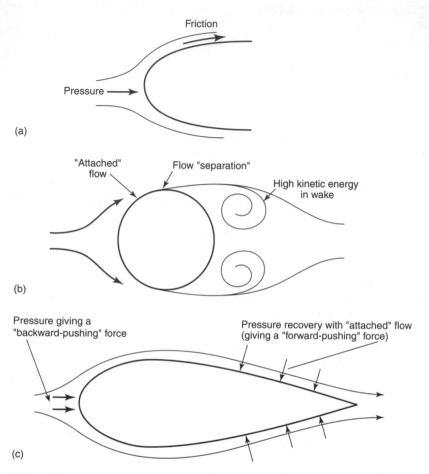

Figure 5.1
Flow around bodies. (a) "Normal" (pressure) forces and "friction" forces; (b) "attached" and "separated" flow around a cylinder; (c) attached flow and pressure recovery on a streamlined body.

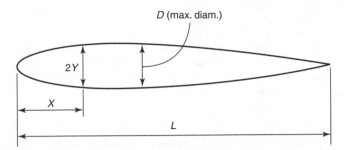

NACA Profile 0020, derived via the formula $Y^2 = a_1X + a_2X^2 + a_3X^3 + a_4X^4 + a_5X^5 + a_6X^6$, where $a_1 = +1.000000$, $a_2 = +0.837153$, $a_3 = -8.585996$, $a_4 = +14.075954$, $a_5 = -10.542535$, and $a_6 = +3.215422$. Length/diameter = 7.00, nose radius/maximum diameter = 0.714, and tail radius/maximum diameter = 0.0143.

Figure 5.2
Low-drag shape: NACA Profile 0020.

Drag coefficient

One aim of aerodynamic experiments on an object is to measure its drag coefficient C_D, defined as the nondimensional quantity

$C_D \equiv drag/(area \times dynamic\ pressure)$.

The drag is the force in the direction of the relative flow (or of the dynamic pressure). The area (A) to be used in the formula is defined later. However it is defined, the product C_DA (which is independent of definition of A) is a very useful number in studies of the drag of bodies. The drag is simply the product C_DA times the dynamic pressure. We list this product later (in table 5.1) for various types of bicycles and other machines. (In chapter 4 this product multiplied by half the average air density was given as the aerodynamic-drag factor K_A.)

The dynamic pressure is the maximum pressure that can be exerted by a flowing stream on a body that forces it to come to rest. At low speeds (say, below 45 m/s or 100 mile/h), the dynamic pressure is closely approximated by

$$Dynamic\ pressure \approx \left(\frac{\rho V^2}{2g_c}\right),$$

in which (for S.I. units) ρ is the air density in kg/m³, and V is the velocity of the air in m/s. The constant $g_c = 1.0$ for S.I. unit systems. It is found in Newton's law of motion:

$$F = ma/g_c,$$

where F is in newtons (N), m is in kilograms, and a is in meters per second squared. In U.S. units, g_c has the value 32.174 lbm-ft/lbf-s^2 when the equation relates m in pounds mass, F in pounds force, and a in feet per second squared. The dynamic pressure versus velocity and altitude is given in figure 5.3. At the time of writing, the HPV speed record was about 35 m/s and was set at an altitude of between 2,500 and 3,000 m, and it can be seen that the dynamic pressure at that speed and altitude would have been over 600 N/m^2.

Different definitions of area and of drag coefficient

The area to be used in the formula can be defined in two alternative ways, each one leading to a different definition and a different value of the drag coefficient C_D. The more usual definition is the frontal area, and unless otherwise stated, the form of drag coefficient that uses this definition of area is the one that we will use in this book. Thus, the drag force is given by

$$Drag = C_D \times frontal\ area \times dynamic\ pressure$$

Another form of drag coefficient is defined in terms of the surface area of the body and is used only for slender and/or streamlined bodies, where the drag is primarily from skin or surface friction, rather than from the eddies coming from bluff bodies. In this book we have given this form of drag the subscript "SA," and it is defined as

$$C_{D,SA} \equiv \frac{drag}{surface\ area \times dynamic\ pressure}.$$

For a given body in a given condition, the surface-area coefficient of drag is smaller than the frontal-area coefficient because the surface area is larger than the frontal area. For a sphere, the ratio of the surface area to the frontal area is 4.0. For a long cylinder of diameter D with spherical ends, the ratio is $4 \times (1 + L/D)$, where L is the length of the straight portion of the cylinder. The measured value of C_D for a rounded-end cylinder aligned with the flow increases with L/D, whereas the value of $C_{D,SA}$ decreases with L/D to compensate for the increasing surface area (figure 5.4).

The significance of not confusing these two definitions can be illustrated by the following anecdote. In the early days of the quest for the Du Pont Speed Prize for the first HPV to reach 29 m/s, 65 mile/h, an MIT student decided that he could win the prize by assembling many pedalers in a line within the same frontal area as one pedaler. He had found that the drag coefficient for a reasonably streamlined single-rider recumbent vehicle

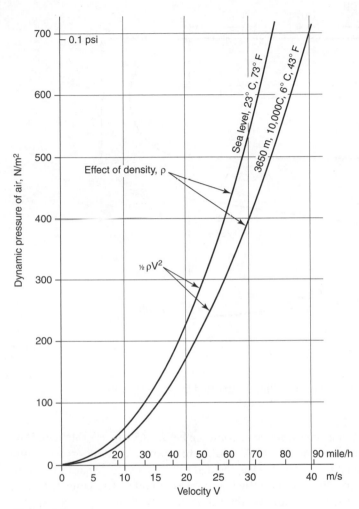

Figure 5.3
Dynamic pressure of air versus velocity and altitude.

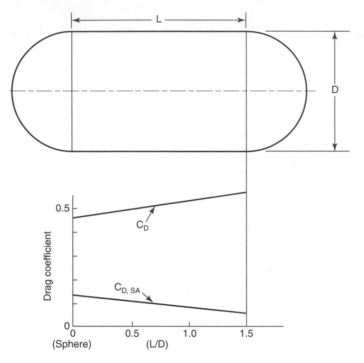

Figure 5.4
C_D and $C_{D,SA}$ for a circular cylinder. (Plotted by Dave Wilson.)

was 0.15 and that the frontal area could be below 0.5 m². He calculated the drag at 29 m/s to be about 38 N, leading to a power required to overcome air drag alone at over 1100 W. He decided to build a vehicle carrying ten to fifteen riders in a line, because the frontal area would be the same, therefore (he thought) the drag would be the same, and the air-drag power required from each of ten riders would be an easily manageable 110 W. He confidently forecast reaching 80 mile/h, 36 m/s.

For various reasons that plagued development, the vehicle was quite slow. But the fallacy underlying the designer's reasoning was that the drag coefficient based on frontal area would not increase as the vehicle was made longer. It would and did, probably quadrupling the drag of a one-person faired body of the same frontal area. It is often preferable when calculating the drag of a streamlined body, therefore, to use the drag coefficient based on surface area. However, either form may be used with confidence so long as the value found experimentally for one configuration is not applied to the analysis of a completely different shape.

The propulsion power ($\dot{W}$) necessary to overcome drag is

$\dot{W} = drag\ force \times relative\ vehicle\ velocity$

(We use W as a symbol for quantity of work, such as joules or ft-lbf, and $\dot{W}$ for the rate of doing work, which is power, in watts or ft-lbf/s or horse-power.) Since the drag force is approximately proportional to the square of the velocity, the power to overcome drag is approximately proportional to the cube of the velocity.

Only in still air is the vehicle velocity the same as the relative velocity used to calculate the drag force. When there is a headwind or a tailwind, the relative velocity is different from the vehicle velocity.

In S.I. units the relationship is

$Power\ (W) = drag\ force\ (N) \times vehicle\ velocity\ (m/s)$

If the drag is measured in pounds force and the velocity is given in feet per second, the power is in ft-lbf/s. This may be converted to horsepower by dividing by 550 (1 hp = 550 ft-lbf/s), or miles per hour (1 hp = 375 mile-lbf/h) may be used:

$$W(hp) = \frac{drag\ (lbf) \times velocity\ (ft/s)}{550[(ft\text{-}lbf/s)/hp]} = \frac{drag\ (lbf) \times velocity\ (mile/h)}{375[(mile\text{-}lbf/h)/hp]}$$

Drag

The drag coefficients of bodies the resistance of which is almost entirely due to pressure drag (e.g., thin plates set normal to the direction of flow) are virtually constant with air speed, once this speed is higher than the "creeping flow" or laminar range (see the discussion of Reynolds number below). But bodies with substantial contributions from the surface-friction drag of the so-called boundary layer of "sticky" or vis-cous flow have drag coefficients that can vary widely in different circum-stances. In general, the flow in this boundary layer can exist in one of three forms:

1. laminar, in which the layers of fluid slide smoothly over one another, as in the foreparts of the three bodies in figure 5.1;

2. turbulent, in which the boundary layer is largely composed of small confined but intense vortices that greatly increase the surface friction, as will most likely be the case at the rearward end of the body shown in figure 5.1c; and

3.   separated, in which the boundary layer, along with the main flow, leaves the surface and usually breaks up into large-scale unconfined jets or eddies, as in figure 5.1b.

If we wanted to produce a low-drag bicycle enclosure, we would prefer that the boundary layer flow be entirely laminar (airplane designers have long tried to arrive at laminar-flow wings). Unfortunately, laminar-flow boundary layers are extremely sensitive. They have a strong tendency to separate from the surface, producing very high levels of pressure drag, because flow separation prevents most recovery of pressure along the downstream part of a body. This pressure recovery gives the body or fairing a forward-pushing force that, if there were no friction, would exactly balance the backward-pushing force at the front of the body.

Turbulent boundary layers have higher surface friction than laminar boundary layers and therefore produce somewhat higher drag; however, they are less likely to separate than laminar boundary layers. Often the lowest levels of integrated drag are produced by forcing the laminar boundary layer on the forward part of a body to become turbulent. At low speeds this may require either the roughening of the surface or the mounting of a "trip" wire at well before the location where separation might otherwise occur. A classic experiment by the aerodynamics genius Ludwig Prandtl showed this effect graphically (Goldstein 1938). Prandtl mounted a smooth sphere in an air stream, measured its drag, and observed the airflow with streams of smoke. The flow separated in so-called laminar separation even before the maximum diameter was reached (figure 5.5, top), and the amount of drag was high. Then he fastened a thin wire ring as a boundary-layer trip to force the boundary layer to become turbulent on the part of the sphere upstream of where laminar separation had previously occurred. The boundary layer indeed became turbulent, and as a consequence the flow remained attached over a much larger proportion of the sphere's surface (figure 5.5, bottom), and the drag decreased greatly, as can be seen from the much smaller wake. Manufacturers of golf balls learned from this and roughen the surface with sharp-edged dimples, producing balls that can be driven faster and farther. (The dimples, combined with top spin, also produce an aerodynamic lift force, which contributes to increasing the ball's range.) In a later section we discuss another possibility for reducing drag: the use of surface suction to pull out the low-momentum inner part of a laminar boundary layer to force it both to stay laminar and to stay attached.

For any one shape of body, the variable that controls the drag coefficient is the Reynolds number (Re), defined in general as

$$Reynolds\ number \equiv \frac{air\ density \times relative\ air\ velocity \times length}{air\ viscosity},$$

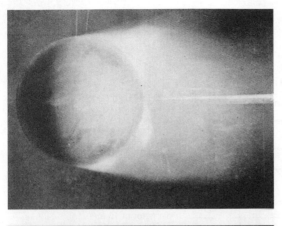

Figure 5.5
Effect of roughness on drag of a smooth sphere (Prandtl's experiment). (From Goldstein 1938.)

where the length has to be specified for each configuration. For a sphere and for a circular cylinder in flow transverse to the cylinder axis, the specified length is the diameter. (One states "the Reynolds number based on diameter.") For streamlined bodies, the length of the body in the direction of the flow is more usually specified. For an aircraft wing, this length is called the "chord." The specified length in bodies like streamlined fairings is more usually the actual length of the body.

For a sphere moving in air at sea-level pressure and 65°F (19°C), this becomes approximately

$$\mathrm{Re} = \frac{2}{3} \text{sphere diameter (m)} \times \text{relative velocity (m/s)} \times 10^5.$$

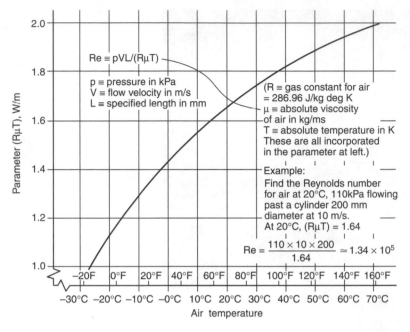

Figure 5.6
Reynolds-number parameter for air. (Plotted by Dave Wilson.)

A more general method of determining the Reynolds number for any pressure and temperature is shown in figure 5.6. Air density is a function of pressure and temperature:

$$\rho = \frac{pressure \text{ (pascals)}}{286.96 \times temperature \text{ (kelvin)}},$$

where the factor in the denominator is R, the "gas constant" for air, 286.96 J/kg-degK. The air pressure can be obtained from the local weather office, but it is always given for mean sea level and may need to be converted for the altitude required by using, for instance, the standard-atmosphere curve of figure 5.7. (The pressure variation with altitude would be useful everywhere on earth; the temperature would vary considerably.) The pressure will probably not be given in pascals (N/m^2) and may be converted using an appropriate part of the following:

1 bar $= 10^5$ Pa $= 0.9869$ atm $= 14.5038$ lbf/sq inch $= 750.062$ mm Hg
$= 29.530$ inches Hg.

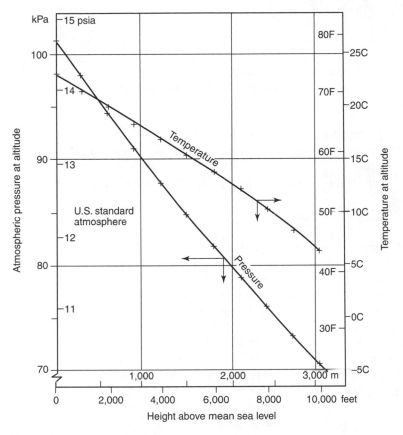

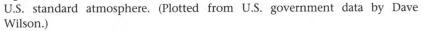

Figure 5.7
U.S. standard atmosphere. (Plotted from U.S. government data by Dave Wilson.)

The temperature in Kelvin is the temperature in Celsius plus 273.15. Sea-level air density is about 1.2 kg/m^3 at 16°C (60°F) and about 1.14 kg/m^3 at 38°C (100°F) for dry conditions. If the humidity is 100 percent, the density drops by about 1 percent at the cooler of these temperatures and by about 2.5 percent at the hotter.

However, it is not strictly necessary to calculate the air density purely to determine the Reynolds number. Since both the density and the air viscosity are functions of temperature, the parameter ($R\mu T$) is just a function of temperature, where μ is the "absolute" viscosity of air in kg/m-s, and T is the "absolute" temperature in degrees Kelvin. The Reynolds number can then be found from the pressure, temperature, velocity, and length alone:

$$\text{Re} = \frac{pVL}{(R\mu T)},$$

where the denominator is the parameter plotted as a function of temperature only in figure 5.6. An example of the use of these charts (figures 5.6 and 5.7) follows.

Example   Find the Reynolds number for air at 20°C, at sea-level pressure (110 kPa), flowing past a cylinder 200 mm in diameter at 10 m/s.

At 20°C the parameter $(R\mu T)$ is 1.64 (from figure 5.6). Therefore

$\text{Re} = 110,000 \times 10 \times 200/(1,000 \times 1.64) = 1.34 \times 10^5.$

Coefficient of drag versus Reynolds number for various bodies

The drag coefficient of various bodies versus the Reynolds number is plotted in figure 5.8. It can be seen that at Reynolds numbers over 3×10^5, even smooth spheres do not need trip wires or rough surfaces to induce turbulence, because a laminar boundary layer will spontaneously become turbulent under these conditions. When the boundary layer becomes turbulent at increased velocity and Reynolds number, the drag coefficient falls sharply from 0.47 to 0.10. (The drop in drag coefficient with increase of velocity or Reynolds number is not usually rapid enough to counteract the need for greater propulsion power, increasing as it does with the cube of velocity. However, hypothetically, certain bodies in certain conditions in which a very rapid reduction in drag coefficient is experienced as the relative velocity V is increased could accelerate by 20–30 percent without any increase in power.) A golf ball about 40 mm in diameter driven at an initial velocity of 75 m/s has a Reynolds number of 2×10^5 at the start and would be in the high-drag-coefficient region if it were smooth. The dimpling shifts the "transition" point to lower Reynolds numbers and gives a low C_D. Thus, paradoxically, a rough surface can lead to low levels of drag.

Compared with a golf ball, a bicyclist travels much slower but has a larger equivalent diameter, so the Reynolds numbers of the two may be similar. A bicyclist using an upright posture may be considered for simplicity as a circular cylinder normal to the flow, a curve for which is shown in figure 5.8. If we take a cylinder diameter of 600 mm to represent an average person, and if we use a speed of 5 m/s, the Reynolds number is 2×10^5, which is below the transition region of about 4×10^5. Therefore there may be some advantage to wearing rough clothing for speeds in this region. Most bicyclists have become aware of the speed penalty that results from converting themselves into smooth but highly unstreamlined bodies (see

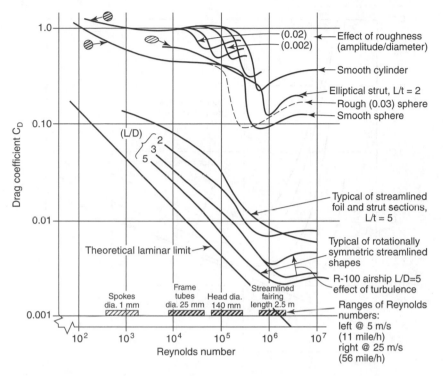

Figure 5.8
Drag coefficient versus Reynolds number for useful shapes. (Plotted by Dave Wilson from data from Hoerner 1965 and other sources.)

figure 5.10 and "Partial and full fairings") by donning a wet-weather cape or poncho, which usually, and somewhat paradoxically, greatly increases wind resistance without increasing cross-sectional area. Perhaps some "trips" woven into the cape material would be beneficial. Even better would be some type of frame that would convert the cape into a low-drag shape. Sharp proposed such a scheme in 1899, and capes with inflatable rims were for sale around that time. (See "Partial and full fairings" for modern variations.)

Most everyday bicycling occurs in the Reynolds-number range of $1-4 \times 10^5$, and the reduction in air drag through the use of some form of practical low-drag shape as an enclosure or "fairing" can approach 90 percent. An even greater reduction in drag can be produced with special-purpose fairings for racing or setting speed records.

Low-drag shapes do not generally exhibit a sharp transition from high drag (separated flow) to low drag (attached flow) as the Reynolds

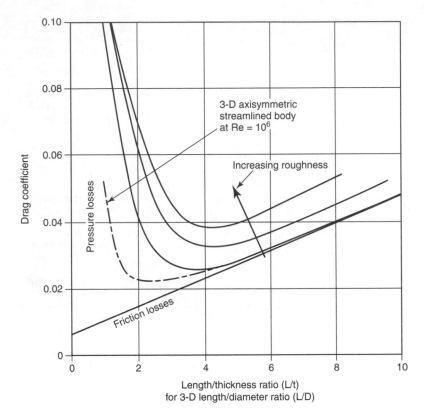

Figure 5.9
Optimum (L/t) of wing and strut sections and of one 3-D streamlined body.
(Plotted by Dave Wilson from data from Hoerner 1965 and other sources.)

number is increased. Rather, the point of transition of the boundary layer
from laminar to turbulent tends to move upstream toward the leading edge
of the body as the Reynolds number is increased. Thus, the drag coefficients
given for streamlined shapes (represented by an airship) in figure 5.8 show
a continuous fall as the Reynolds number is increased in the laminar-flow
region, followed by a moderate rise to the fully turbulent conditions and
then a continued fall.

The Reynolds numbers of streamlined fairings for human-powered
vehicles lie in the interesting transition region between 1.5×10^5 and $1.5 \times$
10^6. The curves in figure 5.9, taken from Hoerner 1959, show that for a
drag coefficient based on maximum cross-sectional (or frontal) area, the
minimum drag coefficient is given by streamlined shapes with a length/
(maximum thickness or diameter) ratio of about four.

Reducing the aerodynamic drag of bicycles

To reduce the wind-induced drag of a bicycle and rider, two alternatives are to reduce the frontal area of rider plus machine and to reduce the drag coefficient that the combined body presents to the air stream. For years, bicyclists have adopted one or other of these alternatives, but only recently have there been concerted attempts to reduce frontal area and drag coefficient simultaneously. The results have been remarkable. A selection of interesting and typical data has been assembled in table 5.1 (Gross, Kyle, and Malewicki 1983; Wilson 1997). The drag coefficients and the frontal areas are given in the first two columns, and the product of the two, C_DA, in the fourth column. Typical values for these three for an "upright commuting bike" are 1.15, 0.55 m², and 0.632 m². Such a bicycle, sometimes called "the British policeman's bicycle," and rider and this set of values are usually regarded as the "base case," to which improvements can be made.

One obvious source of improvement is for the rider to change position. A so-called touring position is used when riding a "road bike" (one with "dropped" handlebars) with the hands on the top of the bars. This reduces the drag coefficient from 1.15 to 1.0 and the frontal area from 0.55 to 0.4 m², giving a reduction in C_DA from 0.632 to 0.40 m². The fifth column of the table shows the power required at the driving wheel to overcome the aerodynamic drag at 10 m/s (22 mile/h), a speed at which aerodynamic drag is becoming dominant on unfaired bicycles. This fifth column shows immediately why ordinary people do not commute on upright bikes at 10 m/s: it requires 345 W (approaching half a horsepower) just to overcome aerodynamic drag. The power the rider puts into the pedals also has to supply losses in the transmission, normally small, and the rolling friction of the tires on the roadway, for which some typical data are given in the last three columns. The total power required to propel the upright bicycle would thus be over 400 W, a level that NASA, testing "healthy men," found could be maintained for only one minute (figure 2.4). Just making the switch to a road bike and using the touring position would reduce the total power required (on level ground in calm wind conditions) to around 275 W, and figure 2.4 shows that a nominally healthy male could keep this level up for about 30 minutes, a typical commuting duration. (It would be atypical to be able to commute for 30 minutes at constant speed, but if the typical male could do that, the distance would be 18 km [11 mile].)

A further dramatic improvement results if the rider uses a racing bike. (A racing bike is little different from the road bike used in the example above, but we have specified a lighter weight and a frontal area that includes the effects of tight clothing and having the hands on the "full-drop"

Table 5.1
Bicycle drag coefficients and other data

Machine and rider	Drag coefficient on frontal area, C_D	Frontal area		$C_D A$	Power to overcome air drag at 10 m/s (22 mile/h)	Power to overcome rolling resistance at 10 m/s for specified total mass (kg) and C_R value		
	C_D	m²	ft²	m²	W	kg	C_R	W
Upright commuting bike	1.15	0.55	5.92	0.632	345	90	0.0060	53
Road bike, touring position	1.0	0.40	4.3	0.40	220	95	0.0045	38
Racing bike, rider crouched, tight clothing	0.88	0.36	3.9	0.32	176	81	0.0030	24
Road bike + Zipper fairing	0.52	0.55	5.92	0.29	157	85	0.0045	38
Road bike + pneumatic Aeroshell + bottom skirt	0.21	0.68	7.32	0.14	78.5	90	0.0045	40
Unfaired long-wheelbase recumbent (Easy Racer)	0.77	0.35	3.8	0.27	148	90	0.0045	40
Faired long-wheelbase recumbent (Avatar Blubell)	0.12	0.48	5.0	0.056	30.8	95	0.0045	42
Vector-faired recumbent tricycle, single	0.11	0.42	4.56	0.047	25.8	105	0.0045	46
Road bike in Kyle fairing	0.10	0.71	7.64	0.071	39.0	90	0.0045	40
M5 faired low racer	0.13	0.35	3.77	0.044	24.2	90	0.003	26
Flux short-wheelbase, rear fairing	0.55	0.35	3.77	0.194	107	90	0.004	35
Moser bicycle	0.51	0.42	4.52	0.214	118	80	0.003	24
Radius Peer Gynt unfaired	0.74	0.56	6.03	0.415	228	90	0.0045	40
Peer Gynt + front fairing	0.75	0.58	6.24	0.436	240	93	0.0045	41
All-terrain (mountain) bike	0.69	0.57	6.14	0.391	215	85	0.0060	50

part of the handlebars; the figures for the rolling drag imply the use of light, supple, high-pressure tires. Loose clothing can increase aerodynamic drag, at speeds of over 10 m/s, by 30 percent.) The drag coefficient goes down to 0.88 (mainly because the head is down in front of the rider's rounded back); the frontal area is 0.36 m², and C_DA drops to 0.32. The power required to ride at 10 m/s is, including tire and transmission losses, about 210 W, which even NASA's healthy man could keep up for almost an hour. People who ride such bikes are more likely to be "first-class athletes," who can be seen from figure 2.4 to be capable of riding at 10 m/s indefinitely, which might be translated as until the need for food, sleep, or other demands of the body must be answered. (The one-hour standing-start distance record for conventional racing bikes was set in 1996 by Chris Boardman at 56.375 km, requiring an estimated average power output of over 400 W.)

Prone, supine, and recumbent positions and bikes
The frontal area presented by a bicycle and its rider can be reduced below that required for a conventional racing bike only by adopting a changed pedaling position. Speed records have been won on bicycles designed for head-first face-down horizontal-body (prone) pedaling, and for feet-first face-up horizontal-body (supine) pedaling, in the strict forms of which a periscope or other viewing device is needed; and for a wide variety of what is known as "recumbent" pedaling. Purists would say that fully recumbent pedaling is supine, and that strictly speaking the position used by the riders of "recumbents" is in fact "semirecumbent." However, the form of bicycle designed to be ridden in such a semirecumbent position has become known in the English-speaking world as "recumbent," or "bent" (and in Europe as *das Liegerad* or *liegfiets*). A well-known successful recumbent, the Easy Racer, is shown in table 5.1 as having a drag coefficient of 0.77, a frontal area of 0.35 m², and a C_DA of 0.27 m², considerably lower than that of the racing bike with the rider in a painful crouch. Therein lies a principal reason for the recumbent's growing popularity at the turn of the millennium: it can be simultaneously fast and comfortable. (These data may not be typical: also given in the table are measurements on a Radius Peer Gynt recumbent, for which considerably higher drag values were measured.)

Partial and full fairings
The organization that controls the rules for conventional bicycle racing, the UCI, has outlawed most measures aimed at reducing aerodynamic drag, including use of the recumbent position, and has even ruled inadmissible the form of racing crouch adopted by Graeme Obree, who beat the one-hour distance record twice in 1993, the second time reaching 52.7 km. However, this book is aimed at giving data helpful to people racing under

all rules (including those of the IHPVA) and to those who just want to use their muscles to travel at either the fastest possible speeds or with the least possible effort at a chosen speed. For these people, the potential for "going recumbent" and/or for using methods of streamlining, including partial or total streamlined enclosures or fairings, is attractive. A fairing also adds weight to a bicycle and makes it bulkier and more difficult to carry and to transport by motor vehicle, and at the present stage of development, it can require a considerable time for the rider to get into and to exit a bicycle equipped with a fairing. Accordingly, many people have devised partial fairings for the front or rear of bicycles.

Data for a bicycle with a partial front fairing, an early model of the Zipper (believed to be the precursor of the Zzipper) on a road bike, are given in table 5.1. This configuration of bicycle and fairing is shown to produce a relatively low drag coefficient and an overall value of C_DA lower than that for a racing bike with the rider in a full crouch. However, when a partial front fairing was fitted to a long-wheelbase Peer Gynt recumbent, both the coefficient of drag and the frontal area increased by small amounts. The notes accompanying the article on these tests (conducted for and published in the German bicycle magazine *Tour*) stated that small variations in the positioning of the fairing produced relatively large changes in drag. The drag coefficients of certain two- and three-dimensional shapes, including some that could be used as front fairings, are shown in figure 5.10 (from Hoerner 1965). The aerodynamic advantages conferred by front fairings have always been somewhat controversial, and research into the flow patterns found with different settings and spacings between the fairing and the rider seems called for.

The very large drag that occurs with a forward-curved half cylinder, which approximates the shape taken up by a poncho or cape on a standard bicycle, suggests the advantages of using stiffeners or other shape-improving means in such clothing. This form of full fairing was briefly mentioned above as an idea put forward by Archibald Sharp in 1899: the use of inflatable tubes to form a poncho or cape or other form of clothing into an aerodynamic shape. Paul van Valkenburgh developed this idea in the Aeroshell. Table 5.1 gives data for the use of this inflatable "suit" plus a skirt to extend the shape to close to the ground. A drag less than half that of the racing bicycle was attained.

The Swiss cyclist Oscar Egg, on a standard bicycle, set a one-hour distance record of 44.247 km in 1914 that lasted for nineteen years. In 1932 he was excited by the high speeds achieved by Faure on the Velocar (see chapter 1) and started experimenting with tail cones to decrease his drag (Mochet 1999) (figure 5.11). Tests at the time showed no improvement in speed. However, it has become popular for the same purpose, particularly in Europe, to fit aerodynamic "tail boxes" behind the seats of recumbent

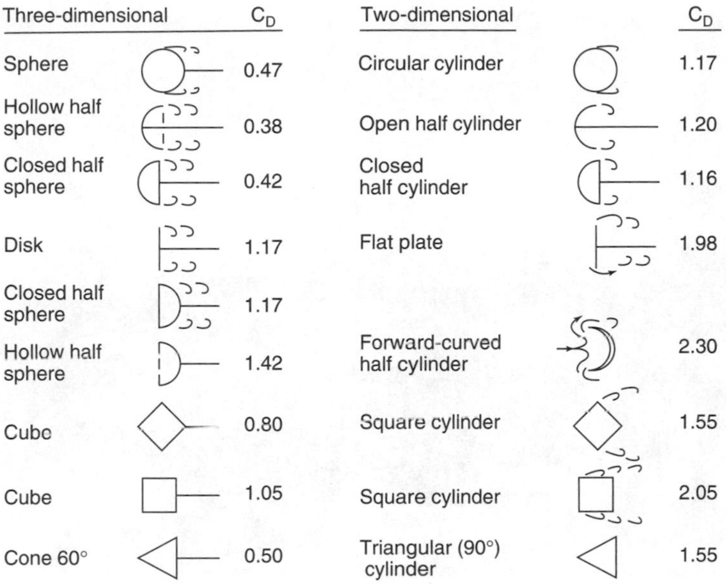

Three-dimensional
C_D
Sphere 0.47
Hollow half sphere 0.38
Closed half sphere 0.42
Disk 1.17
Closed half sphere 1.17
Hollow half sphere 1.42
Cube 0.80
Cube 1.05
Cone 60° 0.50
Two-dimensional
C_D
Circular cylinder 1.17
Open half cylinder 1.20
Closed half cylinder 1.16
Flat plate 1.98
Forward-curved half cylinder 2.30
Square cylinder 1.55
Square cylinder 2.05
Triangular (90°) cylinder 1.55

Figure 5.10
Drag coefficients of shapes for Re = 10^4–10^6. (Plotted by Dave Wilson from data from Hoerner 1965 and other sources.)

VELO FUSEE
Oscar EGG
PARIS

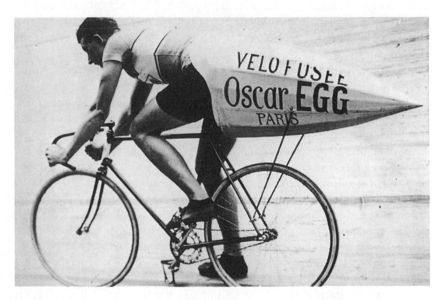

Figure 5.11
Use of tail cone to reduce drag. (From Borge, Le Vélo, p. 116.)

Figure 5.12
Bram Moens with his M5 Low Racer. (Photo courtesy of Bram Moens.)

bikes in order to achieve some pressure recovery. Table 5.1 includes data for a Flux short-wheelbase recumbent fitted with a rear fairing of this type, showing a value of C_DA considerably below that of the unfaired Easy Racer recumbent.

Table 5.1 also includes data for several machines with full fairings, meaning that they come as close to completely enclosing the rider and machine as possible. Chester Kyle's fairing of a road bike had a drag coefficient of 0.10 but a fairly large frontal area, as would be expected of a conventional bike, and the C_DA value was found to be 0.071. Recumbents tend to have higher drag coefficients when these coefficients are based on the frontal area, because the larger surface areas that result from riding in the recumbent position contribute drag, but the resulting C_DA values can be very low. The Avatar Bluebell had a C_DA of 0.056, the Vector recumbent tricycle 0.047, and Bram Moens's M5 Low Racer 0.044 (figure 5.12). The power estimated to be required to overcome rolling drag at 10 m/s is higher in this last machine than that for aerodynamic drag (table 5.1).

Practical fairings
At the present state of the art in fairing construction, full fairings usually must be taped shut over the riders, who then must be released from the

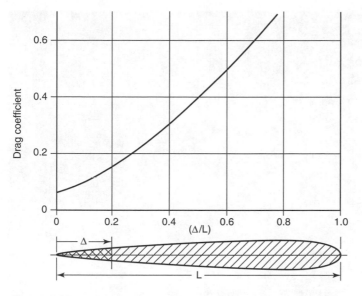

Figure 5.13
Effect on drag of cutoff trailing edges. (Plotted by Dave Wilson from data from Hoerner 1965.)

fairings at the end of their runs. There are many ways in which fairings can be compromised to make them easier to use in normal situations.

One way is to shorten the tail, taking a penalty in reduced pressure recovery, as shown in figure 5.13 (from Hoerner 1965). Many riders prefer to have their heads out of the fairings when using bicycles equipped with them for commuting or recreation and to have gaps in the fairings for access, as on the Lightning recumbents (figures 5.14 and 5.15). It is not possible to predict with accuracy the consequences, in terms of increased drag, that result from making such compromises. We recommend further study of Kyle's work (e.g., Kyle 1995) and of interpolation among the data for relevant machines in table 5.1.

Other aerodynamic phenomena

Boundary-layer suction
A separating flow leaves the surface of a fairing either because it is tripped by some fairly extreme form of roughness or because the boundary layer becomes thick enough for the low-momentum inner layers (those against the fairing surface) to be pushed backward (relatively) by an adverse pressure gradient. Therefore it is reasonable to expect that if these low-

Figure 5.14
Fully faired Lightning X2. (Photo courtesy of Lightning Cycle Dynamics.)

Figure 5.15
Partially faired Lightning F40. (Photo courtesy of Lightning Cycle Dynamics.)

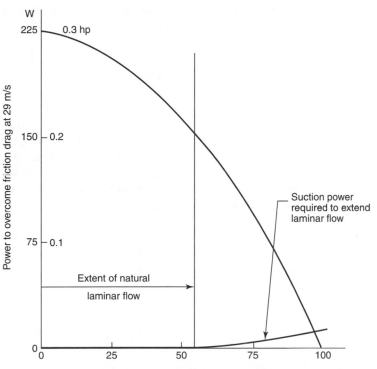

Figure 5.16
Aerodynamic-drag power of typical HPV at 29 m/s, 60 mile/h, versus proportion of laminar flow on fairing. (Plotted by Dave Wilson from data from Holmes 1985.)

momentum layers could be sucked away through holes or slits in the fairing surface, a previously separating flow could be made to remain attached to the surface, and pressure recovery could take place, greatly reducing drag.

Some power is required to suck away the boundary layer, but it is very small compared with the savings in propulsion power, as indicated in figure 5.16 (Wilson 1985, 7). For a typical human-powered fully faired vehicle traveling at 29 m/s (60 mile/h), Holmes (of NASA Langley) calculated that the power required to overcome air drag, were there to be no laminar flow whatsoever, would be 225 W. However, normal "natural" laminar flow would be expected to cover about 50 percent of the vehicle's surface, the drag power of which would be about 160 W. If suction were progressively applied until 95 percent of the vehicle's surface had an attached laminar boundary layer, the propulsion power required to overcome drag would be expected to be under 20 W, and the power required to produce the required suction would also be below 20 W.

This is a tantalizing prospect for anyone planning to break speed records, but caution in this area must be advised. One concern is that any form of bicycle that is pedaled to the maximum of a rider's output is swerving in a sinusoidal-like motion, and at the same time the fairing is being subjected to bumping from the road surface. It is probable that a far higher power than that indicated in figure 5.16 would be necessary to suck away the boundary layer. Also, unless it is possible to supply the suction from the (smoothed) rider's breathing, there will be inefficiencies in the suction-fan internal processes and in the transmission. Figure 5.16 does show why the author has felt that the ultimate speed reachable by a human-powered land vehicle would be on rails, because of the greater steadiness of the fairing that would result (plus the small gaps around the nonsteering wheels, reducing air "pumping," plus the low rolling resistance of steel wheels on steel rails, plus the capability of the rider to use arms and legs to produce power.)

The effects of surface roughness on streamlined bodies

Although, as mentioned above, a rough surface on a poorly streamlined body can give advantages in promoting the transition of the boundary-layer flow from laminar to turbulent, which might permit more recovery of pressure on the aft portion of the body and thus a reduction of drag, there is no doubt that with a streamlined body (which could be defined as one without flow separation), one should strive for as smooth a surface as possible. The effect of simple sand-grain roughness on the skin-friction coefficient, which is defined as the tangential friction drag per unit area divided by the dynamic pressure, is shown in the classical experiments on flow in tubes by Nikuradse (figure 5.17), discussed in Hoerner 1965. The length used in the Reynolds number in Nikuradse's experiments is the length of the tube, and the lines of constant roughness are characterized as the sand-grain diameter divided by this length.

Hoerner also quotes the results of flight tests on the wing of a King Cobra airplane. As received, the wing had surface imperfections. When these were removed, the drag at low-lift conditions (corresponding to the fairing of an HPV) was reduced by 65 percent.

Wind loads from passing vehicles

All bicyclists who have ridden on roads frequented by large, fast-moving motor vehicles have experienced side-wind forces from their passing, but no experimental work concerning the magnitude of the lateral forces

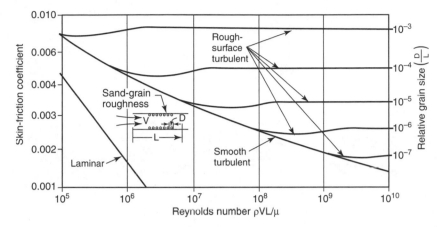

Figure 5.17
Nikuradse's roughness experiment. (Plotted by Dave Wilson from data from Hoerner 1965.)

exerted on actual bicyclists by such vehicles as they pass seems to have been reported. However, Beauvais (1969) has reported valuable work concerning wind effects upon "parked" and jacked-up one-tenth-scale model automobiles. (There is considerable concern in the United States about the safety of jacked-up vehicles at the sides of expressways.) From Beauvais's data we can estimate that a bicyclist may experience lateral forces of typically up to 25 newtons (7 lbf) when overtaken closely by a large vehicle moving at over 30 m/s (70 mile/h). The key word above is "closely": in those cases in which bicyclists are allowed to ride on the shoulders of highways, they should keep as far from high-speed-travel lanes as possible.

Drafting and side-by-side bicycling
A bicyclist is "taking pace" or "drafting" when he travels close behind another moving body, using it to "break the wind." The vortices behind a leading bluff body (see figure 5.1) may indeed help to propel the trailing rider. Drafting is therefore an important part of the strategy in massed-start races. Quantitative data have been gathered on the assistance given by drafting (Kyle 1969).

The second rider ("stoker") of a tandem is drafting behind the leading rider, and therefore incurs little additional drag beyond that which results from the first rider.

When streamlined fairings are used, competitors soon find that there is no benefit in drafting because there are no trailing vortices or large masses of captured air behind an aerodynamically faired shape. In fact,

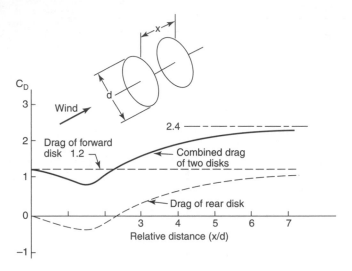

Figure 5.18
Drag interaction between two disks placed one behind the other. (Plotted by Dave Wilson from data from Hoerner 1965.)

some data reported below indicate that there may be a substantial penalty to drafting behind a bicycle equipped with a fairing.

The aerodynamic phenomena involved in drafting are well illustrated by graphs of the drag of pairs of bodies given in Hoerner 1965 and described by Papadopoulos and Drela (1999). The drag of two disks one behind the other (i.e., in tandem) is plotted in figure 5.18. The drag of the forward disk is not affected by the rear disk, which is, however, "dragged along" if it is within 1.5 diameters of the forward disk. This would be the case for riders on conventional tandems.

A better representation of two riders one behind the other is of two circular cylinders (figure 5.19). "When the gap is about two diameters, the lead [cylinder or person] actually experiences a reduction in drag of about 15%. The rear person at that spacing has about zero drag. When the separation increases to four diameters, the lead person loses any benefit, while the rear person's drag is about 25% of the solo value" (Papadopoulos and Drela 1999, 20).

Figure 5.20 offers a similar treatment for streamlined cylinders, which could be regarded as two-dimensional (vertical) fairings for HPVs. When the cylinders are within about one length of one another, the front actually receives a push, whereas the drag on the "drafting" HPV *quadruples*. Presumably the wake from the first fairing causes flow separation over the

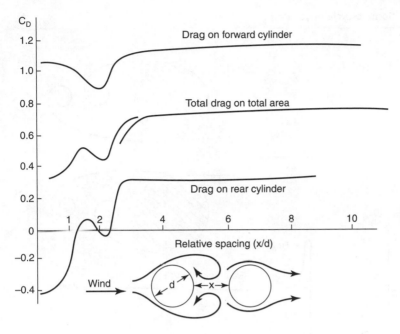

Figure 5.19
Drag coefficients of two circular cylinders, one placed behind the other. (Plotted by Dave Wilson from data from Hoerner 1965.)

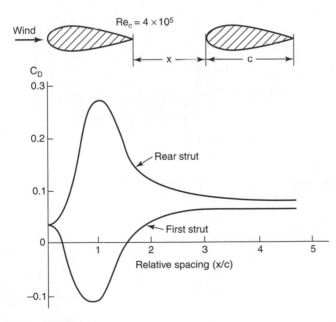

Figure 5.20
Drag of a pair of strut sections, one behind the other in tandem. (Plotted by Dave Wilson from data from Hoerner 1965.)

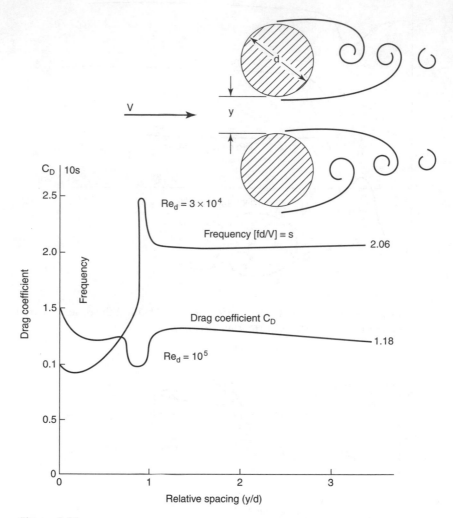

Figure 5.21
Drag and vortex-shedding frequency of a pair of circular cylinders placed side
by side. (Plotted by Dave Wilson from data from Hoerner 1965.)

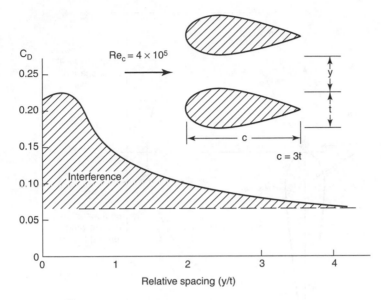

Figure 5.22
Drag of a pair of struts, one beside the other. (Plotted by Dave Wilson from data from Hoerner 1965.)

second fairing. It is no wonder that riders in HPV races do not try to draft one another.

The drag and vortex-shedding frequencies of two circular cylinders side by side are plotted in figure 5.21. When the cylinders are touching, the drag is increased about 25 percent over the solo value. At one-diameter spacing, the drag is reduced about 15 percent over a small range, indicating a sensitive interaction probably related to the high vortex frequencies at that spacing.

When two streamlined cylinders (struts) are side by side (figure 5.22), the drag is greatly increased at small spacings and decreases to the solo values only at relative spacings of over four diameters.

Behavior of faired bicycles in crosswinds

There is little remarkable about the behavior of unfaired bicycles in cross-winds, except for the extraordinary stability they normally display. (It seems extraordinary because if a non-bicycle-riding aerodynamicist were asked to predict the course of a bicycle hit by a sudden gust of wind at, say, 15 m/s (34 mile/h) he would probably estimate either that the bicycle would be unridable in those conditions, or that the rider would be forced

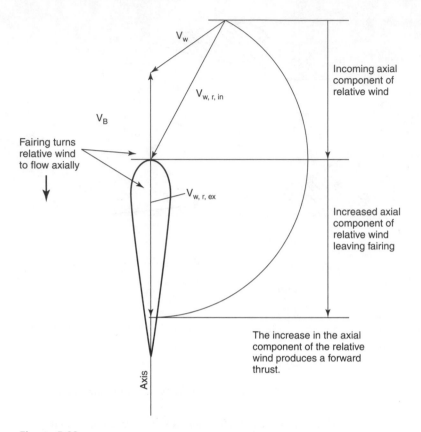

Figure 5.23
How a headwind produces a forward thrust on a faired bicycle. (Sketched by
Dave Wilson from Weaver 2000.)

into a wide swerving path to maintain stability. Yet most riders can ride
fairly precisely (for instance in traffic) in such circumstances.)

It is quite another matter to ride a faired bicycle in crosswinds. Even
the use of front-wheel disks can make riding in crosswinds unpredictable.
The large side area of a full fairing produces transverse aerodynamic forces
in crosswinds that are far larger than those on an unfaired machine. This
topic is too specialized for this text, but the following points can be made
and the references recommended for further guidance.

Milliken (1989) reported simple and effective experiments that
showed that in most cases the aerodynamic center of pressure of transverse flow on a fairing should be ahead of the center of mass to give
good crosswind stability. Fuchs (1998) confirmed this experimental find-

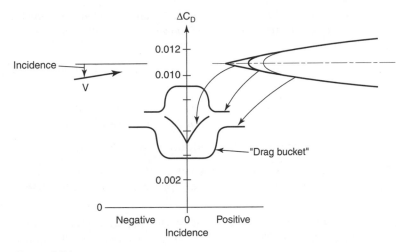

Figure 5.24
Incremental drag (ΔC_D) from leading-edge shape. (Plotted by Dave Wilson from data from Hoerner 1965.)

ing theoretically and arrived at an equation that "allows the designer to trim a single-track vehicle so that it keeps its course in a steady field of crosswind."

Weaver (2000) showed that over a wide range of incidence (the angle the relative wind makes with the direction of motion), the wind acts on a bicycle fairing in much the same way as it does on a sailboat: it provides forward thrust (figure 5.23). The fairing can take flow at a considerable incidence angle (figure 5.24) and turn it almost to leave the fairing in the aft (backward) direction, which (Newton's laws!) provides thrust. Therefore wind has an important effect on records involving bicycling, even if a record-setting ride is made on a circular or oval track, on which, it has been maintained, the negative effects of headwinds would cancel out the positive effects of tailwinds. The net positive thrust that results from the action of wind on a bicycle's fairing can occur even though the fairing has no "camber" or curvature, as does a sail or an airplane wing.

References

Abbott, I. H., and A. E. Doenhoff. (1959). *Theory of Wing Sections*. New York: Dover.

Beauvais, F. N. (1969). "Transient aerodynamical effects on a parked vehicle caused by a passing bus." In *Proceedings of the First Symposium on Road Vehicles, City University of London, U.K.*

Fuchs, Andreas. (1998). "Trim of aerodynamically faired single-track vehicles in crosswinds." Paper presented at Third European Seminar on Velomobiles, Rosskilde, Denmark, August 5.

Goldstein, S. (1938). *Modern Developments in Fluid Dynamics*. London: Oxford University Press.

Gross, Albert C., Chester R. Kyle, and Douglas J. Malewicki. (1983). "The aerodynamics of land vehicles." *Scientific American* 249, no. 9 (December).

Hoerner, S. F. (1965). *Fluid Dynamic Drag*. Bricktown, N.J.: Hoerner.

Kyle, Chester R. (1979). "Reduction of wind resistance and power output of racing cyclists and runners travelling in groups." *Ergonomics* 22, no. 4:387–397.

Kyle, Chester R. (1995). "Bicycle aerodynamics." Chapter 10 in *Human-powered Vehicles*, ed. Allan Abbott and David Wilson, 141–156. Champaign, Ill.: Human Kinetics.

Milliken, Doug. (1989). "Stability? Or control?" *Human Power* 7, no. 3: 9, 14.

Mochet, Georges. (1999). "Charles Mochet and the Velocar." *Recumbent Cyclist News*, no. 52 (July/August).

Papadopoulos, Jim, and Mark Drela. (1999). Some comments on the effects of interference drag on two bodies in tandem and side-by-side. *Human Power*, no. 46:19, 20.

Sharp, A. (1899). *CTC Gazette*, no. 11 (January).

Simons, Martin. (1999). *Model Aircraft Aerodynamics*, 4th ed. Swanley, U.K.: Nexus Special Interests.

Weaver, Matt. (2000). "Body shapes and influence of the wind." *Human Power*, no. 49:21–24.

Wilson, David Gordon. (1985). "Report on address by Bruce Holmes to the second IHPVA builder's workshop." *Human Power* 5, no. 1 (Winter):7.

Wilson, David Gordon. (1997). "Wind-tunnel tests: Review of Tour, das Radmagazin, article." *Human Power* 12, no. 4 (Spring):7–9.

Recommended reading

Fosberry, R. A. C. (1959). "Research on the aerodynamics of road vehicles." *New Scientist 6* (August 20):223–227.

Gertler, M. (1950). "Resistance experiments on a systematic series of bodies of revolution, for application to the design of high-speed submarines." Report no. C297, David Taylor Model Basin, U.S. Navy.

Kyle, C. R. (1974). "The aerodynamics of manpowered land vehicles." Paper presented at Third National Seminar on Planning, Design, and Implementation of Bicycle and Pedestrian Facilities, San Diego, Calif.

Kyle, C. R. (1975). "How accessories affect bicycle speed." Engineering report no. 751, California State University, Long Beach.

Kyle, C. R., V. J. Caizzo, and P. Palombo. (1978). "Predicting human-powered-vehicle performance using ergometry and aerodynamic-drag measurements." In *Proceedings of IMFA Conference, Technical University of Cologne.*

Kyle, C. R., and W. E. Edelman. (1974). "Manpowered vehicle design criteria." Paper presented at Third International Conference on Vehicle System Dynamics, Blacksburg, Va.

Nonweiler, T. (1956). "Air resistance of racing cyclists." Report no. 106, College of Aeronautics, Cranfield, U.K.

Sharp, A. (1896). *Bicycles and Tricycles.* London: Longmans, Green; reprint, Cambridge: MIT Press, 1977.

Van Baak, M. A. (1979). "The physiological load during walking, cycling, running, and swimming and the Cooper exercise program." (Keppel: Krips Repro Netherlands).

Whitt, F. R. (1972). "Is streamlining worthwhile?" *Bicycling* (July):50–51.

6 Rolling: tires and bearings

Introduction

Wheels surely count among the greatest of human inventions. But their ability to convey a load with low resistance depends on their size, the smoothness and firmness of the surface on which they travel, and the properties of tires and suspensions. As was discussed in chapter 4, road irregularities, too, retard motion, by shaking the rider, compressing the bicycle's suspension, or accelerating the bicycle upward. However, this retardation is primarily a question of suspension and will not be discussed here. In this chapter, we delve into the friction and drag of smoothly rolling wheels and turning bearings.

Some historical notes

A wheel's resistance to rolling can increase fiftyfold from pavement to soft soil, far more than resistance increases in walking on those same surfaces. Hence, there was a real incentive to develop paved roads when wheels were adopted for horse-drawn vehicles (figure 6.1). The Roman Empire was the first civilization to put a system of paved roads into use. It is recorded that the times taken to travel across various European routes to Rome were shorter in the Roman era than a thousand years later in the Middle Ages, when the Roman road system had almost vanished through lack of maintenance.

After the Middle Ages, inventions to improve everyday life appeared rapidly. Among these were iron-covered wooden railways, followed by iron wheels and cast-iron rails (1767). These gave rise to the steam railways of Victorian times, which were paralleled by the reappearance of a fair number of paved roads. Thomson (1845) and Dunlop (1888) introduced pneumatic tires that dramatically decreased the impact of forces on the rider due to bumps and thereby made the energy losses for unguided wheels on paved roads similar in nature (though higher in level) to those experienced by the railway wheel. Pneumatic tires also introduced a degree of comfort for those traveling on common roads.

Magnetically levitated and air-supported transportation vehicles have essentially zero friction at normal speeds, but power is required to supply and to control the lift required to enable the vehicles to move. Consequently, hard steel wheels on smooth steel rails require the least power of all systems used to support practical vehicles on land. The intrinsic rigidity of the contacting components means that very little energy can be lost due

Figure 6.1
Replica of Egyptian chariot wheel of 1400 B.C. Note rawhide wrapping to make the tire resilient. (Courtesy of Science Museum, London; reproduced with permission.)

to material deformation as the wheels roll. The average automobile wheel on the best surfaces generally available has ten or more times the resistance to motion of a train wheel on its track, when both carry similar loads. The difference in resistance is due to an automobile tire's intentional deformability, which is necessary to reduce the forces of bumps that the car encounters and thereby to minimize the energy channeled into vertical motion and suspension or occupant vibration.

Rolling resistance

The power needed to pull loaded wheels over a smooth surface depends on the physical properties of both. A great deal of empirical information is available concerning the power requirement for moving all types of wheels on harder surfaces. Wheel-movement requirements under soft-ground conditions have until recently been significant mostly to agricultural engineers and designers of military vehicles but are now of concern also to the designers and users of ATBs. I give some information on soft-ground rolling resistance below.

The term "rolling resistance" as used in this book means the resistance to a wheel's steady motion caused by power absorption in the surfaces of the wheel and of the road, rail, or soil on which it rolls. Rolling resistance does not include bearing friction or the power needed to accelerate or slow the wheel because of its inertia. And it does not include "suspension losses": energy losses in the wheel, suspension, or rider due to impact and vibration. Unfortunately, such losses are inevitable when one is riding on real-world roads.

Table 6.1
Rolling-resistance coefficients (C_R) of four-wheeled steel-tired wagon on 1.5-ton stagecoach

Surface	C_R	Speed	Vehicle
Cubic blocks	0.014–0.022	Slow	Wagon
Macadam	0.028–0.033	Slow	Wagon
Planks	0.013–0.022	Slow	Wagon
Gravel	0.062	Slow	Wagon
"A fine road"	0.034–0.041	4–10 mile/h	Stagecoach
Common earth road	0.089–0.134	Slow	Wagon

Note: Data from Trautwine 1937, 683.

In what follows, the force of rolling resistance is usually represented as a rolling-resistance coefficient C_R times the load carried, just as sliding friction is represented as a friction coefficient times load. This empirical approximation is useful, but its validity is not always borne out by measurements or sophisticated analyses. We recommend that readers interested in rolling-friction theory consult Trautwine 1937, Hannah and Hillier 1962, Reynolds 1876, and Evans 1954 for details about a subject not frequently discussed in textbooks on basic physics. Some empirical rolling-resistance coefficients for a wagon are given in table 6.1.

Bicycle wheels

A conspicuous characteristic of most bicycles is the relatively large-diameter wheels (about 20 percent larger than those of a typical passenger car) turning on ball bearings and shod with tires inflated to two to four times the pressures of passenger-car tires. Even the word "bicycle" acknowledges the importance of wheels to the vehicle it names.

Bicycles' large wheel size benefits bicycle performance in several ways.

- The angle from axle to the point of impact is more nearly vertical in a large wheel than in a smaller one, and so a large wheel can roll over holes or bumps that might completely stop a small wheel. Greater horizontal travel is required before a bump is crested by a large wheel, and so vertical accelerations are gentler. Forces acting to jar the rider are smaller, as are vertical velocities, whose associated kinetic energy is largely unrecoverable.
- There is reduced tire energy loss in smooth rolling on large wheels than on small ones. A large wheel with a tire at high pressure develops a

long, load-supporting contact patch with minimal tire flattening, and so the energy dissipated in the tire structure is lower.

•   Bearing wear is reduced in large wheels compared to small. A larger wheel allows the wheel's bearings to turn more slowly, enabling them to, last longer and contribute less friction.

•   Large wheels reduce the degree of sinkage in soft ground compared to small. Wheels having a large radius (and/or width) resist energy-robbing ground penetration.

•   A large wheel improves the "feel" and stability of steering/balancing.

On the other hand, a large wheel at high speed has higher aerodynamic drag than a small wheel at a comparable speed, and it is difficult to make large, light, slender structures such as large bicycle wheels laterally stiff for precision in steering and strong so that they don't collapse under the weight of radial plus side loads. The light wheels of conventional racing bicycles must be considered marginal in strength. Weak wheels don't usually cause bicycling accidents, but when such an accident does occur, the bicycle's wheels are often destroyed. (However, the rim of a bicycle wheel is a "crush zone" that spares the bicycle and cyclist severe impacts when the tire bottoms out or the cyclist hits a vehicle or wall.)

Comparing the friction of tires and bearings

When a bicycle rolls forward, the tires and the wheel bearings resist the motion to some extent. But while the drag of the tires can be measured or even felt while riding (as long as speed is not too great), the drag of ordinary ball bearings is utterly negligible, as long as they are not adjusted far too tightly (see the calculation below).

Tires

The amount of rolling resistance exerted by a vehicle's tires depends on the load carried and possibly also on the speed at which they are rolling, although the effect of speed on resistance is not ordinarily acknowledged. Tire rolling resistance is usually tested by pressing the vehicle's wheel with a known amount of force against a turning drum[1] and measuring the power required to keep the vehicle in motion. Ideally, the drum will be of considerably greater diameter than the wheel, or else the contact patch will be unrealistically short and wide (i.e., the drum will penetrate deeper into the tire than would a flat road surface), and the drag measurement will be artificially high, as with competition-training rollers. The drum diameter would be less of an issue if identically inflated tires could be pressed together: the contact region would then, ideally, be planar, and the deformation-based drag would be precisely twice that of one tire. However, the sideways motion of the tire against the road within the contact patch, called "scuffing," would also be reduced, so that an artificially low rolling resistance might then be measured. An imperfect wheel, misaligned mounting, or uneven tire construction also may play a role in the amount of drag from the tire, depending on whether the wheel is free to tilt or move vertically as it is while the bicycle is being ridden and whether or not that motion dissipates energy.

Under typical test conditions, the force of rolling resistance F_R is measured via either the operating power or from the coasting (unpowered) deceleration and then is represented as F_V (average vertical force supported) times C_R (coefficient of rolling resistance). For bicycle tires on a smooth hard surface, C_R is usually considered to be between 0.002 and 0.010, depending on inflation pressure, wheel diameter, and tire construction. For a bicycle-plus-rider mass of 80 kg, the total weight carried is 784 N, and the total rolling drag is between 1.5 N and 7.8 N (0.3 lbf to 1.75 lbf). For comparison, aerodynamic drag in low-wind conditions typically ranges between 5 N and 30 N in level riding, depending on speed (figure 6.2).

Tire rolling resistance can also be measured by rolling the tire along flat surfaces,[2] with the following caveats.

•   Slope is highly important. Nominally level indoor surfaces can easily slope 0.001 in places, altering the apparent value of C_R by 10–50 percent. Outdoor variations in slope can be far greater.

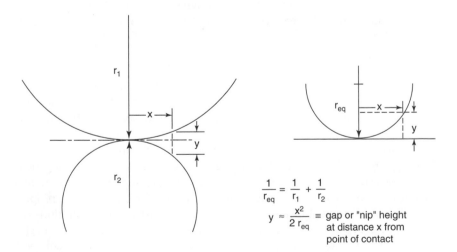

Figure 6.2
Combination of equivalent roller and a plane giving the same y versus x "nip" profile as a two-roller combination.

· Wind is also a concern. Unless the test vehicle has much less frontal area than a normal rider, any wind over 2 m/s will substantially alter the force being measured. In fact, even in windless conditions, air drag must usually be determined and subtracted.

· If the rig on which tests are conducted is not a bicycle skillfully controlled by its rider, outrigger wheels are required to maintain low-speed balance. Any wheel misalignment will add considerably to the drag.

As this book goes to press, an appealing new measurement possibility has appeared. On-bicycle power measurement and averaging (for example, using the PowerTap instrumented rear hub, figure 4.10) makes it possible to ride multiple circuits of a flat course at a constant speed and to determine the average power required. Testing needs to be conducted at low riding speeds (2–3 m/s) to reduce the contribution of aerodynamic drag to about 30 percent of the total, as well as at high speeds (7–10 m/s) to evaluate the aerodynamic drag factor so that the air drag at low speed can be estimated and subtracted. The main problem is that a steady wind adds to the average drag around the circuit (or, for streamlined bicycles, can produce a net thrust). Indoor riding in a large building would eliminate this problem, but then the surface on which the testing was conducted would be different from that of a normal road, and exploring the effect of a range of temperatures would be more difficult.

Preliminary PowerTap measurements on the road and on a home "wind" trainer have suggested significant effects on tire rolling resistance of temperature (with C_R dropping roughly 1 percent for each degree Celsius of temperature rise) and speed (with C_R doubling when wind-trainer speed reaches 5 m/s).

Bearings

Rolling-element bearings use many small balls or rollers to reduce friction and wear. The first widespread use of ball bearings was in bicycles, although the concept of ball bearings was understood prior to their use in bicycles.

Although the rolling-element approach seems "obviously" superior, bearings are highly sophisticated devices, and how well they work depends on various subtle factors. Bearings made of strong materials, properly manufactured and finished with high precision, positioned with the proper configuration and kept clean and lubricated, can last for many millions of revolutions (depending on the load they are required to carry). For bicycle use, the temptation is usually to adopt bearings made of lighter or cheaper materials, reducing the life of the bearings to a tolerable minimum: less than one million wheel revolutions for a bicycle that is not ridden much, and perhaps ten million revolutions for a "serious" bicycle.

The most authoritative contemporary reference on ball and roller bearings is *Rolling Bearing Analysis*, by Tedric Harris (1991). Bearing manufacturers also present basic information in the engineering pages of their catalogs.

Bearing friction

If a bicycle wheel is removed from the bicycle's frame and its axle is turned with the fingers, a small resistance may be felt. This resistance typically is due to the use of a thick grease or of bearing seals: it is not indicative of the friction under load.

If the bearings are adjusted too tightly (preloaded), a better idea of their friction under load may be developed. Low-precision bearings will turn roughly, and with high-quality bearings, it will feel more as if an extra-heavy grease has been added.

The quick-release skewer (the through-axle tension rod used to secure most modern wheels) on a bicycle wheel applies considerable compressive force to the wheel's axle, shortening it by 0.02–0.04 mm and thereby "tightening" the bearing adjustment. The effect of this shortening and the resultant tightening can be felt by placing some washers on the axle to take the place of the bicycle frame and squeezing them with the skewer as if the wheel were installed. This experiment will not simulate any bearing load that might arise from axle bending due to operating loads or preexisting frame misalignment. But even the friction of the bearings as installed in the frame does not reflect the actual friction encountered in riding, which involves a radial wheel load of (say) 450 N (100 lbf).

Perhaps the easiest way to measure the actual bearing friction of a bicycle wheel carrying a bicycle and rider is to find a way to spin the axle while the loaded hub remains stationary. (For example, the axle could be supported "between centers" on an engine lathe.) This approach has the added benefit of eliminating any aerodynamic drag that may be present. In conducting this kind of testing, either the quick-release skewer has to be tightened, or the bearings have to be preloaded in some other way to simulate an actual installation. Weights are attached to the hub or wheel to apply the desired load (e.g., 500 N). When the axle is continuously rotated, torque is needed to prevent the wheel from turning. If the wheel is properly balanced, this torque can be measured with the wheel at any orientation. On the other hand, if the wheel is unbalanced by a known amount, the torque can be measured from the angle through which the wheel turns between the situations of steady forward and steady reverse running. In this kind of test, it is important to avoid excessive rpm, as vibrations occurring as a result of a bent or otherwise imperfect axle can interfere with the measurement, and lubricant will be distributed differently than in a rotating hub.

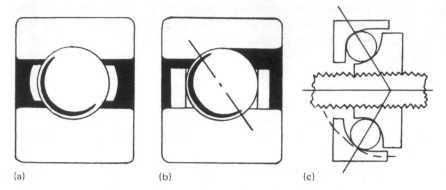

(a)   (b)   (c)

Figure 6.3
Types of ball bearings: (a) annular or radial, (b) 1893 "magneto" (the Raleigh version had a threaded inner race), (c) cup-and-cone (the bearing is self-aligning and can accommodate a bent spindle).

For well-aligned, properly lubricated bearings, approximate friction coefficients (defined at the radius of the circle of rolling elements) are sometimes published by manufacturers. But a more complete treatment may be found in the section on friction torque in *Rolling Bearing Analysis*. In a wheel rolling in a straight line on a flat plane, material hysteresis (Drutowski 1959) may cause friction, but the rolling elements in most bearings also undergo a certain amount of scrubbing motion within their tiny contact areas. For angular-contact (i.e., cup-and-cone) ball bearings like those shown in figure 6.3, the friction coefficient is given as $0.001 \times$ (service load/static load rating)$^{1/3}$. (For needle-roller bearings, or for radial-contact ball bearings, the friction coefficient can be smaller by as much as a factor of five.)

The static load rating is that load that will produce a specified, minuscule, permanent indentation in the ball race under the ball bearing the highest load. For bicycle-sized bearings, the Torrington Service Catalog shows that the static load rating is typically half of the basic dynamic load rating, which is defined as the load at which 90 percent of a group of bearings will last at least one million revolutions. If actual bearing life is taken as eight million revolutions, a conventional bearing-life calculation implies that the service load is also approximately half of the basic dynamic load. Therefore, the friction coefficient should be close to 0.001.

What this means is that a wheel carrying 450 N (100 lbf) would develop a tangential friction force (at the ball circle) of about 0.45 N (0.1 lbf), in addition to any friction from seals, preload friction, etc., that may be present. The finger feel of this friction could be simulated by winding a thread around the axle where its diameter is 16 mm, with a 0.55-N weight at its end, to create 0.004 N-m torque. As this force of 0.45 N acts at the small radius of 0.01 m, it causes a far smaller force at the wheel radius of perhaps 330 mm. Therefore, the drag force due to the bearing friction of one wheel is in the neighborhood of (0.45/33) or 0.014 N and is quite negligible compared with a typical tire rolling resistance of 1–3 N. Clearly, bearings benefit tremendously when the wheel's own radius is much greater than theirs: both drag and wear rate per unit distance traveled are reduced by the ratio of the wheel radius to the bearing radius.

Since wheel bearings contribute so little to overall drag, could we use more economical plain bearings: close-fitting bushings of low-friction metal or plastic? Their bearing radius might be as small as 0.005 m, giving a wheel mechanical advantage of 0.33/0.005 or 66. To add less than 0.001 to the apparent rolling-resistance coefficient C_R (effectively, the difference between an excellent tire and a good one), the plain bearing would have to have a sliding friction coefficient less than 0.07. This is achievable with a modern dry-film lubricant coating. For example, Whitford's Xylan has a friction coefficient quoted as 0.05–0.10 ⟨www.whitfordww.com and also www.garlockbearings.com/du_dx.pdf⟩. A low friction coefficient (0.01–0.07) could be achieved more easily with the addition of a liquid lubricant. Some lubricious solids like PTFE (polytetrafluoroethylene) attain truly low friction coefficients only in certain conditions: at relatively high pressures, with a slight degree of roughness present, and after a certain amount of rubbing (break-in) has built up a film of lubricious material on the mating bearing part.

Appropriate plain bearings therefore appear to offer sufficiently low friction for economy-model bicycles. We do not know whether they have been shunned for reasons that are sound (e.g., messy weeping of lubricant, premature abrasive wear by road dust, need for tapered journals and adjustability to take up wear), or whether they merely suffer from an image problem. (Plain bearings *are* used for derailleur jockey pulleys and for the pinions of internally geared hubs; however, these hubs have a chamber that serves as a reservoir for lubricant. On its cheaper models, Raleigh used plain bearings in pedals and in the lightly loaded upper headset bearings for a short time around 1970. The coefficient of friction was noticeably and annoyingly high. Wear and contamination probably resulted in much worse performance than cited here. Major advantages of rolling bearings are their relative durability and low friction even when poorly lubricated.)

Rolling resistance: theory and correlations

Bicycle-tire rolling-resistance coefficients for smooth surfaces are widely accepted to range between 0.002 and 0.010, making the tires the second-most important contributor, after air resistance, to the level-road drag act-ing on a bicycle. There is considerable uncertainty about precise values of rolling-resistance coefficients in particular cases, and the general effects of factors such as wheel diameter, tire pressure, temperature, and pressure on rolling resistance have not yet been fully explored.

The entire subject of rolling resistance has been treated primarily em-pirically from a variety of perspectives, and much further study is needed. For these reasons, we simply summarize a wide range of published results. One of the most comprehensive available theoretical treatments can be found in chapters 8–9 of Johnson's *Contact Mechanics* (1996).

Tilted ground force due to material inelasticity

The resistance to rolling of a pneumatic tire on concrete, and of a hard wheel on soft ground, have some similar characteristics. In each case, the main resistance to rolling arises from imperfectly elastic deformations of at least one of the materials involved;[3] and larger wheel radius generally reduces drag in both cases. As long as a wheel is not adhesive in any way, the forces exerted by the surface on which it rolls are all compressive. (This condition is sufficient for the net support–plus–resistive force of the ground to act somewhere within the contact patch.) Because the wheel's bearing is low-friction, the force of the ground must act on a line directed through the center of the wheel, and so the wheel's rolling drag is equiva-lent to that presented by the ground forces' being located ahead of the axis and tipped back from vertical to aim at the axle.

If the wheel made only point contact with a geometrically flat sur-face, the force between the wheel and the surface would have to be purely vertical (i.e., no rolling resistance). However, a loaded wheel or ball never makes true point contact with a flat surface: if it did, the contact pressure (force divided by area) would be infinite, and materials failure would occur. In reality, some deformation takes place so as to develop a nonzero area of contact.

The contact pressure is generally not uniform within a particular contact area. However, as long as the pressure's distribution is longitudi-nally symmetrical, the net support force from the road surface remains vertical. Forward rolling causes the pressure to be greater in the leading part of the contact patch than in the trailing part (figure 6.4), which leads to rolling resistance.

The angle through which the support force is tipped is bound to be considerably less than the angle from the wheel's axle to the forward tip of the contact patch. Thus, in otherwise comparable situations, the wheel whose contact patch subtends the smallest angle is likely to have the lowest drag. (However, this implies that wider tires should have lower drag, which is not the case.) If the contact length forward of the axle (i.e., half of the

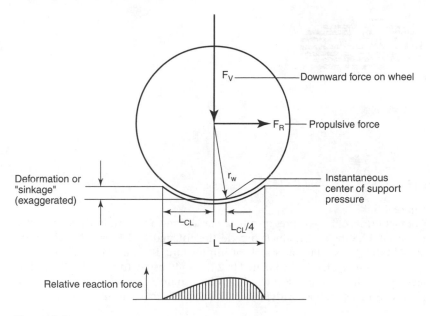

Figure 6.4
Resistance of hard rolling wheel on a soft, elastic surface.

smooth-road contact-patch length) is denoted L_{CL}, then the maximum possible support-force tilt angle in radians is closely approximated by the ratio L_{CL}/r_w, where r_w is the radius of the wheel.

A pneumatic tire on a wheel of large radius produces a long, slender contact patch, whereas a small-radius wheel cannot generate a very long contact patch for a given sinkage, so it has to sink further to create a wider one (figure 6.5). The result is more deformation for each cross section of the tire, plus a larger angle of contact with the surface beneath the wheel.

Form of resistance equation
The foregoing reasoning (combined with some dimensional analysis) suggests a likely form for the force (F_R) that resists rolling of a wheel-supporting force (F_V): $F_R/F_V = f(L_{CL}/r_w)$, where f represents an unknown function increasing from zero.

The ratio F_R/F_V is defined as the coefficient of rolling resistance, C_R. It is often considered to be independent of the load F_V, even though it probably is not (since L_{CL} is affected by F_R). When a single number is given for C_R in the literature, it should be assumed to apply only to specific loading conditions, which unfortunately are not always described.

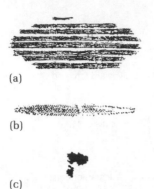

(a)

(b)

(c)

Figure 6.5
Contact prints of bicycle tires on a hard surface and a steel train wheel on a steel track: (a) 12.5-inch × 2.25-inch bicycle tire inflated to 1.8 bar (26 psi) with a 400-N (90-lbf) load (actual length of impression: 100 mm [4 inches]), (b) 27-inch × 1.25-inch bicycle tire inflated to 2.8 bar (40 psi) with a 400-N (90-lbf) load (actual length: 97 mm [3.8 inches]). (c) Steel train wheel of diameter 890 mm (35 inches) on steel track with load of 27 kN (6,075 lbf). (From Whitt 1977.)

The following examples are of simple rolling-resistance analyses or measurements. All contact half-length (L_{CL}) calculations should involve vertical load (F_V), wheel radius (r_w), and a quantity with dimensions of stress, such as modulus (E), inflation pressure (p), or compressive yield stress of the soil (Y). Vertical wheel sinkage y_{WS} is related to L_{CL} through $y_{WS} \approx L_{CL}^2/(2r_w)$. In addition, there may be ancillary geometrical factors that affect the calculation of contact half-length, such as wheel width L_{WW} and radial tire thickness L_{TT} in the case of a rubber-covered cylinder or radius r_T in the case of pneumatic-tire cross section. The drag itself is caused by material energy-loss parameters that are not often tabulated. In the simplest case the energy loss would appear as a multiplicative "loss factor," possibly dependent on speed.

Examples of correlations for different conditions

Firm wheel and firm ground   For a railroad wheel on its track, Koffman (1964) indicates that C_R is proportional to L_{CL}/r_w, or in other words,

$$F_R = F_V K_1 (L_{CL}/r_w),$$

where $K_1 = 0.25$, and is the constant of proportionality.

For an unknown load, $K_1 L_{CL}$ is given as 0.25–0.5 mm, so C_R for a wheel of radius 0.5 m is 0.0005. To estimate that unknown load, compare

$L_{CL} = 9$ mm for 27 kN (6,075 lbf) (figure 6.5) to $L_{CL} = 2$ mm in this case. The Hertzian contact formula for crossed cylinders or for a sphere on a plane, which applies here because of the rounded surface of the rail, implies that F_V is proportional to L_{CL}^3: in other words, F_V in the test may be (27 kN) × (2 mm/9 mm)^3 or 294 N (66 lbf). The precise Hertz formula is

$$L_{CL}/r_w = [3F_V(1 - v^2)/2r_w^2 E]^{1/3},$$

where v is Poisson's ratio for the material (0.270 for cast iron and 0.303 for steel).

For a cylinder rolling on a plane, the *Engineering Encyclopedia* (1954, F532) employs the form $F_R = F_V K_1 (L_{CL}/r_w)$, as in the previous example. For steel cylinders of unknown size carrying unspecified loads (which presumably carry a much wider range of loads than the railroad wheel discussed above), $K_1 L_{CL}$ is given as 0.1 mm to 3 mm.

The eighth edition of *Marks' Handbook* (Baumeister, Avallone, and Baumeister 1978, 3–28) offers a similar correlation for cylinders, with $K_1 L_{CL}$ ranging from 0.005 mm for hard, polished steel to 0.05 mm for ordinary steel and 0.25 mm for rusty steel. These numbers are so low that they may represent only cylinders with small rollers. It is hard to credit the *Handbook*'s assertion that loads varied "from light to those causing a permanent set."

Ignoring surface roughness, the Hertz formula for a long cylinder (or a wide wheel) implies that

$$L_{CL}/r_w = [8F_V(1 - v^2)/\pi r_w L_{WW} E]^{1/2},$$

where L_{WW} is the length of the cylinder. The maximum shear stress is given by

$$[0.045 F_V E/\pi L_{WW} r_w (1 - v^2)]^{1/2}.$$

If the allowed stress level is determined by the strength of a given material and defines the maximum allowed *projected pressure* $= F_V/(2r_w L_{WW})$, the resulting value of L_{CL}/r_w at that load will be a fixed number independent of r_w. Based on the argument that there is no intrinsic way to differentiate the behaviors of a 1-mm roller and a 1-m roller, the same will be true of C_R.

In addition, *Marks' Handbook* provides data for steel agricultural wheels carrying 4.4 kN (1,000 lbf) over concrete: C_R varies between 0.01 and 0.03. It is likely that this resistance is due more to impact with slight amounts of surface roughness than to material energy absorption per se.

Firm wheel and soft ground　The most comprehensive references on soft-ground support and traction are those by M. G. Bekker. In particular, Bekker's *Theory of Land Locomotion* (1956, chaps. 5, 6) applies the classical theories of soil mechanics to the real-world problem of wheel loadings, and *Introduction to Terrain-Vehicle Systems (Part 2)* (1969) updates the earlier work. Both include plentiful references.

As cited in Bekker 1969, one possible power-law fit to results by Grandvoinet for cylindrical wheels could be $F_R/F_V \propto [F_V/(r_w^2 L_{WW})]^{1/3}$. *Marks' Handbook* (Baumeister, Avallone, and Baumeister 1978, 3–28) suggests that C_R for a "properly inflated and loaded" car tire is 0.012 on hard-packed gravel and as high as 0.06 on wet, loose gravel. Wheels on agricultural vehicles loaded with 4.4 kN (1,000 lbf) display values of C_R between 0.05 and 0.09 on sod and between about 0.2 and 0.5 on tilled loam or loose sand.

Simplified calculations for cylindrical and toroidal wheels on yielding ground assume that the soil yields at a fixed compressive stress level Y and does not spring back. (Characterizing soil by a single number is overly simple: see Bekker 1956 for a more general analysis.)

Under those assumptions, the wheel is supported entirely ahead of the axle (i.e., where soil is being indented). C_R is then essentially the ratio

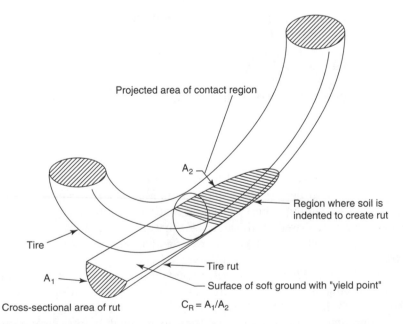

Projected area of contact region

A_2

Region where soil is indented to create rut

Tire

A_1

Tire rut

Surface of soft ground with "yield point"

Cross-sectional area of rut　　$C_R = A_1/A_2$

Figure 6.6
Rolling resistance in soft soil.

between the cross-sectional area of the rut and the vertically projected area of the wheel contact (see figure 6.6). This ratio depends on the load and the soil yield strength (Y), both of which affect the amount of sinkage. The results may be summarized as follows:

1.   For a cylindrical wheel, $F_R = F_V^2/2YL_{WW}r_w$.
•   If wheel radius increases while L_{WW} is fixed, F_R is proportional to $1/r_w$.
•   If both L_{WW} and r_w increase proportionately, F_R is proportional to $(1/r_w)^2$.
2.   For a toroidal wheel, $F_R = (\sqrt{3}/4)F_V^{3/2}/\sqrt{Y}r_T^{1/4}r_w^{3/4}$.
•   If r_w grows while r_T (the radius of the tire cross section) is constant, F_R is proportional to $(1/r_w)^{3/4}$.
•   If both r_w and r_T increase in proportion, F_R is proportional to $(1/r_w)$.

Surprisingly, when F_R is divided by F_V and the expression for L_{CL} is used, both formulas reduce to

$$C_R = L_{CL}/2r_w.$$

For any given shape of wheel-periphery cross section, soft-ground resistance is minimized by reducing the amount of sinkage that occurs. Increasing the wheel's radius always reduces drag, because the support area is greater for a given amount of penetration. Increasing the wheel's width also reduces rolling resistance, although for different shapes of wheels (for example, flat, toroidal, or multiple wheels side by side) the resistance is reduced by different amounts. (Calculations involving wheels of various shapes suggest that for a given load on soft ground, cylindrical wheels, for which the slightest sinkage forms a full-width impression, will have less drag than toroidal wheels of equal width. Perhaps most important, various estimates of drag force involve F_V raised to a power greater than one. Therefore, an additional, properly aligned wheel sharing the load should always reduce the soft-ground rolling resistance of the wheel whose load it shares.)

Users of off-road vehicles refer to many wheels or a large flat support area as "flotation," the extreme example being a "caterpillar" track. Bekker (1969) has shown that in soils with frictional rather than cohesive properties, the best wheel form for a given amount of flotation is large in diameter and narrow.

Among serious cyclists, perhaps the best examples of flotation are the doubled rims and tires of bicycles used in the Alaskan Iditabike race, and the approximately 0.15-m-wide by 0.35-m-diameter tires of the Hanebrink Extreme Terrain Bicycle that was developed for soft sand ⟨http://hanebrinkforks.com/bikes/ext_1.htm⟩.

Soft wheel and firm ground   A great deal of information about pneumatic tires is presented in Clark 1981.

The Evans (1954) equation for a wheel with a solid, cylindrical urethane tire rolling on a rigid plane was tested experimentally by Schael, Thelin, and Williams (1972). The wheels used in Schael et al.'s experiments were 84 mm (3.3 inches) in diameter and 53 mm (2.1 inches) wide, with tires 10 mm (0.4 inches) thick, loaded with 2.1 kN (470 lbf) and then 3.4 kN (770 lbf). Measured values of C_R ranged from 0.006 to 0.046. The speed used in the experiments was not specified.

Schael et al.'s measurements confirmed the sinkage formula, which implies that

$$L_{CL}/r_w = (3F_V L_{TT}/4EL_{WW}r_w^2)^{1/3}.$$

Here, E is the modulus, L_{TT} is the radial thickness, and L_{WW} is the width of the tire.

Schael, Thelin, and Williams showed that $C_R \approx (h/2)L_{CL}/r_w$, approximately twice the predicted value. Here h is a material hysteresis factor. The precise definition of h is not provided in Schael et al.'s paper, but h is apparently related to the fraction of stored elastic energy that is lost in a load-unload cycle of a uniaxial urethane specimen, with values ranging between 0.055 and 0.36. Equivalently, the radian phase shift between stress and deformation at 11 Hz was used, which presumably is valid for one particular rolling velocity only. See McClintock and Argon (1966).

IRC tire-testing data (Brandt 1998) were commissioned by the tire manufacturer Avocet (see figure 6.7) and represent several sew-up and clincher[4] tire models, from each of two tire suppliers, tested over a range of pressures. The wheel was loaded by a force of 490 N and rolled against a smooth drum of unknown diameter, at unspecified speed and temperature. The C_R values obtained seem a little high, perhaps because a drum was used, which is equivalent to testing a wheel of smaller diameter.

These very clean data are significant in showing that C_R does not approach zero as tire pressure increases (one way to see this is to plot C_R versus $1/p$). Therefore, a purely power-law theory cannot hold.

Brandt (1998) has noticed that the curves for the sew-up tire models in figure 6.7 cross those for the clincher tire models, as if sew-ups had intrinsically lower drag, as one might expect, but that the rim adhesive was adding a constant offset.

An approximate analysis of a slender pneumatic bicycle tire on a hard road can be performed for a contact-patch length ($2L_{CL}$) considerably longer than the tire width ($2r_T$). The treatment is similar to Rotta's (1949) analysis, as outlined by Bekker (1956). The approach is to calculate the amount of sinkage at each position along the contact patch, from which

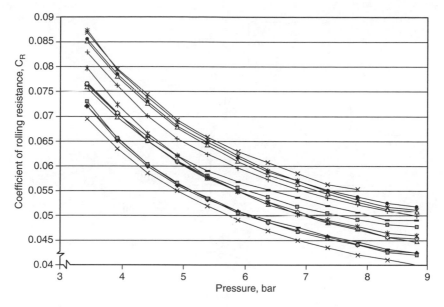

Figure 6.7
Tests of miscellaneous tires against a drum showing effects of inflation pressure.
(Data from Brandt 1998.)

contact width at that position can be determined. The total calculated
contact area multiplied by inflation pressure is then equated to F_V. The
resulting equation is

$$L_{CL}/r_w = (3F_V/4K_2pr_w^2)^{1/3}.$$

The coefficient K_2 represents the ratio of the contact half-width to the
sinkage of the tire cross section (figure 6.8), which is approximately constant for well-inflated tires in normal use (i.e., when sinkage is not excessive). Tire width per se doesn't enter the expression, but K_2 does
depend somewhat on the ratio of tire width to rim width (i.e., rim-flange
separation).

F. R. Whitt carried out tests by inking a wheel with known loads and
pressing it against paper, then measuring the rim's sinkage. This experiment showed that the behavior of real tires departs from that predicted by
the simple model for several reasons. Perhaps the most important is tread
pattern and thickness: only with a tire with unpatterned tread of uniform
thickness is pavement pressure equal to inflation pressure and contact area
equal to F_V/p. A variable-thickness tread permits contact zones in which the

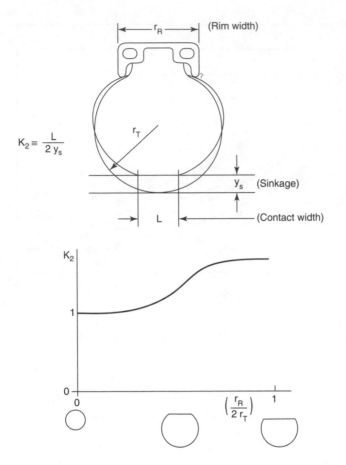

Figure 6.8
Contact mechanics of slender thin-tread tire.

taut tire fabric at the zone edges is not perfectly horizontal. A band in which the tire's tread is worn flat will increase the tire's contact area without increasing supporting force. Contrariwise, a raised tire rib will be pressed to the ground not only by the inflation pressure of the tire, but also by tension in the tire fabric, which can bulge down toward the ground without quite meeting it. On one worn, relatively thin-tread tire, K_2 was approximately 1.05, but contact width also had a no-load nonzero value because of a flat wear band on the tire.

Another factor that causes the behavior of tires to depart from the model's predictions is that squashing (which was not modeled) just beyond the contact-patch ends causes contact length to be shorter than predicted, in a behavior similar to that of the center rib. In Whitt's experiments, the actual length of the contact patch was just 84 percent of that calculated from measured sinkage. This compares well with the factor of 85 percent found by Smiley and Horne (as cited by Clark 1981).

The data fitted well to $L_{CL}/r_w = (F_V/0.95pr_w^2)^{3/8}$. This is a slightly different power law than the one given above, but at a typical load and pressure, it is within 7 percent of the theoretical calculation when multiplied by 0.84 to correct L_{CL}.

The energy loss of a slender-tire cross section ought to be related to its maximum deformation, which is determined by wheel sinkage (y_{WS}). Wheel sinkage can be approximated as $y_{WS} \approx L_{CL}^2/2r_w = (r_w/2) \times (L_{CL}/r_w)^2$, which is theoretically proportional to $1/p^{2/3}$.

The drag force acting on a tire is really the energy dissipated per unit length of tire when a cross section of the tire is deformed by the full amount of sinkage (y_{WS}). Two contributions to this dissipation seem likely. The first is from viscoelastic bending loss in the rubber of the tire, a linear phenomenon dominant at low pressure. It should probably have a magnitude proportional to y_{WS}^2 (or to $1/p^{4/3}$ when load is given) and be affected by time of deformation (L_{CL}/V, where V is the bicycle velocity) and also by temperature. The second contribution is from friction loss between the fibers of the tire cords, a nonlinear phenomenon significant at higher pressures. This should be roughly proportional to y_{WS} (which characterizes the amount of tire slip) times inflation pressure (p), which relates to the pressure's squeezing fibers together. Therefore, this loss should be proportional to $p^{1/3}$.

It is also possible that the tire tread continually undergoes a slight slippage or rubbing (for example, where it lifts away from the road at the tail of the contact patch). This seems like a reasonable explanation of the tread wear that takes place as the tires are ridden, but we have no analytical model for it.

If we try to fit the IRC data to $C_R = K_3/p^{4/3} + K_4 p^{1/3}$, the results are surprisingly good. (K_3 and K_4 are constants.) To assess the quality of the fit, we divide through by $p^{1/3}$ and see if the quantity $C_R/p^{1/3}$ is a straight-line function of $1/p^{5/3}$. But an even straighter line occurs with $K_3/p^{0.9} + K_4 p^{0.3}$. In this last case, a typical fit is $C_R = 0.001[(23.7 \text{ bar}/p)^{0.9} + (p/0.611 \text{ bar})^{0.3}]$. Such a formula implies that rolling resistance would reach a minimum around 20–30 bar (300 psi) and climb thereafter, if the tire could sustain such pressures.

Unfortunately, there is no unique "correct" fit to the IRC data. Other simple formulas such as $C_R = K_3/p^{5/4} + K_4$ provide almost as good a fit. Whether such curve fitting can actually distinguish the contributions of hysteresis and friction at various pressures, via the magnitudes of K_3 and K_4, may be learned only through considerable further experimentation. A particularly worthwhile step would be to measure the effects on one wheel of load, speed, temperature, roughness, and drum radius.

Another useful test would be to vary the tire's test load and inflation pressure in strict proportion. In that case, the shape of the loaded tire would not alter much in the course of the test. Drag due to material energy losses in bending would therefore not change, whereas drag due to frictional rubbing would increase in proportion to load and equally to pressure. If a straight line resulted when the drag force was plotted versus the load, then the two phenomena could be separated.

To conclude, one of the main ideas from both theory and the few available data is that attempts to plot C_R should be based on a conception of C_R as a function of F_V/p, or even of F_V/pr_w^2. The precise power law or other equation form employed for the purposes of plotting might change, but this combined quantity has a good chance of correlating effects of all three variables. In particular, if pneumatic-tire rolling resistance has been evaluated over a range of pressures, this concept permits at least an educated guess about the possible effects of F_V and r_w.

Bicycle tire diameter and road roughness

Virtually our only measurements of the effects of bicycle-tire diameter and road roughness came from Frank Whitt and were presented in earlier editions. Unfortunately, with his passing, occasional uncertainties or inconsistencies in Whitt's measurements now cannot be resolved. We present the important data Whitt gathered, along with critical commentary, and hope that future work will eventually supplant it.

Whitt measured low-speed rolling resistance at a variety of pressures for a 27-inch wheel on "smooth" and "medium-rough" surfaces and for a 16-inch wheel on a "medium-rough" surface. The load was 90 lbf. Smoothed curves for the rolling-resistance measurements are presented in figure 6.9. Road roughness seems to have increased C_R in Whitt's experiments by about 44 percent.[5]

When contact length was measured, the rough-surface results were fit reasonably well by $C_R = 0.075(L_{CL}/r_w)$ for both sizes of wheel. Leaving out any empirical length correction, the previous section showed that L_{CL}/r_w is equal to $2y_S/L_{CL}$ and proportional to $(F_V/pr_w^2)^{1/3}$ or $(F_V/pr_w^2)^{3/8}$. Therefore an alternative version of this correlation would be $C_R = 0.068(F/pr_w^2)^{1/3}$.

It seems that Whitt took the proportionality of C_R to L_{CL}/r_w to imply that C_R would be inversely proportional to R_W. However, the formulas just given imply either $C_R \propto 1/r_w^{3/4}$ or $C_R \propto 1/r_w^{2/3}$, which changes this proportionality. The difference arises because L_{CL} also depends somewhat on r_w.[6]

The lowest curve, representing Whitt's smooth-surface results for a 27-inch wheel, seems to be incompletely plotted. When supplemented with tabular information from the previous edition of this book, it seems

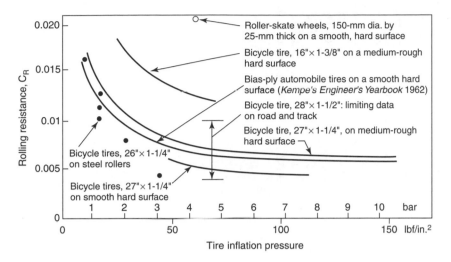

Figure 6.9
Low-speed rolling resistance of bicycle and automobile tires versus inflation pressure. (Unattributed data from Whitt 1977.)

that a reasonable power-law fit over the normal range of pressures might be $C_R = 0.0046(100 \text{ psi}/p)^{0.44}$, or equivalently $0.048(F_V/pr_w^2)^{0.44}$.

The wheel on which the measurements in table 6.2 were based, a 51-mm (2-inch)-wide bicycle tire at 1.2 bar (18 psi) with various loads, fits reasonably well to $C_R = 0.076(F_V/pr_w^2)^{0.60}$, if the wheel radius is taken to be 330 mm (13 inches). From the car-tire data in Carr and Ross (1966), with a 3.2-kN (720-lbf) load, $C_R = 0.013[1 + (0.59 \text{ bar}/p)]$.

Speed had a negligible effect in Whitt's experiments over the reported range of 8–20 m/s (30–50 mile/h). C_R on concrete pavement was found to be about 0.001 higher than on asphalt (see figure 6.10).

Kyle and Edelman (1974) found an effect of speed in some of their tests of bicycle tires. They give rolling resistance as $C_R(1 + V/V_{DD})$, where C_R is the rolling resistance at low speed, and V_{DD} is the speed at which rolling resistance is extrapolated to be twice its low-speed value (i.e., drag is doubled), typically about 17 m/s. Their results are shown in table 6.3.

Kyle and Edelman's data, considered along with those in *Kempe's Engineer's Year Book* (1962), allow C_R to be calculated as

$$C_R = 0.005\{1 + (2.1 \text{ bar}/p)[1 + (V/29 \text{ m/s})^2]\}.$$

It would be interesting to compare the V^2 tire term in this equation to the wheel's aerodynamic drag.

Table 6.2
Rolling resistance of wide bicycle tires

Cross section		Load		Speed		Inflation		Rolling resistance		
inches	(mm)	(lbf)	(N)	(mile/h)	(m/s)	(lbf/inches^2)	(bar)	(hp)	(W)	C_H
2	51	120	53	20	8.9	10	0.69	0.1	75	0.016
2	51	120	53	20	8.9	18	1.2	0.07	52	0.011
2	51	120	53	20	8.9	30	2.1	0.05	37	0.008
2	51	150	66	20	8.9	18	1.2	0.1	75	0.013
2	51	180	80	20	8.9	18	1.2	0.12	89	0.013

Source: Patterson 1955.

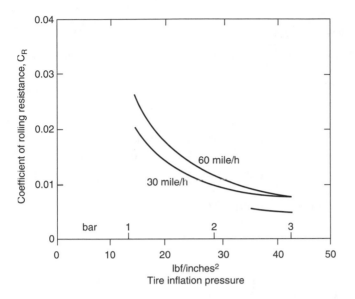

Figure 6.10
Effect of inflation pressure on rolling resistance of automobile tires. The two upper curves are for bias-ply tires; the lowest curve is for radials. (From Bekker 1956.)

Table 6.3
Rolling resistance of bicycle tires

	Pressure				
	lbf/inches^2	kPa	Vehicle	C_R	V_{DD} (m/s)
Vittoria imperforable Seta 27-inch tubular	105	724	Bicycle	0.0029	17.6
Criterium 250 27-inch tubular	105	724	Bicycle	0.0039	22.7
Clement Criterium Seta Extra 27-inch tubular	105	724	Tricycle	0.0019	Very high
Hutchinson 27 × 1$\frac{1}{8}$-inch clincher	60	414	Bicycle	0.0047	16.1
Hercules 26 × 1$\frac{3}{8}$-inch clincher	40	276	Bicycle	0.0066	—
United 21 × 2$\frac{1}{4}$-inch clincher	40	276	Tricycle	0.0061	—

Source: Kyle and Edelman 1974.

Table 6.4
Values of C_R for low-drag tires

C. R. Kyle and P. Van Valkenburgh (1985) Add 10–35 percent for asphalt.	0.0016–0.0032	sew-ups on linoleum
	0.0023–0.0029	clinchers on linoleum
	0.0017	track sew-up on concrete
C. Kyle (1986) p. 134	0.0016–0.0026	track sew-ups
	0.0028	Moulton clincher (17)
	0.0033–0.0037	road sew-ups
	0.0039	road clinchers
Avocet data (*Kempe's Engineer's Year Book* 1962)	0.0039–0.0049	road tires, 120 psi (Note: drum used.)
Senkel (1993) (Scott 1889)	0.0016–0.0042	
Lafford (2000)	0.0043	top 700C clincher, 100 psi

Bekker (1956, 208) and Ogorkiewicz (1959) offer the following alternative correlation (for automobile tires):

$$C_R = 0.0051\{1 + (1.09 \text{ bar}/p)[(1 + F_V/3 \text{ kN}) + (1 + F_V/30 \text{ kN})(V/39 \text{ m/s})^2]\}.$$

A key question is whether (for 27-inch wheels, say) the lowest achievable C_R is 0.004, 0.003, 0.002, or even less. The data in table 6.4 have been published for low-drag tires.

Increase of speed due to a reduction in C_R

Using the equations and methods of chapter 4, it is not hard to analyze a specific situation to determine how a given change in rolling resistance, or equally in road slope, will alter speed at a fixed power level. What is more difficult is to develop simple, generally applicable conclusions.

Our approach is to assume a base C_R of 0.004 with system weight of 700 N (158 lbf) and present the effect of subtracting 0.001 from C_R. If the change in C_R is actually 0.002, the changes in speed should approximately double. Or if drag is added, the effect on velocity should be read as a *decrease*. Two extreme cases of aerodynamic-drag factor K_A will be plotted: 0.19 for a crouched rider in racing clothes, and 0.39 for a bolt-upright rider in loose winter clothes. The results are presented in figure 6.11.

For the given reduction in C_R, we can see that the low-drag rider attains a speed increase of 0.2 m/s at the low end of the normal speed range, decreasing to about 0.1 m/s at the high end.[7] The high-drag rider experiences about half as much increase in speed as the low-drag rider. These results may be approximated by

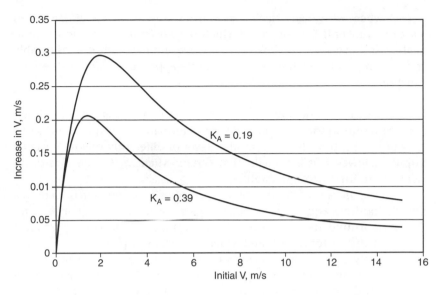

Figure 6.11
Increase in speed with reduction in C_R from 0.004 to 0.003 (system mass = 71.4 kg). (Plotted by Jim Papadopoulos.)

$$\Delta V/V \approx -\Delta F_R/3K_A V^2,$$

where the change in F_R is just weight times the change in C_R.

For reference purposes, reducing wheel diameter from 27 to 16 inches is anticipated to add about 40 percent to C_R, using the two-thirds-power formula, so for our base case, we should add 0.0016 to C_R. Similarly, halving the inflation pressure should add 26 percent, or about 0.001 in our base case. Perhaps the greatest improvements in C_R will derive from purchasing the proper tire.

Tire construction to minimize energy loss

Tires have a variety of features that reduce the energy loss in smooth-surface rolling (i.e., that lower rolling resistance). An introduction to some of these can be found in Shearer 1977a and 1977b.

Although a "solid-steel" tire has lower rolling resistance than any pneumatic tire on a smooth surface, on a normal road it would have to leap over every little pebble it encountered. The great superiority of pneumatic tires when no steel rails are available is that they simply "swallow" minor bumps, with almost no change in force. Therefore, no shocks are applied to the rider, and suspension losses occur only when roughness is severe.

A wheel with many small steel springs around the circumference in place of a rubber tire (examples of which have been developed in the past) might also have very low loss of power over bumps, but at considerably greater complexity. So the following points apply to pneumatic-tire construction only:

· The special fabric forming a tire is generally not interwoven, as interweaving results in thread rubbing during deformation. Instead, there is one layer of parallel threads above another layer of differently oriented parallel threads. (However, it is thought that certain Michelin Hi-Lite tires do in fact have an interwoven sidewall.)

· Tire fabric is used "on the bias": instead of radial and circumferential threads, positive and negative helix angles are used, making it easy for the fabric to undergo circumferential stretching or compressing when it is formed into a torus, and also when the tread is pressed close to the rim. (Bias-ply construction increases lateral stability because of the greater length of rim supporting the contact patch, and because of triangulation of support for the contact patch, but it also increases scuffing owing to "Chinese finger puzzle" effect when loaded (a tube of interleaved biased fibers that tightens on a finger when stretched). Some radial bicycle tires were not accepted in the market.)

· Thin layers bend and spring back more easily than thicker ones. The very thinnest possible threads are used, glued together in the thinnest possible layer. Thread thickness may be characterized by "thread count" (number of threads per inch). Higher-strength fibers are ideal for bicycle tires, as long as they resist abrasion. A thin tread and inner tube are also desirable, consistent with the desired length of life of the tire. The development of tubeless tires for bicycles would help reduce drag encountered in riding them.

· The strength of pneumatic bicycle tires is maximized to allow inflation to high pressure. Hard tires deform less than softer ones on a smooth surface but still have a long enough contact region to "swallow" many pebbles. However, they are more susceptible than softer tires to minor changes in road level, which lead to large bump forces.

· Interior layers such as thorn deflectors are avoided in bicycle tires or constructed integrally (rather than constituting a separate, slideable layer).

· The materials from which bicycle tires are constructed, especially the rubber, are selected for good rebound (i.e., low energy loss due to deformation). Air is ideal for rebound, as its pressure hardly varies. However, it is known that low energy loss in a tire's tread can mean poorer traction in slippery conditions (see Bowden and Tabor 1951, 1964, 1973).

Notes

1. The usual procedure for two rollers in contact with radii (r_1, r_2) is to equate the combination to an "equivalent" roller of radius r_{eq} that, when touching a flat plane, leads to the same dependence of gap or "nip" height on fore/aft distance (see figure 6.2). The formula is $(1/r_{eq}) = (1/r_1) + (1/r_2)$. This formula suggests that a wheel with a diameter of 660 mm (26 inches) pressed against a training roller with a diameter of 100 mm (4 inches) should have drag similar to that of wheel with a diameter of 89 mm (3.5 inches) pressed against a plane.

2. Although it seems attractive to evaluate drag from the force required to restrain or "tow" the wheel on a powered treadmill, such a technique must account for the effect of the treadmill's soft belt, which will create added resistance.

3. Some literature suggests that a train wheel on a rail also shares this characteristic, and this perspective was presented in the second edition of this book. However, it now seems that the primary cause of train wheel resistance is frictional slippage within the contact patch: because of the cone angle of the wheel, the contact patch includes different wheel radii moving at different velocities. With that caveat, we have retained the main results for train wheels presented in the second edition, since for some reason they nicely fit the two-dimensional analysis being presented.

4. The more accurate names for these tire types are "tubular," for the sewn-up casing containing an inner tube that is glued to a special rim, and "wired-on," for the open casing containing wire or cord beads that seat on the rim usually having a separate inner tube.

5. In ordinary on-road determinations, it is not possible to distinguish the effects of roughness on C_R (defined at constant axle height) from the effects of roughness that lead to vibration and impact (and hence a loss of energy principally within the rider's body or bicycle suspension). The latter will be highly speed-dependent and affected by the springiness and damping inherent in the load being carried.

6. Part of the difficulty in evaluating the effect of wheel diameter is that tires for smaller wheels have often been made very differently from those for the more usual large wheels. They are mostly for children's sidewalk bicycles, for which puncture resistance and low cost are primary considerations and high drag may also be advantageous, from a safety perspective! On the other hand, the designer and engineer Alex Moulton has been able to develop a smaller-diameter tire whose rolling resistance rivals that of tires of normal size; tires of this type were used on the General Motors SunRayce solar car. Given the difficulty of obtaining tires with differing R_W but identical cross section and construction, the best way to determine the effect of R_W might be to test one given wheel on a variety of drum diameters. The combination of drum and wheel can be related to an "equivalent" wheel radius, as shown earlier. Not only can wheel-radius effects be explored in this way, but several points from data so obtained should permit reliable extrapolation to a drum of infinite radius (i.e., a flat surface).

7.   Low-drag tires may be of more than academic interest to racers: 0.1 m/s is about 1 percent of speed, suggesting thirty-six seconds saved in an hour-long event.

References

Baumeister, T., E. A. Avallone, and T. Baumeister III, eds. (1978). *Marks' Standard Handbook for Mechanical Engineers*, 8th ed. New York: McGraw-Hill.

Bekker, M. G. (1956). *Theory of Land Locomotion*. Ann Arbor: University of Michigan Press.

Bekker, M. G. (1969). *Introduction to Terrain-Vehicle Systems* (Part 2). Ann Arbor: University of Michigan Press.

Bowden, F. P., and D. Tabor. (1951). *Friction and Lubrication of Solids*, vol. 1. London: Oxford University Press.

Bowden, F. P., and D. Tabor. (1964). *Friction and Lubrication of Solids*, vol. 2. London: Oxford University Press.

Bowden, F. P., and D. Tabor. (1973). *Friction: An Introduction to Tribology*. New York: Anchor-Doubleday.

Brandt, Jobst. (1998). Internet posting (includes Avocet data plus some discussion).

Carr, G. M., and M. J. Ross. (1966). "The MIRA single wheel rolling resistance trailers." Report, Motor Industries Research Association, Nuneaton, U.K.

Clark, S. K., ed. (1981). Mechanics of Pneumatic Tires. U.S. Department of Transportation National Highway Traffic Safety Administration, DOT HS 805 952 (formerly National Bureau of Standards monograph 122). Washington, D.C.: Superintendent of Documents, U.S. Government Printing Office.

Drutowski, R. C. (1959). "Energy losses of balls rolling on plates." *Trans. ASME* (Series D) 81, no. 233:311.

Engineering Encyclopedia. (1954). New York: Industrial Press.

Evans, I. (1954). "The rolling resistance of a wheel with a solid rubber tire." *British Journal of Applied Physics* 5:187–188.

Hannah, J., and M. J. Hillier. (1962). *Applied Mechanics*. London: Pitman.

Harris, Tedric A. (1991). *Rolling Bearing Analysis*. New York: Wiley.

Johnson, K. (1996). *Contact Mechanics*. Cambridge: Cambridge University Press.

Kempe's Engineer's Year Book. (1962). Vol. 11. London: Morgan.

Koffman, J. L. (1964). "Tractive resistance of rolling stock." *Railway Gazette* (6 November):889–902.

Kyle, Chester R., and W. E. Edelman. (1974). "Man powered vehicle design criteria." Paper presented at Third International Conference on Vehicle System Dynamics, Blacksburg, Va.

McClintock, F. A., and A. S. Argon. (1966). *Mechanical Behavior of Materials*. Reading, Mass.: Addison-Wesley.

Ogorkiewicz, R. M. (1959). "Rolling resistance." *Automobile Engineer* 49:177–179.

Patterson, P. D. (1955). "Pressure problems with cycle tires." *Cycling* (April):428–429.

Reynolds, O. (1876). "Rolling friction." *Philosophical Transactions* 166:155–156.

Rotta, J. (1949). "Zur statik des luftreifens (Statics of a pneumatic tire)." *Ingenieur Archiv* 17.

Schael, G. W., J. H. Thelin, and B. L. Williams. (1972). "The load-deflection and rolling characteristics of polyurethane wheels." *Journal of Elastoplastics* 4 (January):10–21.

Shearer, G. R. (1977a). "The potential for improvement in tire response." Paper no. 770871, Society of Automotive Engineers, Warrendale, Penn.

Shearer, G. R. (1977b). "The rolling wheel: The development of the pneumatic tyre." In *Proceedings of the Institution of Mechanical Engineers*, U.K. 191 (November).

Trautwine, J. C. (1937). *The Civil Engineer's Reference Book*, 21st ed. Ithaca, N.Y.: Trautwine.

Whitt, Frank R. (1977). "Tyre and road contact." *Cycle Touring* (February–March):61.

7 Braking

Introduction

The friction of dry solid substances

Experiments have shown that when two surfaces are pressed together with a force F_V, there is a limiting (maximum) value F_F of the frictional resistance to motion. This limiting value is a definite fraction of F_V, and the ratio F_F/F_V is called the coefficient of friction, μ. Therefore, $F_F = \mu F_V$. For dry, rigid surfaces, μ is affected little by the area of the surfaces in contact or the magnitude of F_V.

When surfaces start to move in relation to one another, the coefficient of friction falls in value and is dependent on the speed of the relative movement. For steel wheels on steel rails, the coefficient of friction can be 0.25 when the wheels are stationary and 0.145 at a relative (sliding) velocity of 18 m/s (40 mile/h). Polishing the surfaces lowers the coefficient of friction (one cause of brake fade), as does wetting. The coefficients of metal-to-metal dry friction are about 0.2–0.4 (down to 0.08 when lubricated); for leather to metal they are 0.3–0.5. (These coefficients are for stationary conditions and decrease with movement.) Brake-lining materials against cast iron or steel have a friction coefficient of about 0.7, and this value decreases less with movement than for other materials. Elastomers (rubbery materials) deform under load, which causes their friction to be highly variable. In contrast with that of dry, rigid surfaces, the friction of elastomers is affected by contact area, increasing with greater area. Thus, such measures as "dimpling" brake rims can be counterproductive. The frictional resistance of elastomers is at a maximum when the material is made to "creep" along a surface. As true sliding begins, the coefficient of friction falls, decreasing with increasing relative velocity.

The variability of friction with contact area and relative motion, coupled with the flexibility of brake mechanisms that can change the contact area as the load increases, often leads to a "stick-slip" sequence that, occurring repeatedly and rapidly, gives rise to brake squeal.

Bicycle brakes

Two places where solid-surface friction occurs must be considered in normal bicycle braking: the brake surfaces and the road-to-wheel contact. (Track bicycles are braked by resisting the motion of the pedals, the rear cog being fixed to the wheel hub without a freewheel.)

Five types of brakes have been fitted to regular bicycles for ordinary road use. The plunger brake is used on some present-day children's bicycles

Figure 7.1
Plunger brake on Thomas Humber's safety bicycle. (Reproduced with permission from Nottingham Castle museum.)

and tricycles and was used on early bicycles such as the ordinary or high-wheeler and on pneumatic-tired safeties up to about 1900 (figure 7.1). Pulling a lever on the handlebars presses a metal shoe (sometimes rubber-faced) against the outer surface of the tire. Plunger brakes were and are used on solid and pneumatic tires; their performance is affected by the amount of grit taken up by the tire from the surface on which it is riding, which fortunately increases braking effectiveness and wears the metal shoe rather than the tire. Such brakes are very poor in wet weather because the tire is being continuously wetted (see below for the effects of water on braking).

The internal-expanding hub brake is similar to automotive hub brakes, but it is less well protected from water incursion than the automotive version, and therefore its performance varies in wet weather. Hub brakes used to be popular on medium-weight "roadsters" in the 1930s, but they lost favor mostly because of a high weight compared to rim brakes. They have been reintroduced in an improved form by Sturmey Archer (figure 7.2). Hub brakes are popular on the rear wheels of tandems and on various other human-powered vehicles to eliminate the rim and tire heating that rim brakes produce, which is particularly serious on long descents in mountainous areas. (See comments on this topic later in the chapter.)

The backpedaling or "coaster" hub brake brings multiple disks or cones together when rotation of the cranks is reversed (figure 7.3). These brakes operate in oil and are entirely unaffected by weather conditions. They are very effective on a bicycle's rear wheel; they cannot be fitted to the front wheel because the actuating force required is too great to be applied

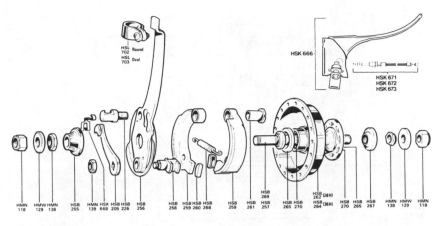

Figure 7.2
Exploded view of Sturmey-Archer internal-expanding hub brake. (Courtesy of Sturmey-Archer Ltd.)

Figure 7.3
Exploded view of Bendix backpedaling hub brake. (Courtesy of Bendix Corp.)

by hand. (However, see the Calderazzo patent described later in the chapter (figure 7.10) for a possible way of using coaster brakes on a bicycle's front wheel.) They cannot be used with derailleur gears, and if the chain breaks or comes off the sprockets, all braking is lost.

The disk brake, having become the preferred form of brake in motorcycles, automobiles, race cars, and aircraft, is becoming accepted as the optimum system for all-terrain bicycles, and this acceptance will undoubtedly extend to other types of bicycles as well. Disk brakes can be operated either by cable or hydraulically from normal hand levers (figure 7.4). The effective braking diameter is normally at less than half the wheel diameter, which requires a higher braking force than for rim brakes but keeps the braking surfaces away from the wheel spray in wet weather. Disk brakes normally have good wet-weather performance. Those found in cars are generally made of cast iron, and those on bikes have usually been steel. In aircraft, "carbon-carbon" (carbon fibers in a graphite matrix) brake disks have been used: a typical duty is for them to absorb 3 MJ/kg, versus less

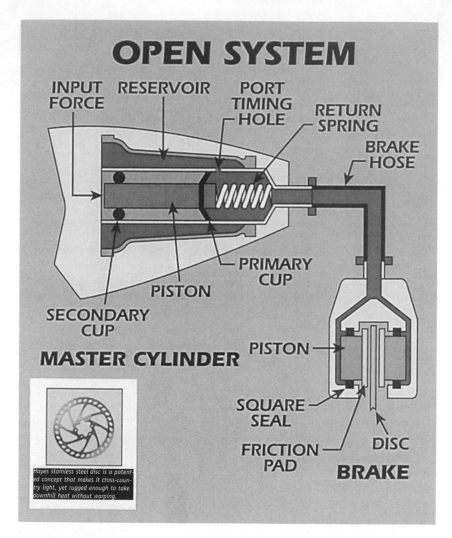

Figure 7.4
Disk-brake system. (Courtesy of Hayes.)

than 1 MJ/kg for steel disks. Motorcycle disk brakes employing carbon-carbon have further increased the permissible loading by incorporating ceramics into the disks, and we expect that this development will propagate into other applications, including bicycles.

The rim brake is the most popular type of bicycle brake around the world. A pad or "block," usually of rubber-composition material, is forced against the inner or the side surfaces of the wheel rims, front and rear. Because the braking torque does not have to be transmitted through the hub and spokes, as with the preceding three types, and because the braking force is applied at a large radius, these brakes are intrinsically the lightest of the types discussed here and result in the lightest bicycle design. Rim brakes are, however, very sensitive to water (the coefficient of friction with regular combinations of brake pads and steel wheel rims has been found to fall, when the brakes are wet, to below a tenth of the dry value) and can suffer rapid wear. They require continual adjustment (provided automatically in a very few designs) and pad replacement more frequently than every 3,000 km (2,000 miles). Automobile brake shoes on disk brakes, with heavier duty, last around 80,000 km (50,000 miles).

Bicycle manufacturers in Western countries have solved the wet-weather-braking problem by switching from steel to aluminum wheel rims, the friction coefficient of which does not fall nearly as catastrophically when the rims are wet as does that for steel. The aluminum alloys used in such rims are also much softer and wear much faster than does steel. Particles of grit can become embedded in brake pads, thereby scoring the rim surface with potentially deep grooves, or the rims can just become generally thinner because of overall wear. The high pressures used in modern tires (up to at least 12 bar, 1.2 Mpa, 170 lbf/in^2) can then cause the rims to explode outward, with a high likelihood of locking the wheel (Juden 1997). (This has happened, with relatively low-pressure tires, four times to the author.) This is a very serious event if it occurs in the front wheel. Some aluminum-alloy rims can be supplied with a flame-sprayed ceramic coating, which greatly reduces the rate at which they wear.

Power absorption of brake surfaces

Drum brakes for modern motor vehicles can be designed by allowing a certain power to be absorbed per unit area (about 7–12 MW/m^2) of braking surface (*Kempe's Engineer's Year Book* 1962). Another measure, used for disk brakes for aircraft, is the energy that can be absorbed in a single braking action: under 1 MJ of energy per kilogram of disk material can be dissipated in a steel disk, and 3 MJ/kg in carbon-carbon disks, as mentioned above.

The amount of power that needs to be absorbed by a vehicle's brakes depends upon the speed and mass of the vehicle and on the desired deceleration rate. For a typical bicycle of 12 kg (26.5 lbm) and rider of 75 kg (165 lbm), let us determine the power loading required at the brake pads for strong braking on the level and on descending a steep hill. We will specify that these pads have a total area of 2,500 mm² (3.9 in²) and that the retardation be −0.5 g (half gravitational acceleration) from 9 m/s (20 mile/h) when on the level, and from twice this speed when on a downhill of 15 percent slope. Gravitational acceleration (g), is 9.81 m/s² (32.17 ft/s²), and expressing braking decelerations as proportions of g is useful, because it gives directly the proportion of the vehicle's and rider's weight that must be applied as braking force. The time t for a retardation a is given by

$$V_2 = V_1 + at,$$

where $V_2 = 0$ and V_1 is the initial velocity. Therefore, $V_1 = -at$, and so

$$t = \left(\frac{V_1}{a}\right) = \left(\frac{9 \text{ m/s}}{-0.5 \times 9.81 \text{ m/s}^2}\right) = 1.835 \text{ s}$$

for the level-road case and twice this, 3.67 s, for the downhill stop from twice the initial speed. The stopping distance is

$$S = \frac{V_1 + V_2}{2} t = 9\left(\frac{1.835}{2}\right) = 8.26 \text{ m}$$

for the first case and four times this, 34.04 m, for the second case. The initial kinetic energy is

$$\frac{mV^2}{2g_c} = \frac{(77 + 12)}{2}(9)^2 = 3,604 \text{ J}$$

and is also four times this value, 14,416 J, for the second case. The power dissipation falls from a peak (at initial applications of the brakes) to zero (when the bicycle comes to rest).

Determining brake duty (largely a function of surface heating) requires the mean power dissipation over the time (t), which is given by

$$3,604/1.835 = 1.96 \text{ kW}$$

for the first case. Thus, the power absorbed per unit of brake-block area is

$$(1.96/2500) \times 10^6 = 0.79 \text{ MW/m}^2.$$

For the second case, the potential energy of the bicycle and rider must also be dissipated in the brakes. This is

$$34.04 \times 0.15 \times (77 + 12) \times 9.81 = 4{,}457 \text{ J}.$$

The total power dissipated is $(14{,}416 + 4{,}457) = 18{,}873$ J. The mean power dissipated is 5.14 kW, and the power absorbed per unit of brake-block area is 2.06 MW/m^2.

This is about one-quarter of the loading allowed in automobile-brake practice. Therefore, the rim surface area is more than adequate for braking. However, many riders in mountainous country have learned, to their dismay, that the thermal mass of and the heat transfer from a wheel rim are small. Rim brakes can cause the rim's temperature to rise quickly to the point at which the rubber cement holding tire patches, or even the tire itself, softens, and the tires will deflate or (in the case of "stick on" tubular tires) come off the rim. When these failures occur at speed on the front wheel, serious accidents are possible. We discuss this further later in this chapter.

The adequacy of a vehicle's braking surface is, of course, only one factor in determining the distance required to stop the vehicle. It is necessary in addition to be able to apply an adequate force to the brake system to bring the vehicle to a halt. Bicycle brakes are often deficient in this respect, especially in wet weather (when the coefficient of friction is greatly reduced) and especially for the front wheel (where most of the braking capacity is available because of weight transfer during deceleration).

Friction between tire and road

If we assume that an appropriate amount of force can be applied to the vehicle's brakes and that the brakes' pads or linings have been proportioned so that they will not "fade" (suffer a decrease of coefficient of friction) on account of heating, the stopping capacity of the brakes depends directly upon the grip (or coefficient of friction) of the tires on the road. For pneumatic-tired vehicles, this grip varies from 0.1 to 0.8 times the support force between tire and road, according to whether the surface is, for example, dry concrete or wet ice.

Longitudinal stability during braking

The weight of a bicycle and its rider does not divide itself equally between the bicycle's two wheels. A typical value for the weight distribution on a road bicycle is 40 percent front, 60 percent rear, a proportion we examine in the example following. This distribution applies on level ground when

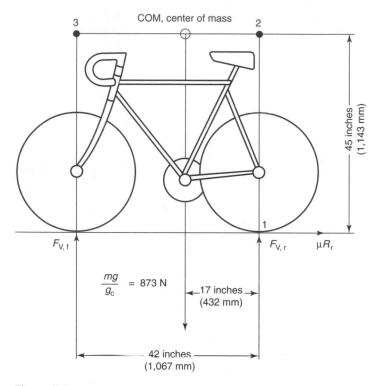

Figure 7.5
Configuration specified for braking calculations.

the bicycle is either at rest or is moving at constant velocity. The distribution can change dramatically, particularly during strong acceleration ("doing a 'wheelie'") and during braking (possibly leading to "taking a header"). To determine whether or not the braking reaction is important, let us estimate the changes in wheel reactions when the typical bicycle and rider above brakes at half the acceleration of gravity.

If the wheelbase is 1,067 mm (42 inches) and the center of gravity of rider and machine is 432 mm (17 inches) in front of the rear-wheel center and 1,143 mm (45 inches) above the ground (figure 7.5), we can calculate the front-wheel reaction $F_{V,f}$ when the bicycle is stationary or when the rider is riding at constant speed by equating moments about point 1 in figure 7.5:

$$F_{V,f} \times 1{,}067 \text{ mm} = 89 \text{ kg} \times 9.81 \text{ m/s}^2 \times 432 \text{ mm};$$

therefore, $F_{V,f} = 353.5$ N, and

$F_{V,r} = 873 - 353.5 = 519.5$ N.

(The total weight is $89 \times 9.81 = 873$ N)

During the 0.5 g braking, a total braking force of $0.5 \times 873 = 436.5$ N acts along the road surface. The front-wheel reaction $F_{V,f}$ around point 2 in the figure is now

$F_{V,f} \times 1,067$ mm $= 873$ N $\times 432$ mm $+ 436.5$ N $\times 1,143$ mm.

Therefore $F_{V,f} = 821$ N, and by subtraction, $F_{V,r} = 52$ N.

Thus, the rear wheel is in only light contact with the ground. Only a slight pressure on the rear brake will cause the rear wheel to lock and skid. The front brake therefore has to provide over 90 percent of the total retarding force at a deceleration of 0.5 g, even if the tire-to-road coefficient of friction is at the high end of the typical range (0.8). Therefore, brakes that operate on the rear wheel only, however reliable and effective they may be in themselves, are wholly insufficient to take care of emergencies.

Another conclusion from this calculation is that a deceleration of 0.5 g (4.91 m/s^2) is almost the maximum that can be risked by a crouched rider on level ground before he risks going over the handlebars. We can calculate the maximum possible deceleration as a proportion of g by setting $F_{V,r} = 0$ in the above case. Then, taking moments of force (torques) around point 3 in the figure, we have

873 N $\times (1,067 - 432)$ mm $= (a/g) \times 873$ N $\times 1,143$ mm,

yielding 0.56 g, or 5.45 m/s^2.

Riders of tandems and recumbent bicycles and drivers of cars do not have this limitation on deceleration: if their brakes are adequate, they can theoretically brake to the limit of tire-to-road adhesion. If the tire-to-road coefficient of friction of their vehicle is 0.8, they are theoretically capable of a deceleration of 0.8 g, which is over 40 percent greater than that of a seated bicyclist with the best possible brakes. For this reason (and many others) bicyclists should never tailgate motor vehicles.

Skilled riders increase their deceleration capability when descending steep slopes by crouching as low and as far behind the bicycle's saddle as possible.

Minimum braking distances for stable vehicles

If it is assumed that the slowing effect of air resistance is negligible in braking, a relatively simple formula can be used to estimate the minimum

Table 7.1
Coefficients of adhesion and rolling resistance for automobiles

Surface	Coefficient of adhesion	Coefficient of rolling resistance
Concrete or asphalt (dry)	0.8–0.9	0.014
Concrete or asphalt (wet)	0.4–0.7	0.014
Gravel, rolled	0.6–0.7	0.02
Sand, loose	0.3–0.4	0.14–0.3
Ice	0.1–0.2	0.014

Source: Carr and Ross 1966.

stopping distance S (m) of a vehicle fitted with adequate braking capacity and having a center of gravity sufficiently low or rearward in relation to the wheelbase for there to be no danger of the rear wheel(s) lifting during braking:

$$S = [V_1 \ (\text{m/s})]^2/[20(C_A + C_R)],$$

where C_A is the coefficient of adhesion and C_R that of rolling resistance. (C_A is the value of the coefficient of friction [μ] of a rolling wheel just before skidding occurs.)

Table 7.1 gives typical values for the adhesion and rolling-resistance coefficients for automobile tires to enable minimum stopping distances to be calculated. In practice, greater distances are needed for braking than those based on the formula and on a high adhesion coefficient. Test data gathered by Hanson (1971) for rim brakes are given in table 7.2.

Rear-wheel-only braking

Let us see what braking distance we may expect if the same rider and bicycle studied earlier, starting from 9 m/s (20.1 mile/h), brake with the rear brake only to the limit of tire adhesion. We assume that the rear brake is strong enough to lock the wheel if desired, and that the coefficient of friction (μ) between the tire and the road surface is 0.8. Then the maximum retarding force is $0.8 \times F_{V,r}$, where $F_{V,r}$ is the perpendicular reaction force at the rear wheel. This rear-wheel reaction force $F_{V,r}$ is somewhat less than the value during steady level riding or when the bicycle is stationary, because deceleration results in more reaction's being taken by the front wheel. Let us take the moments of forces about point 3 in figure 7.5. Under the assumed static conditions the machine is in equilibrium:

Table 7.2
Test data on initially wetted rim brakes

Point	Braking force		Coefficient of friction μ
	lbf	N	
1 (wet start)	22	97.9	0.17
2 (prerecovery)	22	97.9	0.17
3 (recovering)	26	115.7	0.20
4 (recovering)	31	137.9	0.24
5 (recovering)	35	155.7	0.27
6 (recovering)	39	173.5	0.30
7 (recovered)	44	195.7	0.34
Turns of wheel before onset of recovery		30	
Turns of wheel during recovery		20	
Total turns to recovery		50	

Source: From Hanson 1971, 32.

$$F_{V,r} \times 1{,}067 \text{ mm} + \mu F_{V,r} \times 1{,}143 \text{ mm} = 873 \text{ N} \times (1{,}067 - 432) \text{ mm},$$

$$F_{V,r} = 279.8 \text{ N } (62.9 \text{ lbf}) \quad \text{for } \mu = 0.8.$$

Then the deceleration (a) as a ratio of gravitational acceleration (g) is given by Newton's law:

$$F = ma/g_c, \quad \text{therefore } a = Fg_c/m = -\mu F_{V,r} g_c/m,$$

$$(a/g) = -\mu F_{V,r} g_c/mg = 0.8 \times 279.8 \text{ N}/873 \text{ N} = 0.256.$$

So the retardation with rear braking is less than half the value at which, using the front brake to the maximum safe limit, the rider would be about to go over the handlebars ($0.56g$).

The time taken for this deceleration is given as before by

$$V_1 = -at,$$

$$t = (-9 \text{ m/s})/(-0.256 \times 9.81 \text{ m/s}^2) = 3.58 \text{ s},$$

and the stopping distance is given by

$$S = (V_1 + V_2)t/2 = 9 \text{ m/s} \times 3.58 \text{ s}/2 = 16.1 \text{ m } (52.9 \text{ ft}).$$

Therefore, the minimum stopping distance is about twice that for reasonably safe front-wheel braking. In practice an even longer stopping distance than the minimum is likely to be required, because a deceleration level sufficiently below the limit where skidding starts would normally be chosen.

Wet-weather braking

Wet conditions affect, usually adversely and often to a considerable extent, both adhesion of bicycle tires to the road on which they are riding and the grip of rim brakes on the rim of bicycle wheels. We shall show below that stopping distance can in some cases increase in wet weather to over ten times the dry value. On the other hand, motorcycles and cars fitted with shielded disk or drum brakes are little affected by wet weather unless the brake is for some reason temporarily submerged in water.

Braking distances for bicycles equipped with conventional rim brakes on steel rims are approximately quadrupled in wet weather. Hanson (1971) and Armstrong (1977) used laboratory equipment to simulate wet-weather braking of a bicycle wheel: their separate tests yielded the following significant findings.

For brake pads of normal size and composition running on a regular twenty-six-inch (equivalent to 650 mm) plated steel wheel, tests at the Massachusetts Institute of Technology (MIT) (see Hanson 1971 and Armstrong 1977 and figures 7.6 and 7.7) showed that the coefficient of friction when the pads were wet was less than a tenth of the dry value. Moreover, the wet wheel would turn an average of thirty times with full brake pressure applied before the coefficient of friction began to increase, and a further twenty turns were necessary before the full dry coefficient of friction was attained (table 7.2). The dry coefficient was not recovered if water was being added to the brake pads or rims after brake application, as might occur during actual riding in very wet conditions.

The MIT tests were conducted on wheels with steel rims because of the severe drop-off in braking efficiency when rims of this material were used with any of the brake pads then (1971) available. Since that time there have been several developments in wet-weather braking, spurred partly by the aim of the International Standards Organization (Technical Committee TC/149) and of the U.S. Consumer Product Safety Commission to formulate generally acceptable safety standards for the performance of bicycle brakes in wet weather (see below).

The use of coefficients of friction to measure vehicle stopping distances is frowned upon by some investigators who believe that the notorious variability in measured values of these coefficients makes bicycle stopping distance from a standard speed of 15 mile/h (6.7 m/s) on an

Figure 7.6
Test setup (at MIT) for brake-pad materials in Hanson's 1971 experiments. The spring allowed the test pad to follow an inevitably uneven rim without large variations in force. Strain gauges in the support allowed measurement of normal and tangential forces. (Courtesy Allen Armstrong.)

actual or simulated bicycle the only valid measure of stopping distance. However, such stopping-distance tests have tended to confirm the validity of the MIT results.

Several different brake-block materials were investigated in the tests at MIT, and the results of these investigations are shown in figure 7.7 and table 7.3. Although many of the brake materials employed in the investigations are designated only by numbers, it can be seen that what were at that time (and still are in most of the world) standard bicycle brake pads ("B rubber") have the highest dry coefficient and the lowest wet coefficient of friction of all materials tested. Attempts to improve the wet friction of these pads by cutting grooves of various types in the pads or by using "dimpled" steel rims were unsuccessful. Similar findings have been reported by others.

Jow (1980) found that braking distances for wheels with aluminum-alloy rims were highly dependent on the material of the brake blocks or pads. In most cases the stopping distances when the wheels were wet were two to three times longer than under dry conditions (figure 7.8). However,

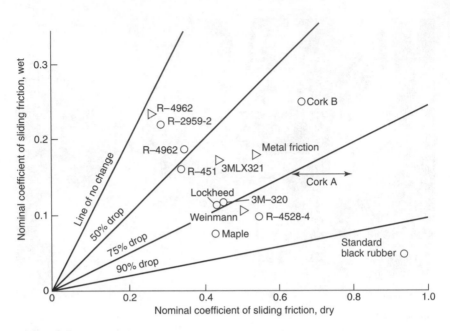

Figure 7.7
Friction coefficients for wet and dry braking. Rim materials: (O) nickel-chromium-plated steel; (▷) aluminum alloy. (Data from Hanson 1971 and Jow 1980.)

one block, the Shimano Dura-Ace EX, held the wet stopping distance to under 20 percent longer than when dry.

Other data using an aluminum rim were gathered by Armstrong (1977) on a lathe setup at MIT. Armstrong tested six brake-block materials:

1. a standard Weinmann red-rubber block;
2. a proprietary U.S. pad of the time, "Metal Frictions";
3. an experimental 3M "wet or dry" polymer pad;
4. a pad of Raybestos R-451 material;
5. a pad of Raybestos R-4962 material; and
6. a pad of Raybestos R-2959 material.

The R-451 and R-2959 damaged the (uncoated) aluminum rim by picking up small bits of aluminum that became embedded in the block and then scored the rim. The R-4962 did not damage the rim during Armstrong's tests, which involved a few hundred turns of the wheel at most. The R-4962 also had another remarkable property: nearly identical wet and dry coefficients of friction. (We wonder if the Shimano Dura-Ace EX

Table 7.3
Data gathered at MIT on brake-pad materials at equivalent speed of 4.5 m/s (10 mile/h)

Friction material	Nature of run	Average μ_{dry}	Average μ_{wet}	$\dfrac{\mu_{wet}}{\mu_{dry}}$	Turns to recovery	Remarks
R-451	dry	0.33	—	—	—	$\mu = 0.39$ at 120°F (48.9°C)
R-451	wet-dry	0.34	0.17	0.50	50	
B rubber	wet-dry	0.95	0.05	0.05	55	Erratic recovery
R-4528-4	wet-dry	0.55	0.10	0.18	54	
Maple	wet-dry	0.44	0.09	0.20	42	$\mu_{max} = 0.56$ during rec'y
Lockheed	wet-dry	0.45	0.12	0.27	25	
R-451	wet-dry	0.34	0.17	0.50	53	
Cork A[a]	dry	0.63	—	—	—	
Cork A	wet	—	0.26	0.41	—	
Cork A	dry	0.79	—	—	—	
Cork A	wet	—	0.19	0.24	—	
Cork B[b]	dry	0.67	—	—	—	
Cork B	wet	—	0.19	0.28	—	
Cork A	wet[c]	—	0.16	—	—	
Cork B	wet[c]	—	0.25	—	—	
R-451	dry	0.43	—	—	—	
R-451	wet-dry	0.37	0.17	0.46	70	

Source: Hanson 1971, 34.

[a] Orientation A: Layers parallel to friction face.
[b] Orientation B: Layers perpendicular to friction face.
[c] After a 48-h soak.

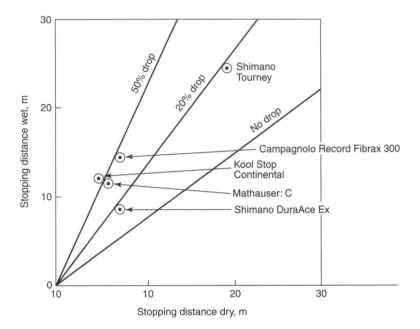

Figure 7.8
Stopping distances wet and dry with aluminum-alloy rims and various brake-pad materials. (From Jow 1980.)

pad tested by Jow was of similar material. Manufacturers are understandably reluctant to disclose information about the composition of their products.) For applied forces of 40, 80, 120, 160, 200, and 240 lbf, the frictional forces given by R-4962 when dry were 13, 22, 35, 47, 58, and 69 lbf, respectively, giving apparent coefficient-of-friction values of 0.32, 0.27, 0.29, 0.29, 0.29, and 0.29, respectively; and by the same material when wetted, the frictional forces were 20, 37, 46, 58, and 65, respectively, giving apparent coefficient-of-friction values of 0.30, 0.25, 0.31, 0.29, and 0.27, respectively.

The same promising material, R-4962, used as a brake pad on steel showed a 2:1 ratio of dry to wet coefficient of friction. The Weinmann standard red-rubber pad had a 0.08 wet coefficient of friction on a steel rim and 0.12 on aluminum.

The (UK) Road Research Laboratory (Anonymous 1963) found that rim-brake wet-weather performance (presumably that on steel rims) can be improved by the use of brake pads longer than the usual 2 inches (51 mm). Softer pads than are common these days are also desirable, along with more rigidity in the brake mechanism and in the attachment of the brake to the frame of the brake itself. This finding agrees with the earlier observations that although static dry friction may be independent of the areas of the surfaces in contact, when there is relatively high-speed sliding, a higher effective coefficient of friction is given if the area of contact is increased.

A consequence of the MIT work with steel rims was the development of a brake that could use aircraft brake-pad materials found by Hanson (1971) to suffer very little drop in friction coefficient in going from dry to wet conditions. The friction coefficient was too low to be used in a regular caliper brake, because too large a squeeze force would be required. It was not possible to strengthen a regular caliper brake and then to increase the leverage, because a consequence of increased leverage is decreased brake-pad motion. (Bicycle wheels of present construction cannot be relied upon to run true, so that a considerable brake-pad gap must be allowed.) Therefore, a brake with two leverages was developed. When the brake lever is initially squeezed, the pads are moved under very low leverage (low force, large movement). As soon as the pads contact the rim, a slider in the brake mechanism locks up, and further movement has to take place through a high-leverage, high-force action. The brake therefore has the additional advantage that it automatically takes up pad wear without further adjustment. The dual-leverage brake was redesigned by Positech, Inc. and tested. Used on the front wheel only, with a regular caliper brake on the rear, it regularly achieved stopping distances of less than 25 percent of those given by regular brakes on steel rims in wet conditions (3.5 m from 6.7 m/s, instead of the usual 15–20 m). However, it was not taken up commercially. It was described a little more fully in the second edition of this book.

Transmission of braking force

The forces generated by hand-operated brakes in early bicycles were transmitted along rods and levers. The invention of "Bowden cables" (flexible steel tension cables inside flexible steel compression housings) in 1902 offered simultaneously a saving in weight and in manufacturing cost coupled with freedom to design both the bicycle's frame and its brakes in different ways. Unfortunately, designers apparently forgot that the laws of sliding friction apply inside a Bowden cable just as they do at braking surfaces. The force transmitted by the inner cable is continuously reduced, particularly around bends, according to the formula

$$(F_1/F_2) = e^{\mu\theta},$$

where $e = 2.718$, and μ is the coefficient of sliding friction. The total angle (θ radians) through which a brake cable is bent along its whole length should be used in calculations involving this formula. The cradling or squeezing action of the outer cable on the inner increases the apparent coefficient of friction by a small extent, in the same way as the friction of a V-belt is increased by the squeezing action of its pulley.

Perhaps fortuitously, the front brake cable on regular bicycles has a smaller total bend angle than does the rear, and it is easy to get a greater braking force at the front, where it is needed. (The rider must develop the requisite skill to apply the brake in such a way that the point at which a "header" is precipitated is not reached.) Additional friction in the large total bending angles of the rear cable can decrease the force applied to the rear pads by 20–60 percent compared to that applied at the front by the same braking action. (Unlubricated brake cables often rust internally, reducing the transmitted brake forces to unsafe levels.) Brake-cable "casings" with linings of low-friction plastic, such as PTFE, have been developed, and it is highly desirable, because of the additional friction in the rear cable, that such casings become standard. However, it has been pointed out above that the rear brake of a bicycle requires less actuating force than does the front if locking (skidding) is to be avoided. Although bicycle brakes with self-adjusting wear take-up mechanisms have been offered commercially, these were not successful. Virtually no present brakes that are currently available allow adjustment without wrenches through the whole range of brake-block wear, which can lead to extremely dangerous conditions in bicycles ridden by less mechanically able persons.

At the time of writing, hydraulic actuation of rim and disk brakes is becoming increasingly popular on the more expensive all-terrain bicycles. Force transmitted hydraulically is entirely unaffected by bends in the hydraulic tubing, and the amount of friction generated is negligible. Moreover, the amount of friction stays low during the life of the brake. The brake pads in some hydraulic rim brakes are attached to the pistons of the "slave" cylinders and so move linearly in toward the wheel rims during braking. Such linear motion offers the significant advantage that, as the pad wears, there is no danger of its going into either the tire sidewall or the spokes, as can happen with the pads on some rim brakes as they wear.

Other developments in bicycle braking

It has been claimed that leather, which was first used for bicycle brakes in the late 1800s because of its good resistance to wear, coefficient of friction, and ability to conform to the profile of the wheel's rim, also possesses outstanding wet-braking properties when used against a chrome-plated surface. It is stated that this is true for chrome-tanned leather, but not, apparently, for leather tanned by the older "vegetable" process. Chrome-tanned leather gives a ratio of wet to dry friction of between 0.5 and 1.0, for

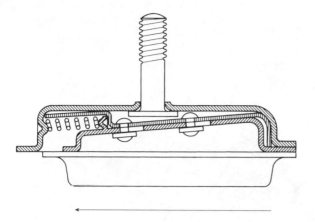

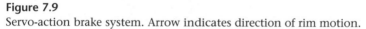

Figure 7.9
Servo-action brake system. Arrow indicates direction of rim motion.

reasons not fully understood but connected with the porosity between the fibers of the leather and their affinity for water. Fibrax Ltd. brought out a brake block in a leather reputed to be from buffalo hide. It was reported to give outstanding performance, with wet stopping distance no more than 30 percent greater than the dry. In 1980, Fibrax introduced a leather block for use with aluminum rims.

Rather too often, brakes are developed in which the braking forces themselves supply part of the actuating force ("servo-action" brakes). A typical system (figure 7.9) incorporates angled ramps within the brake shoes, so that the brakes, in being pulled forward by the wheel rim during braking, are also forced inward to give a stronger squeeze (but only if there is significant friction in the cable that the hand lever is not merely pushed out). A disadvantage of such positive-feedback arrangements is that they magnify the differences between dry and wet friction coefficients. Brakes with such arrangements may give strong braking action with a light actuating force when dry but provide insufficient braking even with a maximum squeeze action when wet.

What is needed, rather, is an added negative feedback stage to limit braking force in dry conditions to less than the amount that would result in the rider's being projected over the bicycle's handlebars. A braking system incorporating such a combination of positive and negative feedback (figure 7.10) was developed by Calderazzo (Hopgood 1979). Only the rear-wheel brake is actuated by the rider. The rear brake is mounted on a lever pivoted near the wheel axis so that it is carried forward during braking. In moving forward, it actuates (through a cable or hydraulic line) the

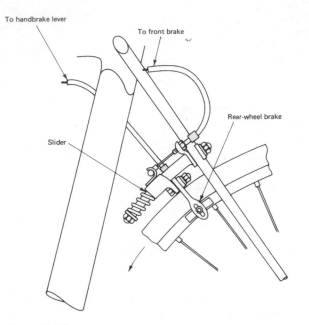

Figure 7.10
Calderazzo feedback brake system. When the hand brake is operated, the rear brake is carried forward on a slider against a spring, actuating the front brake through the cable simultaneously. If the bicycle starts to pitch forward, the rear wheel is no longer rotated by the road surface, and the front brake is released. (From Hopgood 1979.)

front brake, with any reasonable desired degree of force multiplication. Accordingly, little effort need be required for strong braking to be obtained. At the point at which the rear wheel would start skidding, braking at the front wheel is automatically limited. In hundreds of tests with this system, in which testers made "panic stops" from high speeds on different surfaces, never did a rider even begin to go over the handlebars. (The front forks of the test bicycle eventually failed through fatigue: testimony to braking effectiveness and to the inadequacy of the design of the fork.) This promising system apparently died in patent litigation. As mentioned earlier, it could well be used to actuate other types of brakes (e.g., a coaster brake) in the front wheel.

Rim temperatures reached during downhill braking

Wilson (1993) studied rim temperatures attained during steady downhill braking of the type that is required in cycling on mountain roads. His

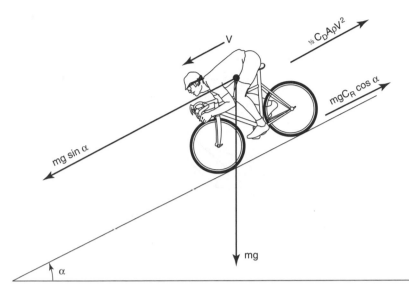

Figure 7.11
Forces on a slope.

results showed that temperatures reached are likely to be dangerously high for standard road bicycles that rely on rim brakes alone, and even higher for bicycles with smaller front wheels and for tandems. It is therefore highly desirable that all tandems be equipped with at least one brake that does not heat the wheel rim, such as a drum or disk brake. Wheels with deep-section, streamlined rims will run cooler than those with narrow, unstreamlined rims that produce separated air flow, which has little cooling effect. Wide rim strips used under the tires and tubes can insulate the tube somewhat from the heated rims. It is also important in downhill cycling that braking be applied to both wheels fairly evenly, but with a bias in favor of the rear wheel, because of the extreme danger of a front-tire blowout at speed. The model indicates that it is better to go either slower or faster than about 10 m/s to limit the increase in rim temperature that results from braking. However, although the steady-state model shows that the faster the bicycle is allowed to travel on a downhill, the lower will be the rim temperature, if one has to brake suddenly from high speed at, for instance, a sharp bend in the road, one will produce a very high transient temperature, and therefore, the danger of tire failure will increase sharply.

The model is based on the forces on a vehicle descending a steep hill at steady speed (figure 7.11). The power dissipated in the brakes is the net downslope force multiplied by the vehicle speed.

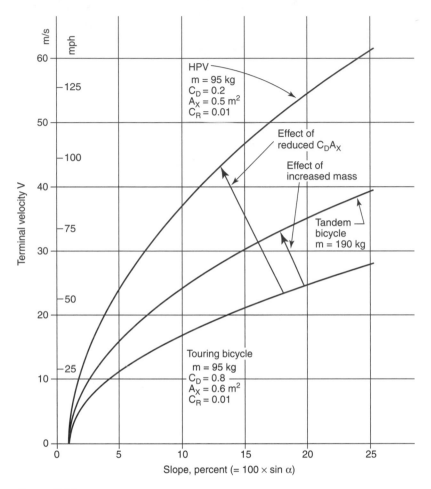

Figure 7.12
Terminal velocity versus slope. (From Wilson 1993.)

$[mg \sin \alpha_S - mgC_R \cos \alpha_S - C_D A V^2]V.$

The terminal velocities, for which the net downslope force is zero, predicted from this equation for various slopes are shown for various bicycles in figure 7.12, together with the specifications used for different bicycles. The thermal model showing, principally, the area from which the rim heat can be dissipated in convection heat transfer, is presented in figure 7.13. Radiation and conduction of heat through the spokes and the tire have been ignored in order to err on the conservative side. The heat-transfer

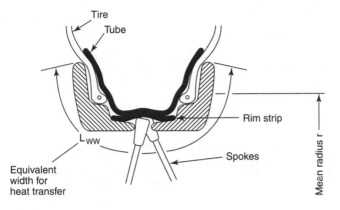

Figure 7.13
Thermal model. (From Wilson 1993.)

model uses one developed for turbine disks by Bayley and Owen (1970) to show general trends rather than to predict absolute values.

The resulting expression for the difference in temperature between the rim and the air is, for typical sea-level values of properties for air:

$$\Delta T = \frac{9.81m(\sin \alpha_S - C_R \cos \alpha_S) - 0.6C_D A V^2}{100rL_{WW}[0.5 + 1.125(1 - 0.0632rV)]}$$

where

m is the mass of the rim (kg);

α_S is the slope of the hill;

C_R is the rolling-resistance coefficient of the tires;

C_D is the drag coefficient of the bicycle and rider;

A is the frontal area of the bicycle and rider (m^2);

V is the speed of the bicycle (m/s);

r is the mean radius of the rim (m); and

L_{WW} is the effective width of the rim (figure 7.13) (m).

(See Wilson 1993 for an equation incorporating values of properties of air at other than sea level.)

Rim-temperature increments above ambient temperature are shown in figure 7.14 for various machines. The increments are zero at zero speed

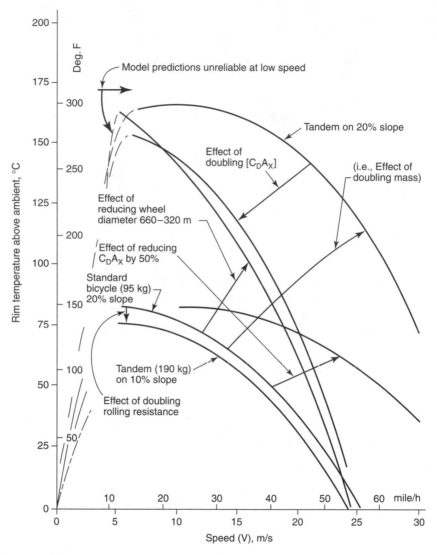

Figure 7.14
Temperature rise of rims during long braking on hills. (From Wilson 1993.)

and at terminal speed. Between zero and terminal speed, there is a speed at which the incremental rim temperature will be at a maximum. The thermal model is unreliable at low speed, so that the estimation of the speed at which maximum rim temperature will be reached is particularly imprecise. However, the overall shape of the curvers must be close to reality. It makes the dilemma of downhill racers very clear: going fast avoids heating the rim unless emergency braking has to be applied, in which case the danger of overheating the rim is sudden and serious.

References

Anonymous. (1963). *Research on Road Safety*. London: HMSO.

Armstrong, Allen E. (1977). "Dynanometer tests of brake-pad materials." Positech, Inc., Lexington, MA.

Armstrong, Allen E. (2000). Personal communication to author regarding further tests made in 1977.

Bayley, F. J., and J. M. Owen. (1970). "The fluid dynamics of a shrouded disk system with a radial outflow of coolant." *Journal of Engineering for Power*; Paper no. 70-GT-6, ASME, New York, N.Y.

Carr, G. W., and M. J. Rose. (1966). *The MIRA Single-Wheel Rolling Resistance Trailers*, MIRA Research Report 1966/5. Motor-Industry Research Association, Nuneaton, U.K.

Hanson, B. D. (1971). "Wet-weather-effective bicycle rim brake: An exercise in product development." M.S. thesis, Massachusetts Institute of Technology, Cambridge.

Hopgood, R. C. (1979). Personal communication on the Calderazzo patent case.

Jow, Richard. (1980). "Bicycle workshop: Stopping power. Here's a comparison test of 18 brake blocks ..." *Bicycling* (May):110–116.

Juden, Chris. (1997). "How thin may the braking rim of my wheel get?" (technical note). *Human Power* 13, no. 1 (Fall):20.

Kempe's Engineer's Year Book. (1962). Vol. 11. London: Morgan.

Wilson, David Gordon. (1993). "Rim temperatures during downhill braking." *Human Power* 10, no. 3 (Spring/Summer):15–18.

8 Steering and balancing

Introduction

Balancing a bicycle is possible at rest only with a special technique known as a "track stand," but is easy when moving forward. Like walking on stilts, balancing a bicycle derives from an ability to steer the support points to a position under the center of mass. Many bicycles are capable of making the necessary steering adjustments automatically, without any rider input.

Unfortunately, the mathematics purporting to describe bicycle motion and self-stability are difficult and have not been validated experimentally, so design guidance remains highly empirical. The most significant design detail is a geometric quantity called "mechanical trail."

This chapter will discuss some simple steering-related observations and the rapid steering oscillation known as "shimmy." The important subject of tire behavior ("slip angles" and "scrub torques") will be introduced, and its connection with bicycle handling will be described. Finally, the most difficult area of human factors, involving human perceptions and adaptability, will be explored briefly. This topic has been studied in greater detail for aircraft piloting, which is an easier problem to solve than balancing a bicycle, because the pilot is far lighter than the airplane.

The most visible "mystery" in balancing a bicycle is that the bicycle can be balanced on just two points of support. Indeed, there's a sensation that it would be impossible to fall down even if one tried.[1]

An important quantity frequently mentioned in this chapter is "steering torque": the turning effort the rider must apply to the bicycle's handlebars to steer as desired, and particularly to hold the bicycle in a steady turn. It is defined as a force times a lever arm and measured in units such as newton-meters or pounds-foot. Not only is steering torque significant as an indication of the muscular effort required to steer a bicycle, but more importantly, it indicates how the handlebars would initially tend to reorient (i.e., in opposition to the steering torque) if released.

Special characteristics affecting bicycle steering

The geometry and mass distribution of a bicycle's steering mechanism play a significant role in handling, but scientific study of such matters has been relatively inconclusive. Part of the reason for this lack of conclusive research is that the concepts involved in a largely self-balancing vehicle are fairly subtle, and the relevant equations are complex (i.e., hard both to derive and to interpret). But far more important is the central role of the bicycle's "pilot": unlike the pilot of an airplane or even the rider of a motorcycle, the rider is by far the heaviest part of the system in bicycling and is able to use all kinds of body motions (sometimes called "body English"), largely unconsciously, as control inputs. Furthermore, the handling behavior that "feels good" to a rider is always changing, conditioned by adaptation and affected by fatigue.

There is a comprehensive "received wisdom" about the design features that supposedly make for good bicycle handling in a given situation (e.g., high-speed cornering, or negotiating a slippery trail), and for all we know, the prescriptions offered by this received wisdom may often hold true. At this juncture, what science can prescribe remains far more limited.

Wheeled-vehicle configurations

A wide variety of human-powered vehicles has been built: two types are shown in figure 8.1. Among them are circus unicycles, large-wheel monocycles with the rider inside the rim, bicycles (one or more riders in line, or a side-by-side couple on a "sociable"), dicycles (two wheels side by side), and tri- or quadracycles.

Though modern bicycles and adult tricycles appear very similar from the side, their characteristics are distinctly different.

- Bicycles must be balanced, requiring some skill. Tricycles are innately stable, even at rest.
- When traction is good, bicycles can easily corner at high speed. Balance is maintained by leaning. Side forces on the wheels are relatively small, so the wheels need only low stiffness and strength against lateral forces. Most tricycles cannot lean, so fast cornering is possible only by "hiking" one's body to the inside of the turn to avoid rollover (figure 8.2). Wheel side loads are then large, except in the case of a special construction that tilts the wheels in the direction of the center of the turn. A vehicle with this special construction is sometimes called a "leaning trike."
- When traction is poor, bicycle balance can be lost, and in a crash, the wheels can be subjected to large side loads. Tricycles are not particularly affected by slippery conditions, except when cornering.
- A bicycle's narrow width and single track makes it far easier than a tricycle to thread between obstacles and to carry up stairs.
- Even though a bicycle must operate in a state of balance, this balance is easily achieved even with offset mass, because the lean angle can always be adjusted to place the center of mass over the support line.[2] However, offset mass generally causes a steering torque. Tricycles are largely unaffected by offset mass unless the center of mass falls outside the triangle formed by the contact points of the three wheels.

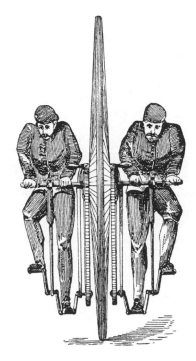

Figure 8.1
Unicycles, dicycles, and tricycles. (From Sharp 1896, 184, and Harter 1984, 25, 29.)

- A side slope or cambered road has an effect on both types of vehicle that is quite similar to that of a steady turn. It has almost no effect on a bicycle but gives rise to an annoying steering torque on a tricycle and side force on its rider—or in extreme cases, the risk of rolling over.
- With more than two wheels, or even on a dicycle, which has two wheels on a single axle, misalignment may cause one wheel to direct the vehicle slightly to one side, against the resistance of one or more others, with the potential for substantial rolling resistance and wear. These may be prevented either by very accurate alignment, even when the vehicle is deformed by rider weight, or by a self-aligning caster arrangement.

In examining diverse bicycle constructions from the point of view of handling, it can be helpful to consider each wheel, front and rear, as part of a separate "assembly," and then to ask the following (see figure 8.3).

- To which assembly is the rider joined?
- Where is the rider's center of mass?
- Where does the line of the steering axis fall?

Figure 8.2
Cornering tricycle. (From *Cycling and Sporting Cyclist*, September 14, 1968, p. 19.)

Broomstick analogy

A bicycle balances when its center of mass (COM) is "over" its support. At rest or in steady horizontal motion, "over" means vertically above. But in horizontally accelerated motion, such as a steady turn (imagine sitting in a fast-turning merry-go-round) "over" means at an angle, such that the combination of downward gravitational pull and horizontal centripetal force aims directly from the bicycle's COM to the point at which it is supported, as in figure 8.4.

An examination of the simple exercise of balancing a broomstick upright in the palm of the hand can elucidate many important aspects of bicycle balancing. The key rule is that *unstable balance of an unstable rigid[3] body requires an accelerated support*. Whether its support point is at rest or moving steadily, a broomstick inverted and placed on the palm of the hand

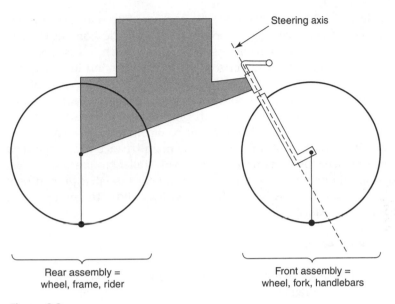

Steering axis

Rear assembly =
wheel, frame, rider

Front assembly =
wheel, fork, handlebars

Figure 8.3
Definitions of front and rear assemblies.

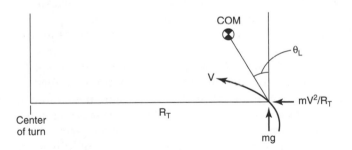

Figure 8.4
Definition of "the center of mass, COM) being over the support" when a bicycle travels around a curve. Traveling in a circle of radius R_T at speed V involves an inwarad (centripetal) acceleration of V^2/R_T. The ground force must be composed of a vertical component supporting the weight and an inward component creating the acceleration. For balance, the COM must be "over" the support point, where "over" is defined as an angle $A_L = \tan^{-1}(V^2/gRT)$. (This is a simplified description ignoring several minor secondary effects.) To prevent skidding, the road should be cambered (tilted) to be more nearly perpendicular to the bicycle.

is unstable and will simply fall over. (A gyroscopically stabilized top is a quite different case.) Balancing a broomstick, or a bicycle, consists in making the small support motions necessary to counter each fall as soon as it starts, by accelerating the base horizontally in the direction in which it is leaning, enough so that the acceleration reaction (the tendency of the center of mass to get left behind) overcomes the tipping effect of unbalance. The base must be accelerated with proper timing to ensure that the rate of tipping vanishes just when the balanced condition is reached. Even more sophisticated control is needed to maintain balance near a specified position, or while moving along a specified path. Taller broomsticks fall less quickly than shorter ones (the time it takes an object to fall is proportional to $[y_{CM}/g]^{1/2}$, where y_{CM} is the height of the COM above the support) and so are easier to balance.[4]

How bicycles balance

A rider balances a bicycle in the left-right direction by *steering it while rolling forward* so as to accelerate the support of the bicycle laterally.[5] Restraining a bicycle's steering makes it unrideable, a fact that is put to good use in steering locks for deterring bicycle theft. Surprisingly, the small steering motions necessary to right a bicycle after a disturbance can take place *automatically*, even with no rider, as can be demonstrated by releasing a riderless bicycle to roll down a gentle hill and then bumping it.[6]

It is widely believed that the angular (gyroscopic) momentum of a bicycle's spinning wheels somehow supports it in the manner of a spinning top. This belief is absolutely untrue. *Gyroscopes can react against (i.e., resist) a tipping torque only by continuously changing heading.* For example, a tilted top can resist falling only by continuously reorienting its spin axis around an imaginary cone. Locked steering on a forward-rolling bicycle does not permit any wheel reorientation, and the bicycle will fall over exactly like a bicycle at rest, no matter how fast it travels, or how much mass is in the wheels. To be sure, bicycle wheels actually are changing heading continuously whenever the steering is turned, but their mass is too small to be of importance: the resulting gyroscopic support moment is tiny compared to the "mass times acceleration times center of mass height" moment that predominantly governs bicycle balancing.

Still, there is an extremely interesting gyroscopic aspect to bicycle balance: the angular momentum of a bicycle's front wheel urges it to steer (i.e., to precess) *toward the side on which the bicycle leans,* as can be demonstrated by lifting a bicycle off the ground, spinning the front wheel, and briefly tilting the frame. In other words, the gyroscopic action of the front wheel is one part of *a system that automatically assists the rider in balancing the bicycle.* If the angular momentum of this gyroscopic action is canceled

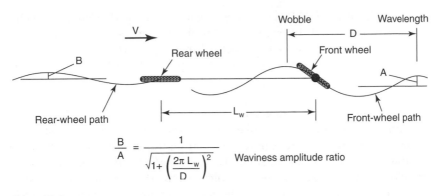

Figure 8.5
Front-wheel track compared to that of the rear. Front-wheel track is wavier.

(as Jones [1970] did with an additional, counterrotating, front wheel), considerably more skill and effort are needed for no-hands riding.

The broomstick analogy presented above is only partly applicable to a bicycle. A bicycle is actually supported at two distinct points that generally accelerate somewhat differently. A low-speed slalom maneuver after riding through a puddle demonstrates that the front wheel travels a much wavier path than the rear, which also lags in phase (figure 8.5). Only in a steady turn do the contacts of both wheels with the ground follow paths of comparable curvature.[7] Therefore, lateral acceleration (equal to rolling-velocity squared divided by path radius of curvature, or equivalently to rolling velocity times rate of change of wheel heading [in radians/s]) is generally greater for the front contact than for the rear. In fact, only at the front of the bicycle frame[8] can lateral acceleration be brought about relatively rapidly, by accelerating the steering angle and by maintaining a rate of increase of steering angle. The steering angle must settle to a steady value before the front acceleration reduces to match the rear.

One implication of the delayed and reduced lateral acceleration of a bicycle's rear contact is that mass over the front contact is far more easily balanced than that over the rear. If mass is to be carried over both contacts, keeping that in the rear lower than that in the front will allow the front-support acceleration to exert more control over balance.[9]

The basic means by which bicycles are balanced and controlled involves *vehicle supports that travel only in the direction in which they are pointed* (implying that they can support loads perpendicular to that direction), the front being steerable. These steering functions can be performed not only by conventional large-diameter bicycle wheels, but also by small-diameter wheels, as on a foldable scooter, skates on ice, skis or runners on snow, and fins or foils in water.[10] It is even possible to tip a four-wheeled motor vehicle up on the wheels of one side and to balance and steer it like a motorcycle for as long as desired. In each of these cases, the required steering torque may differ, and the particular sideslip, inertia, or flex may affect the feel, but all are essentially bicycles.

Countersteering to generate lean

An unstable balanced object like a broomstick or bicycle must have the appropriate (say) leftward lean to maintain significant acceleration leftward of the center of mass. In other words, *the support point must first move to the right of the system center of mass to create the lean.* The motion of the support point can be hard to observe while riding, because it happens so quickly and unconsciously. To see it most clearly, one can ride a bicycle along a painted line on the right edge of a road and watch the front wheel's position while making a quick maneuver to change lanes rightward. One will notice a brief leftward deviation of the front wheel's path, caused by briefly steering leftward before settling into a sustained rightward steer angle (figure 8.6).

But in fact every cyclist knows so-called countersteering very well, unconsciously: it is the only possible way to maneuver the bicycle, or to stop with the right degree of lean to allow whichever foot the cyclist chooses to touch the ground. That everyone who knows how to ride a

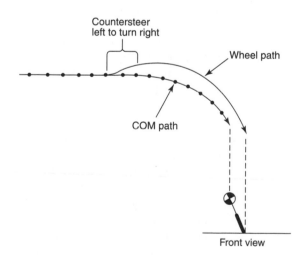

Figure 8.6
Brief leftward "countersteering" to generate the rightward lean necessary for rightward turning.

bicycle already unconsciously understands the method becomes clear when we are riding close to the edge of a curb or a slight drop-off. Riding closer than about 125 mm makes us feel nervous and "trapped": we know that it will be necessary to turn *toward* the danger in order to steer *away* from it. If there's no room, we sense that trying to escape will take us over the edge. Nonetheless, it is useful to practice abrupt, forced countersteering, for use in emergency maneuvering (see, for example, Forester 1993).

Countersteering is also needed when one encounters a side-wind gust or when one is pushed by a neighboring rider. Whenever any force pushes us to the right, we must briefly turn right to generate leftward lean, so as to counter that force steadily.

Incidentally, all principles of unstable balance apply to runners too. To accelerate, a runner leans forward, and to decelerate quickly, he leans back, by getting the feet out in front. Running straight and then turning to turn leftward suddenly requires a step to the right, off the path, to generate lean.

Basic bicycle-riding skills

At the height of the 1890s bicycle boom, bicycle-riding schools sprang up in major cities. But for most of us who acquired our bicycling skills at a later time, learning to ride was typically a trial-and-error process conducted near home. Does the study of bicycle balance offer any insight into the process?

•   The common advice to "turn toward the lean" is good. A quick method for teaching this is described by U.S. patent 5,887,883 to Joules.
•   It's hard to see how training wheels can inculcate any of the desired balancing habits, unless they are off the ground (i.e., used only when at rest). (Such a positioning frequently occurs as the training wheels wear and the supports bend).
•   Unintended upper-body motions probably act as disturbances. Perhaps a temporary back support would be helpful. (Most recumbent bicycles have permanent back supports.)
•   A heavier front wheel and more trail (discussed later in this section) should exert a stabilizing influence on a bicycle and hasten its rider's awareness that the steering has an automatic tendency to perform the balancing task for him.

Beyond these thoughts, we are attracted by the commonsense idea of having those learning to ride a bicycle adjust the bicycle's seat low enough to plant their feet on the ground and practice by coasting down gentle, grassy slopes. Also, a scooter is an excellent tool for learning to balance, almost free of the risk of a fall, as stepping off onto the ground is easy.

Once basic balancing of a bicycle is mastered, important cornering techniques can be developed.

- Paying attention to surface conditions that provide poor traction (e.g., loose gravel or wet leaves). These are also impediments to secure balancing in straight-line riding. A small slippery patch in a turn may be negotiated by briefly steering wider to reduce horizontal contact force, though unfortunately this technique will increase the lean angle. Allowing one's body to bend easily relative to the bicycle's frame when entering and leaving the turn will minimize wheel sideforce during the transient maneuvers involved in a turn.
- Adopting the gentlest possible turn radius (i.e., starting wide then grazing the apex). The rider must be prepared to brake hard before entering a turn if he finds his speed to be too high.
- Gripping the bicycle's handlebars more tightly when cornering hard at high speed, to stiffen the arms and to reduce instability.
- On a bicycle with a free-wheel, holding the inside pedal in a raised position during hard cornering, so that it does not strike the pavement. The limit of cornering traction on good dry pavement is typically 40° or more, but the inside pedal typically strikes the pavement at 30–35°. Holding the bicycle more upright than the upper body places more stress on the wheels but may allow continued pedaling through a corner without slowing. On a fixed-gear bicycle (one without a free-wheel), this technique is essential.

Cyclists often extend the inside knee in executing a turn. This practice offers the seemingly marginal benefits of making the bicycle's frame slightly more upright, keeps a little more of the tire tread on the pavement, and counters the tendency of a steered wheel's "mechanical trail" (defined later in this section) to shrink because of lean. Snapping the leg inward momentarily decreases sideforce and may enable recovery if traction is lost during the turn.

Grooves or ridges (like streetcar tracks) that can trap a wheel are particularly dangerous for bicyclists, because balance depends on being able to steer left or right. Slippery conditions make even shallow ruts dangerous because there's too little friction to climb out of them.

Effect of bicycle configuration on steering and balancing

Some of a bicycle's behavior is explicable in terms of basic geometry: the placement of the wheels, the line of the steering axis, and the centers of masses of the front and rear assemblies.[11] In figure 8.7, two wheels, possibly of differing radii, are separated by a horizontal distance called the *wheelbase* (L_W). The steering axis is drawn as a dotted line (typically it is

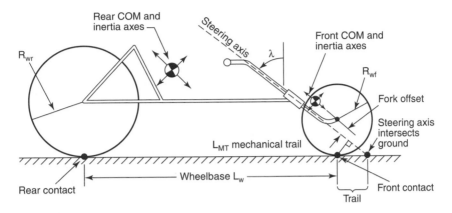

Figure 8.7
Bicycling parameters related to handling and stability. Each wheel has a spin moment of inertia that causes gyroscopic phenomena.

tipped back from vertical by an angle λ and intersects the ground ahead of the front-wheel contact). The line of the steering axis commonly passes below the front axle, that is, the fork is bent forward.

The perpendicular distance by which the front axle is ahead of the steering axis is called the *fork offset* (L_{FO}) (a more common but confusing name is *fork rake*). The horizontal distance by which the front contact is behind the imaginary point where the steering axis intersects the ground is called *trail* (similar to automotive caster). A more significant quantity is the *perpendicular distance* from the front-wheel contact to the steering axis (positive when the axis is ahead of the contact point), which has been called *mechanical trail* (L_{MT}).[12] (This lever arm has such a great effect on handling and stability that we have tabulated its value for a number of different bicycles in tables 8.1 and 8.2). Within this skeletal bicycle geometry, it is usually necessary to know the position of the mass center of the front assembly and that of the rear to pass judgment on stability.

For many analytical purposes, the radii of a bicycle's wheels are not significant. A simple model with the same contact points and steering axis involves just tiny "zero-radius" wheels (see figure 8.8). Such an approximation effectively freezes the mechanical trail at a fixed value, whereas on an actual bicycle, L_{MT} is somewhat reduced during hard cornering.

A variety of observations from this simple model can easily be understood.

•   Riding straight while bending the torso to the left side of a bicycle's frame requires the frame to lean rightward to maintain balance (i.e., COM

Table 8.1
Mechanical trail for typical bicycles

Bicycle type	Angle of steering axis with vertical, radians (angle with road, degrees)	Fork offset, mm	Mechanical trail, mm
Touring	0.314 (72)	50.5	55.5
Touring	0.314 (72)	50.7	55.2
Touring	0.314 (72)	47.5	58.5
Touring	0.297 (73)	57.9	42.3
Road racing	0.297 (73)	57.4	42.8
Road racing	0.279 (74)	50.0	44.5
Road racing	0.279 (74)	66.9	27.6
Road racing	0.271 (74.5)	55.1	36.5
Track racing	0.262 (75)	52.1	36.7
Track racing	0.262 (75)	65.4	23.4

Note: All bicycles included in table have a wheel radius of 343 mm.

Table 8.2
Trail of specific bicycles

Specific bicycle	Trail, mm
Raleigh sports utility, three-speed	34.9
Peugeot touring, 650B tires	38.1
Dahon folding bicycle	39.7
Gitane Hosteller	47.6
Raleigh International	47.6
Holdsworth Italia touring	50.8
Mineapolis Wonder (original fork)	50.8 (approx.)
Viking Tour of Britain (touring/racing)	57.1
Kvale racing	57.1
Raleigh-made Huffy touring	60.3
Mineapolis Wonder (replacement fork)	63.5
Ideor track racing	63.5
Holdsworth Italia road racing	69.8
Schwinn Paramount track racing	69.8
Elliott track racing	76.2

Source: Forester 1989.

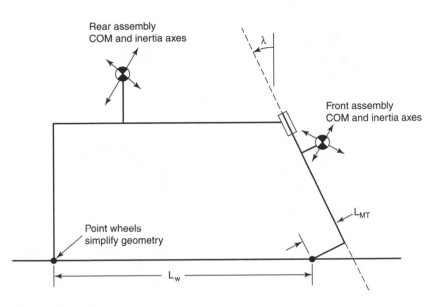

Figure 8.8
Bicycle model with point wheels. If the bicycle is held level and the steering is turned, front wheel will lift off the ground.

"over" the line joining the contact points). The vertical support force on the front contact will have a component perpendicular to the wheel; this component acts through the lever arm of the mechanical trail and tends to turn the handlebars to the right, as can easily be felt. The small effect of handlebar weight simply adds to this torque (see figure 8.9).[13]

- While walking beside a bicycle with one hand on the saddle, one way to steer left is by briefly applying a leftward twist (i.e., a yawing torque) on the saddle. This torque creates a rightward sideforce on the front assembly's trailing contact, steering it leftward. After the turn has been established, a steady rightward torque is needed to prevent the steering from turning further to the left. Alternatively, leaning the bicycle leftward allows the offset weight of handlebar and wheel center to turn the wheel left, slightly assisted by ground pressure on the front contact (figure 8.10). The effect of rapidly jerking the saddle to either side is more complex.

- When riding a bicycle at low speed (e.g., 2.5 m/s) in a circle, being careful to keep your torso in the plane of the frame, and controlling the handlebar position with just one finger, it is clearly necessary to restrain the bicycle from sharpening the turn. This characteristic is primarily a reflection of *the system's potential energy being at a maximum in the upright,*

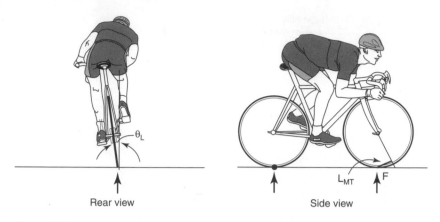

Figure 8.9
Steering torque from frame tilt when riding straight. The COM will be over the support line connecting the wheel contacts. The vertical force at each wheel can be resolved into a component parallel to the wheel plane and a component perpendicular to the wheel plane. The ground force (F) supporting the front wheel tends to turn it leftward, with moment $FL_{MT} \sin \theta_L$. In addition, the scrub torque at the front contact and the handlebar weight also promote leftward steering.

straight configuration. When the handlebars are turned,[14] the center of mass falls by an amount proportional to the mechanical trail (equivalently, with the mass center at a fixed height, turning the handlebars would lift the front contact off the ground). The resulting torque cannot be demonstrated at rest because of tire friction . As will be discussed below, the tendency to sharpen a turn is part of bicycle self-stability. On theoretical grounds, it was expected theoretically that front-wheel gyroscopic effects would eliminate this characteristic at high speeds, but new research suggests that tire properties might prevent such effects from accomplishing this.

▪   If a cyclist is supported perfectly vertically by an assistant, turning the bicycle's steering to the right places the center of mass to the right of the support line, creating a strong tendency to tip to the right. (See the rear view in figure 8.11.) In this case, it is helpful to suppress the instinctual tendency to turn toward the side of leaning.

▪   Low-speed turns to the right place the front contact to the left of the frame plane; to retain balance the frame must therefore lean left. In low-speed turns, therefore, *the frame leans to the outside.* Only above a certain minimum speed (defined approximately by $V = \sqrt{gL_{MT}x_{CM}/y_{CM}}$ does the bicycle frame actually lean toward the center of the turn.

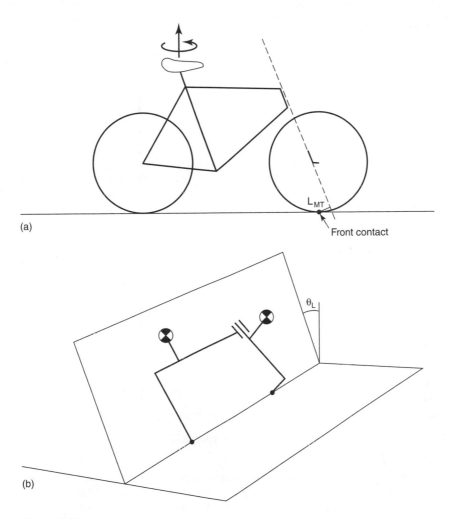

(a)

Front contact

(b)

Figure 8.10
Two methods of steering a bicycle held only by its saddle: (a) A twisting moment tending to turn bicycle leftward generates a rightward sideforce from road to front contact. Because of L_{MT}, this steers the front wheel leftward. A reverse twisting moment is needed to straighten the steering again, because COM is lowered by the handlebar turn. (b) Vertical ground force at front contact and weight of front assembly at its COM both develop leftward-turning tendency on conventional bicycles.

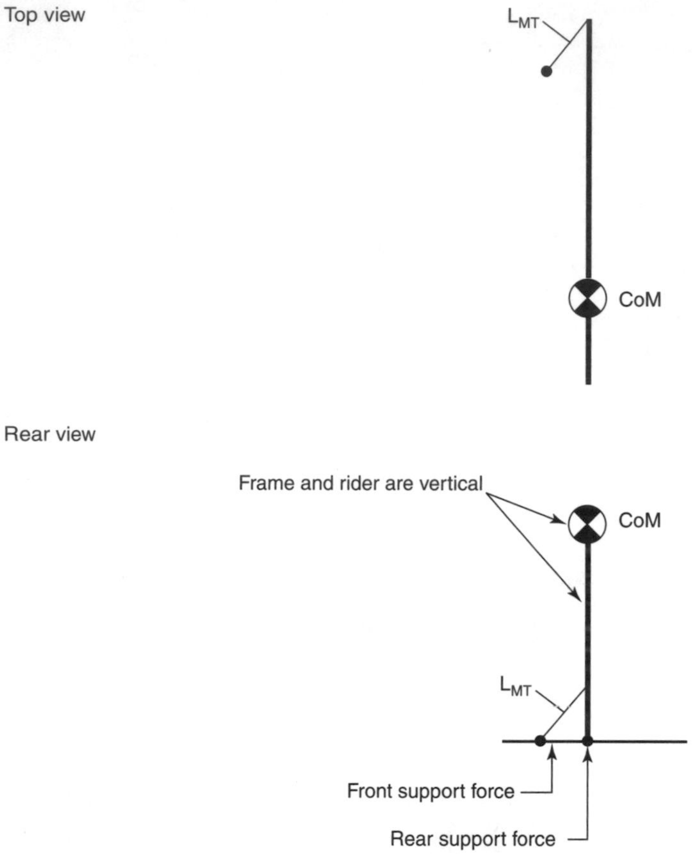

Top view

Rear view

Figure 8.11

Importance of keeping front wheel straight when being held upright before a race. Rear view of point-wheel bicycle with steering turned rightward; front contact is now to the left of the frame. COM is no longer over the support line, and bicycle tends to tip rightward, sometimes severely.

In no-hands riding, and if gyroscopic torques and special tire-contact-patch torques are ignored, the center of mass is at its maximum height in the "balance plane" defined by the center of mass and the two ground contacts. The handlebar torque is zero (i.e., the steering is in the "balanced" orientation). The tendency of the bicycle to turn to one side, or equivalently the need for torso lean to travel straight, is in this case a sensitive indicator of various asymmetries. At moderate speeds, no-hands handlebar orientation can be predicted qualitatively by the principle that the steering can achieve equilibrium only by turning *against* any disturbing torque. Here are some examples.

• Applying a clockwise (rightward) torque bias to the steering (e.g., with a taut rubber band from the seat-post pulling on a string wrapped around the steering axis) ultimately leads to the steering's being turned counterclockwise (i.e., a leftward turn). Alternatively, the rider's torso must lean to the right of the frame, so frame lean creates a countering leftward torque.

• Intentionally misaligning the front wheel relative to the forks (say, bottom displaced to the left of the rider, and top to the right) also creates a steering torque initially tending to turn the handlebars (rightward, in this case). In equilibrium they are therefore turned *leftward*: the bicycle curves to the left. Misaligning the rear wheel so that its forward edge is moved rightward also generates leftward steering.

• Torso lean to the left of the frame tilts the frame to the right, generating a torque tending to turn the steering right. The equilibrium configuration therefore involves steering to the left. Relative to the frame's midplane, the cyclist leans in the direction of the intended turn and then straightens the torso to return the bicycle upright.

In theory, these expected turn directions are reversed above a so-called "inversion speed" (see "Nonoscillatory instability"). However, the existence of this phenomenon has recently been called into question by the recognition that tire properties may suppress it, at least when the rider's weight creates a substantial contact-patch length and resulting turn-sharpening torque when leaning.

Experiments with a riderless bicycle
The foregoing experiments are often easier and more revealing with a riderless bicycle. Other riderless investigations have also been conducted:

• Jones (1970) varied trail and also installed an off-ground counter-rotating front wheel as an "anti-gyro;" and

- Hand (1988) added front-wheel mass, reversed the fork, and experimented with sideforce and steering-torque disturbances. (Hand's experiments are analogous to Milliken's [1989] string-pulling experiments with a ridden bicycle.)

To perform steady-state riderless experiments, it is essential to have low-friction steering, a condition of initial alignment that allows the bicycle to travel straight, and a design that affords intrinsic stability at the test speed. It is then possible to engage in bicycle activities similar to a game of catch (rolling the bicycle to a partner) or kite flying (propelling and leaning the bicycle by pulling on an attached string).

Nonlinear determination of balance speed and steering torque

Steering torque is the foundation for a variety of other approximate or exact nonlinear computations to determine, for example, the lean of a bicycle's front wheel, or the turn radius generated by a given steering angle θ_S and lean angle θ_L. As a bicycle with a given steering angle is leaned further and further, the front contact point moves forward and the rear moves rearward: a steered bicycle lying on the ground would virtually touch at its foremost and rearmost points. At this extreme juncture the turning radius is less than the wheelbase! This extreme example is given merely to make the point that turning radius depends somewhat on lean angle.

   Mechanical trail has been identified as an important geometric quantity related to steering: it is the moment arm allowing lateral contact forces to affect steering torque and effectively gives the front wheel a caster action. But since the contact point migrates forward around the wheel, mechanical trail can decrease significantly. When a conventional bicycle is first leaned, then steered slightly toward the lean, the forward pitch increases. At the same time the front contact point moves toward the point where the steering axis cuts the ground. When those points blend, the pitch is at its greatest, and the trail is zero. Further steering moves the contact point forward of the steering axis and again reduces the pitch. A similar motion of the front contact point is found if the steering is fixed and the lean angle is varied. The lean angle for vanishing trail can easily be calculated for any steering angle, either exactly or approximately.

   Considering the approximate balance condition in a turn, $V^2 = gR_T \tan \theta_L$ (this approximates all the bicycle mass as directly over the rear contact point and ignores tire thickness) for a typical bicycle. Figure 8.12 illustrates the turning conditions under which mechanical trail vanishes. As one example, with a 45° lean angle, trail vanishes at a turn radius of 5.2 m (this is a speed of about 7 m/s). A reduction (or even elimination) of mechanical trail can play a significant part in rider perception of stability.

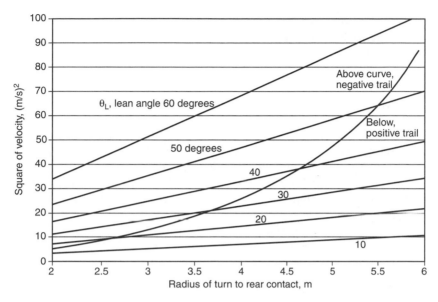

Figure 8.12
Trail is reduced or eliminated in hard cornering. (Plotted by Jim Papadopoulos.)

When mechanical trail becomes negative, a bicycle becomes unstable, and the front wheel rapidly turns to its limit unless actively restrained. In many bicycle crashes, the front wheel ends up turned as far as it will go because of this phenomenon.

The exact nonlinear statics calculation required to determine the balanced lean angle of an ideal bicycle riding in a tight circle is relatively messy—difficult even for an advanced-dynamics university class. The complicated configuration multiplies the difficulty of summing up gyroscopic torque vectors and accounting for the little-known tendency for a high-front, low-rear mass distribution to press harder on the bicycle's front wheel, altering the steering torque. A solution for rigid tires has been published by Kane (1977).

However, the rigid-tire approach leaves out the potentially important contribution of tire scrub torque: because of the finite size of its contact patch, a vertical wheel traveling in a circle requires a steering torque to keep it going around the turn.[15] Man and Kane (1979) later accounted for more realistic tire properties, but their account was not credible, because it allowed any desired turn radius once lean and steer angles were fixed. A more believable treatment may be found in Cossalter, Doria, and Lot 1999, the analysis in which contains the most important comparison between steering torque calculated with rigid tires and that calculated with tires able to scrub. The result of Cossalter, Doria, and Lott's analysis, for a motor-cycle, was to suppress the high-speed steering-torque reversal calculated theoretically for the rigid case.

Correcting the straight-line steering torque of an imperfect bicycle

A bicycle that is not perfectly symmetrical generally requires an annoying steering torque to travel straight, or an upper body lean, when ridden with no hands. It is conventional to check a bicycle's wheel alignment by plac-ing a straightedge at two points of the rear wheel near ground level (i.e., parallel to the intersection of the wheel plane and ground plane) and determining whether the front wheel grazes the same straight-edge when turned parallel to it. But this test alone cannot indicate whether straight-line riding on the bicycle will require a steering torque.

Straight-line riding requires only that the ground traces of the bicy-cle's wheels (i.e., the line of intersection of each wheel plane and the ground plane) be parallel. In that condition, with the bicycle in balance, steering torque arises whenever the steering axis does not pass directly above the front contact or does not pass directly below the front-assembly mass center. Any of a number of factors (load imbalance, wheel tilt or off-set, steering-axis misplacement) can give rise to steering torque, and any other of those same factors can be altered to reduce or cancel that torque.[16]

Bicycle dynamics (unsteady motions)

Consideration of steady riding conditions can teach us only so much. We must also consider dynamics: how does the bicycle enter or leave a turn? How does it respond to a wind gust, a handlebar twitch, or an upper-body lurch? Such brief, unsteady situations are the key to bicycle stability, con-trollability, and the management of disturbances. A useful compilation of motorcycle-dynamics references is available at ⟨http://www.mecc.unipd.it/~cos/DINAMOTO/bibliography/references_moto.html⟩.

An understanding of dynamic behavior requires recourse to equa-tions governing bicycle motion. Over the last century, scores of analyses have been attempted, some comprising more than a hundred pages filled with complicated equations. Unfortunately, most analyses are incorrect, either because of faulty methods or because of errors in algebra.

In an exhaustive study by Hand (1988), the following treatments (plus just a few others not cited here) were found to be relatively complete, correct, and useful: Whipple (1899), Carvallo (1900, 1901), Klein and Sommerfeld (1910), Doehring (1955), Sharp (1971), and Weir (1972). (In an otherwise masterful and comprehensive monograph, Neimark and Fufaev [1972] derived incorrect equations by omitting the change in po-tential energy due to steering angle.) The study by Roland (1973) at

CALSPAN (Cornell Aeronautical Labs), while impressively ambitious and thorough, unfortunately lacked some of the supporting detail needed to validate it. The elegant and economical treatment by Franke, Suhr, and Riess (1990) may be questionable because of the steering equation it employs.

Most of the published equations of bicycle motion are relatively hard to study or use because of complexities in the derivation and notation. The linearized equations developed by Papadopoulos (1987) were formulated specifically to be as simple and symmetric as possible. A basic introduction was published by Olsen and Papadopoulos (1988). We strongly advocate that any future derivations conscientiously demonstrate consistency with known equations, as a way of catching basic errors.

Far less attention has been paid to measurements of actual riding dynamics. An outstanding exception is Roland 1973. On the motorcycle side, some interesting techniques and results are presented in Weir and Zellner's (1979) experimental measurements of steering transients.

Linearized equations of motion, ignoring tire width

The tire's behavior due to nonzero contact-patch length and consequent lateral deformation of the casing is described below, but in rolling forward with no sideslip or scrub torque, a coasting bicycle with a rigid, no-hands rider has just two important degrees of freedom: the lean angle (θ_L) of the frame (let's choose rightward) and the (leftward) steer angle (θ_S) of the fork relative to the frame. (Variation in forward velocity is also a degree of freedom, but it can be ignored in most analyses.) Just two equations of motion are significant: one based on leaning (rolling) moments, and one based on steering moments. Additional equations are needed to incorporate tire sideslip or rider sway in response to maneuvering forces.

For mathematical tractability, "linearized" equations can be developed that are accurate for small angles of lean and steer. A complete set of system parameters for such an analysis follows:

- three for the entire system: wheelbase (L_W), mechanical trail (L_{MT}), and steering-axis rearward tilt (λ);
- for each assembly (i.e., front and rear), seven parameters:
- the COM coordinates (x_{CM} and y_{CM}) relative to the wheel contact;
- the mass (m), plus the moments and product of inertia (I_{xx}, I_{yy}, and I_{xy}) taken at the center of mass;
- the ratio ($K_{H/V}$) of the wheel's spin angular momentum to its rolling velocity; and
- in addition to these seventeen geometric and inertial parameters, the bicycle's rolling velocity (V).

The resulting equations can be written in the simplified symbolic form described by Papadopoulos (1987), in which the coefficients on the constants depend on the above seventeen bicycle parameters and the C_{--} and K_{--} coefficients also involve a velocity that is assumed to be constant:

1. The lean-moment equation: $M_{LL}\ddot{\theta}_L + M_{LS}\ddot{\theta}_S + C_{LS}\dot{\theta}_S + K_{LL}\theta_L + K_{LS}\theta_S = 0$ (i.e., the C_{LL} term is zero). The M, C, and K are abbreviated representations of bicycle parameters. This equation is derived by equating the moments of forces and inertial reactions around the ground line at which the frame plane intersects the ground. If the bicycle frame had outrigger support wheels, the right-hand side would be a "supporting moment."
2. The steering-moment equation: $M_{SL}\ddot{\theta}_L + M_{SS}\ddot{\theta}_S + C_{SL}\dot{\theta}_L + C_{SS}\dot{\theta}_S + K_{SL}\theta_L + K_{SS}\theta_S = T_Q$ (steering torque). The derivation is more complex, but the right-hand side is the steering torque exerted by the rider.

Mathematically, these are described as coupled constant-coefficient linear differential equations for the lean angle and steering angle, in terms of steering torque T_Q (which is set to zero for no-hands riding). The single dots above θ represent "rate of change of angle" (i.e., angular velocities). The double dots above θ represent "rate of change of angular velocity" (i.e., angular accelerations). Angles are measured in radians (one radian is about 57 degrees) because this eliminates a conversion factor in the equation. (In addition, the value for a small angle in radians is very nearly equal to the slope of that angle.)

In principle these equations can predict behavior such as changes in bicycle lean angle due to a given time-varying steering angle and what steering torque T_Q will be required, or how the bicycle will behave when ridden with no hands (i.e., when $T_Q = 0$). But although these equations can be mined for insight, their usefulness is limited, because they do not include the large effects of undefined rider body motions. Their main value is for riderless bikes. Nevertheless, they are useful in the absence of something better. Some brief examples should suffice to illustrate their use.

· In one case, consider riding straight and upright, with no hands. A side-wind gust suddenly pushes the rider to the right, essentially taking the bicycle out of balance. In the early stages of the ensuing motion, only the angular accelerations are large, so angular velocities and angles themselves may be neglected. With zero steering torque, the second equation gives a relation between lean acceleration and steering acceleration. The first equation makes it possible to eliminate either lean or steering angle and solve for the other to find a lean acceleration away from the wind, as well as a steering acceleration direction that depends entirely on the details of rider mass distribution, front-assembly mass distribution, and mechanical trail. At the end of a process of wobbling and readjustment, all the angular accelerations and velocities are zero, and the same process may be used to determine the steady-state lean (toward the wind) and steady-state steering angle (away from the wind). Perhaps more interesting than initial and steady response is the maximum deviation into the wind before the rider is back in control. Determining this deviation requires solving both equations simultaneously, along with a supplementary relation giving bicycle lateral motion from the steering angle.

• In the case of a tricycle, the right-hand side of the first equation must be written to account for a supporting moment from the rear wheels. But as long as that moment is not of great interest, only the second equation, in which the lean angle and angular velocity and acceleration are all zero, is needed. What is left will be recognized by any physics student as the equation for a "mass-damper-spring," although the coefficient signs are no longer automatically positive. If the steering angle is prescribed, the torque can be determined. Otherwise, if the torque is prescribed, a time history of the steering angle can be determined.

Fuchs (1998) has adapted the above equations to the steady motion of a streamlined bicycle in a crosswind; his study includes additional valuable references.

Stability

One of the questions related to bicycling that can be studied via equations involves a bicycle's *intrinsic stability*: when does an uncontrolled bicycle automatically tend to ride straight and upright?

In the field of dynamics, stability has a precise meaning. For our purposes, a steady motion (such as rolling straight and upright) is stable if, after it is disturbed, it eventually settles down to being steady again. In the technical sense, a well-aligned[17] riderless conventional bicycle with freely turning steering bearings is stable over a range of speed that depends on its design (say, 3–6 m/s). If it is bumped while it is in motion, it will soon return to straight upright running. The nature of its stability is defined by a settling time (how quickly the disturbance dies away) and possibly a frequency, if the system (e.g., the steering) tends to oscillate while settling down to steady motion.

The presence of a "dead" no-hands rider, rigidly joined to the rear assembly, does not alter the picture much. The system's mass is then distributed differently, so the resulting equations have slightly different coefficients, and the stable speed range is typically raised to 6–8 m/s.

Unfortunately, hands-off stability with no rider input of any kind does not seem to have much to do with a bicycle that "feels" stable. For example, all standard (uncontrolled) bicycles and motorcycles lose stability both at low speeds (when they execute greater and greater weaving oscillations) and also theoretically at high speeds (when they fail to recover from a turn and instead progressively increase their lean and turn angles in a spiraling crash). But no competent rider has much cause for complaint when riding a typical bicycle at speeds between 2 and 15 m/s. We might speculate that a bicycle whose instability grows too quickly, or whose recovery oscillation is a little too fast to track, will always be hard to ride. But that notion requires further investigation. This is the domain of *human factors* (see Sheridan and Ferrell 1974; Weir 1972; Roland 1973; and the 1990 military (MIL) standard on airplane piloting qualities).

Experienced cyclists actually seem to redefine the technical term "instability" to mean "oversensitivity to small input torques." That is, a bicycle could be perfectly stable at a certain speed with a no-hands rigid rider and yet might seem too skittish, or even unsafe, if each little shift of body weight or hand pressure caused a large steering deviation (Ruina 1987).

Having offered some caveats about the limited significance of technical stability studies, we still find it interesting to ask when an uncontrolled bicycle is stable. It's an intriguing scientific question and may help in identifying more important issues regarding bicycle handling. (See two important collections of motorcycle-related stability papers from the Society of Automotive Engineers [1978, 1979].)

The bicycle-dynamics equations presented above can reveal no-hands uncontrolled stability by any of several routes.

•　　By direct simulation (i.e., instant-by-instant numerical solution of the differential equations) to calculate the motion of a bicycle starting from a small initial lean. By inspecting the results, it can be determined whether the bicycle straightens up or crashes.[18] (It doesn't really matter what the small initial disturbance is: an unstable bicycle will *almost always* wobble or fall, and a stable one will *always* straighten up.) The disadvantage of this approach is that it is hard to determine general rules from specific cases (see Roland 1973).

•　　If the constant-coefficient equations are solved exactly by standard algebraic methods, stability is revealed by the eigenvalues (exponential growth factors). These are generally complex numbers, their real parts (x-coordinates) reflecting growth tendency, and their imaginary parts (y-coordinates) indicating oscillation frequency. If their real parts are all negative, then steering disturbances decrease over time, whereas if any one eigenvalue has a positive real part, then its corresponding pattern of disturbance is predicted to grow infinitely. This approach has disadvantages similar to those of the previous one: the algebraic eigenvalue formulas are too complex to use, so eigenvalues are generally determined numerically for specific cases of interest (see Doehring 1955 and Weir 1972).

- If the main interest is in identifying a simple criterion of stability, and not the details of a bicycle's motion as it either straightens up or crashes, then the Routh-Hurwitz stability criteria may be employed. If four specific algebraic quantities (functions of velocity and the bicycle parameters) are all positive, the bicycle will be stable. If a given bicycle is stable at a certain speed, then altering the design or the speed may destroy that stability. Loss of stability is revealed by monitoring just two of the four quantities: the most complicated one that reveals oscillatory instability, and the simplest one that reveals nonoscillatory exponential growth. The simpler algebra allows some concepts to be explored or proved relatively easily (see Papadopoulos 1987). Some conclusions of this approach follow.

Nonoscillatory instability

The simplest criterion for establishing a bicycle's stability is just the condition that in a steady turn, it should try to sharpen its steering angle. In other words, the steady-turn handlebar torque required of the rider must be such that it *restrains the steering from turning further.* If an uncontrolled bicycle lacks this property, it never picks itself out of a turn but gradually increases its lean and steering angles while following a tighter and tighter spiral, a phenomenon referred to in the motorcycle-stability literature as "capsizing."

In a conventional bicycle, turn-sharpening behavior is afforded at low speeds by the steering geometry and the front-assembly COM position, which together tend to increase any steering angle if the bicycle is balanced. But at high speeds, gyroscopic "stiffening" effects reduce the geometry-based turn-sharpening tendency of the steering.[19] Finally, at a critical "inversion" speed, there is theoretically no need for steering torque or upper-body lean to hold any turn. (Above the inversion speed, the steering will tend to self-center, thus reversing the ordinary sense of required handlebar torque or torso lean.)

In principle, all conventional bicycles and motorcycles possess a steering-torque inversion speed, and above this speed they will display the mild nonoscillatory instability described above (i.e., they will capsize).[20] For typical ridden bicycles, this speed is in the range of 5–8 m/s. But in actual bicycle riding, torque reversal and instability are not very apparent. However, these statements contrast with results in Zellner and Weir's (1978) paper on motorcycle-handling experiments.

The tendency of an uncontrolled bicycle to capsize at high speed is not a matter for concern to most riders. The instability that results from this tendency is so slight that it takes many seconds to develop, and slight unconscious upper-body motions of the rider probably suffice to compensate for it.

However, at low speeds, violation of the turn-sharpening criterion for stability through poor design causes an uncontrolled bicycle to capsize far more quickly. Since gyroscopic effects are then negligible, the requirement is for maintaining stability in a poorly designed bicycle essentially static, and may be stated in either of two equivalent ways.

- The upright, straight bicycle must be at an absolute maximum of potential energy with regard to any combination of reasonable steering and lean angles.
- A stationary, balanced bicycle, if its handlebars are turned *while balance is maintained*, must lower its center of mass, or equivalently, must generate a steering torque that tends to increase the steering angle.

A bicycle with a vertical steering axis and negative fork offset to produce a trailing front contact does not satisfy these requirements. With the bicycle at rest, steering the front wheel to the left, and tilting the entire bicycle to the right to bring it into balance, *raises* the center of mass.

The turn-sharpening stability criterion can be given as a simple design formula, but only with the help of several symbols (see Papadopoulos 1987). For simplicity, we will take the tack of ignoring the normally small front-assembly mass offset forward of the steering axis, which tends to turn the steering if the stationary bicycle is held in balance with its steering turned. The criterion for automatically straightening up is then:

$$0 < (x_{CM}/y_{CM})(L_{MT}/L_W) < \sin \lambda.$$

That is to say, mechanical trail must be positive, and the rearward tilt of the steering axis must be at least a small positive value depending on the mechanical trail.[21] This last relation can be shown geometrically (figure 8.13):

- draw a line from the rear contact through the system center of mass;
- call the point on this line vertically above the front contact (P); and then
- the steering axis must intersect the ground ahead of the front contact and pass below P and above the front contact. In fact, with normally large bicycle wheels, it is usual for the steering axis to pass below the front-wheel center, which is to say that the front fork has positive offset (L_{FO}).

Who was the genius who thought of tilting a bicycle's steering axis? And was this tilting valued for its stability benefits, or only for something

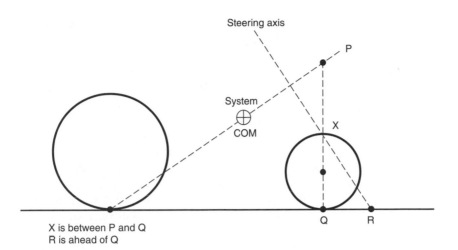

Steering axis

System

COM

P

X

Q

R

X is between P and Q
R is ahead of Q

Figure 8.13
Geometric stability requirement for negative turns. (The requirement restricts the position of the steering axis.)

more mundane like minimizing hand-force steering disturbances during stand-up pedaling[22] or preventing rearward bending damage from striking a pothole (Brandt 2000)? The development of a tilted steering axis is one of the major mysteries of bicycle evolution. John Allen (2001) writes:

In the early days of the safety bicycle, the handlebars were placed close to the cyclist, as had been traditional and necessary with high-wheelers, with their very serious pitchover problem. High-wheelers had little or no forward angling of the front fork: it would not have been practical because it would have prevented the cyclist from standing over the pedals, and would have placed the force vector from pedaling too far from the steering axis, making steering difficult. Bicycle evolution involved innumerable experiments, but the answer is most likely mundane: the fork was angled forward in order to keep the handlebars close to the cyclist, and for the front wheel to clear the feet, in spite of what intuitively would seem to be a stability reduction. This development occurred *before* the discovery [by Major Taylor?] that a greater distance to the handlebars improved both power production and aerodynamics. A longer stem also greatly improves stability when riding with one hand on the handlebar, an important side benefit which would not accrue simply by lengthening the top tube and keeping the fork vertical.

Oscillatory instability
Bicycles can also go unstable in an oscillatory fashion.[23] For example, an uncontrolled conventional bicycle rolling at just below a stable speed will

steer too far and overcorrect a lean, thus weaving back and forth at increasing amplitude. Unfortunately, the mathematical expression defining this instability is too complex to be included here. While it is *not* possible to guarantee that this instability will be prevented by adequate trail and angular (gyroscopic) momentum at higher speeds, these factors usually help and seem to be essential to the stability of conventional designs. However, if we trust the equations, it is also apparently possible to have unconventional bicycles that are intrinsically stable over an immense speed range even though they lack trail and gyroscopic effects altogether.

In the second edition of this book, considerable space was given to Jones's stability parameter u (1970). It is now possible to put Jones's appealingly simple analysis into perspective. He did correctly identify the tendency of a leaned bicycle's front end to rise and fall vertically in proportion to steering and lean angles, and hence the torque tending to turn the steering toward a lean.[24] When this torque is divided by the load borne by the front contact, the result is just L_{MT}, plus a small quantity related to front-assembly weight and its offset forward from the steering axis, which can be ignored if we assume a lightweight front assembly. To determine the parameter u, Jones then nondimensionalized L_{MT}: $u = L_{MT}/2R_W$. It is unfortunate that Jones took this last step, because the wheel radius has little to do with the behavior being investigated. To sum up, Jones's experiments were very revealing, but his premise was faulty. Nonetheless, u[25] can be recognized as a very important stability ranking quantity among bicycles with two wheels of the same size.

Tricycle stability

The stability of some systems similar to bicycles has been studied: a skateboard with rigid rider, a rolling hoop or disk, a towed trailer with a flexible hitch, a shopping cart, and a riderless tricycle. As outlined at the beginning of this chapter, tricycles act very differently from bicycles in numerous ways. In fact, the sudden countersteering unconsciously used by bicyclists to create lean for in-balance cornering poses additional risk to a tricycle of rolling over or collapsing a wheel. However, handling and stability are far easier to analyze for tricycles than for bicycles because with a tricycle, there is no lean angle to worry about, just the steering angle. It is far simpler to derive the single equation of motion. Alternatively, it is possible just to set the lean angle (and its rates of change) to zero in the bicycle-dynamics equation related to steering torque.

The resulting equation for the steering angle of a tricycle is analogous to the classical mass-damper-spring equation. The mass term is always positive. The damper term is simply proportional to velocity. It can be made negative, causing either growing oscillations or divergence, by front-assembly mass behind the steering axis or negative mechanical trail.

Finally, the spring term is roughly comparable to that for a conventional in-balance bicycle. For normal bicycle geometry, the spring term will be negative at low speeds, tending to decenter the steering. For tricycles, however, in contrast to the case for bicycles, this decentering is a cause of *instability* and no advantage. At higher speeds the spring term is generally positive and proportional to the square of velocity.

A tricycle may make an ideal test bed for demonstrating various kinds of instability, since it need not necessarily crash to signify success. The effective mass (inertia), damping, and springiness of a tricycle's steering can all be modified easily by adding simple mechanical hardware. Measuring tricycle stability with precision may allow heretofore unknown properties of the tricycle's front tire to be determined, as discussed below.

With unusual tricycle configurations, it appears theoretically possible to provide stability at all speeds and also to eliminate the steering torque of riding a crowned road, for example, by keeping mechanical trail small and placing the right amount of front-assembly mass behind the steering axis. As shown in figure 8.14, it is possible to determine a "fixed line" for the front assembly of a tricycle at rest, such that points on that line lie neither left nor right when the front wheel is turned. Mass added to the front assembly ahead of the fixed line increases the torque required to travel in a curve. The rear assembly also has a vertical fixed line with the same property.

Shimmy

Shimmy is an unnerving bicycle instability that can sometimes cause an inexperienced rider to crash. When a bicycle undergoes shimmy, the steering oscillates right and left several times per second, with growing amplitude. Similar wheel vibrations are well known in airplane nosewheels (as described briefly by Den Hartog [1985; see his equation 7.39 and problems 213–216] and still an area of active research), shopping-cart casters, and motorcycles (where a violent occurrence of shimmy is termed a "tank slapper": see Society of Automotive Engineers 1978).

Before outlining an explanation for shimmy, it is worth considering what to do if it happens. Shimmy presents a danger when the cyclist panics and attempts actively to return the handlebars to center. The shimmy frequency is so high that the muscular response occurs too late, accelerating the handlebars when they are already well on their way to the other side and increasing the oscillation. (In aircraft, this dangerous phenomenon is known as pilot-induced oscillation.) As long as the problem is not compounded by active intervention, any of several different methods seems to stop the oscillation at once:

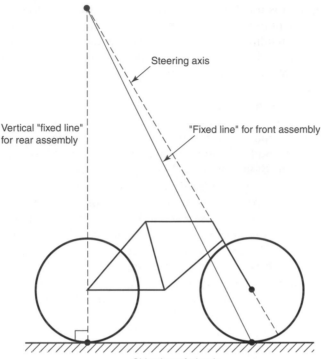

Steering axis

Vertical "fixed line"
for rear assembly

"Fixed line" for front assembly

Side view of tricycle

Figure 8.14
Mass added to the front assembly ahead of the "fixed-line" confers no-hands
recovery from a turn.

- reducing weight on the saddle (by standing slightly) deprives the
vibrating system of a key restraint and adds considerable damping;
- clamping the top tube of the frame between the rider's knees tremendously alters the vibrating mass and also adds damping; and
- lightly using the hands in a passive "resisting" or "motion-reducing"
mode also increases damping.

Shimmy theory

Very limited experience with a given bicycle and body position suggests
that no-hands shimmy appears at a critical speed and grows to a final
steady amplitude at any higher speed, with greater amplitudes for greater
speeds. The frequency of shimmy is relatively unaffected by speed.

Speculative causes of bicycle shimmy include an untrue wheel, loose
bearings, or gyroscopic effects. Although these factors may sometimes

come into play, they are clearly not essential to the phenomenon. A key bit of evidence for this conclusion is the constancy of shimmy frequency at different bicycle speeds. In contrast to this constancy, an untrue wheel will create a steering disturbance once each wheel revolution, and the gyroscopic precession of a rotating wheel involves wobbling twice per revolution, as can easily be verified with an unattached spinning wheel.

Shimmy is evidently a *self-excited oscillation*: there is no alternating force turning the handlebars back and forth except that generated by their motion. In instances of shimmy, the equation of motion shows a negative number for vibration damping, causing vibrations to grow rather than decay. The vibration energy is provided by the moving bicycle.

It is not our intention to present a detailed dynamic analysis of shimmy, and indeed many aspects of the phenomenon remain in question. But it is both appropriate and feasible to present a simple system with bicycle-like features and explain how shimmy arises in the case of that system.

A castered wheel (like the front wheel of a bicycle), or more generally, any trailed rolling system, such as a trailer, is capable of surprising energy interchanges with the unit that is towing it. In a situation in which its tongue (or pivot axis) oscillates back and forth laterally, details of the distribution of its mass affect the sideforce it imposes on the ground. Because of the angular deviation of the trailer from the straight-line path of the towing vehicle, the wheel's sideforce on the ground has a fore-aft component that will either propel the towing vehicle (as does the tail of a fish) or retard it. Imagine a person sitting at the back of a pickup truck that drives along a straight road, and imagine that the person is holding a trailer hitch and swinging it side to side to make the trailer follow a sinusoidal path. If the result is to propel the truck forward (i.e., do work on the truck), then the side-to-side motion will require effort (power) from the person moving it. On the other hand, if the result is to retard the truck, power from the truck's engine will flow into the person's hand, and the hitch will try to "run away" to either side.

A very simple distribution of mass that is easily seen to pump energy into the vehicle hitch is simply a large polar moment of inertia centered at the axle. If the system is traveling forward fast enough that the hitch length is far smaller than a wavelength of the oscillation on the road surface, then the angle of the trailer towbar is essentially aligned with the path of the wheels. It can be seen that the rotation rate of the trailer is maximum rightward at the far left crest of the oscillation and maximum leftward at the far right crest. Maximum angular acceleration occurs as the trailer crosses the center line of the oscillation. What is important is that the force required to cause angular acceleration is *opposite* to the hitch velocity: as the hitch moves from left to right, the force of the hand holding the hitch is directed to the left. In other words, the trailer pumps power into the arm of the person holding it, trying to increase the speed of its lateral motion in both directions, right and left. Given that a towed wheel (or trailer) with appropriate inertial properties is capable of pumping energy into lateral oscillation, then such oscillations are to be expected. When energy-absorbing devices (dampers) are present, it is to be expected that a higher speed must be attained before the power delivered at the hitch can overcome the damping tendencies. A simple system analogous in several ways to a bicycle's front end viewed from the side and from above is shown in figure 8.15. The system has the following elements:

- a wheel or contact point towed a trailing distance (L_{MT}) behind a hitch point (analogous to the front-wheel contact point's being towed behind a "hitch point" low on the steering axis);
- a significant polar moment of inertia (I_{zz}) of the towed wheel: a conventional bicycle wheel has much of its mass quite far away in comparison to the trail;
- the mass (m) of the bicycle head-tube area, whose lateral inertia force is transmitted simultaneously to the rider and the steering axis (or hitch) (i.e., a series loading); a similar situation arises when a mass is supported at the midpoint of a rigid horizontal bar whose ends are supported with springs. The rider's mass, much greater than that of the head-tube area, is assumed not to move laterally during the vibration. If the head-tube area mass is moved laterally while the wheel is not permitted to steer, it will experience a composite stiffness (k), derived mainly from the torsional flexibility of the frame, including forks; and
- a damper c in series with k to represent the energy-absorbing connection between the frame and rider (i.e., slip at the saddle). A firm connection is modeled by a large value of c. Somewhat counterintuitively, a firm connection corresponds to very little energy dissipation.

In this basic system, shimmy is predicted when velocity exceeds

$$V = \frac{kL_{MT}}{c}\left(1 + \frac{I_{zz}}{mL_{MT}^2}\right),$$

or approximately

$$\frac{kI_{zz}}{mcL_{MT}}.$$

This relationship implies that it is important to keep stiffness high, damping constant low (paradoxically, increasing energy absorption), head-

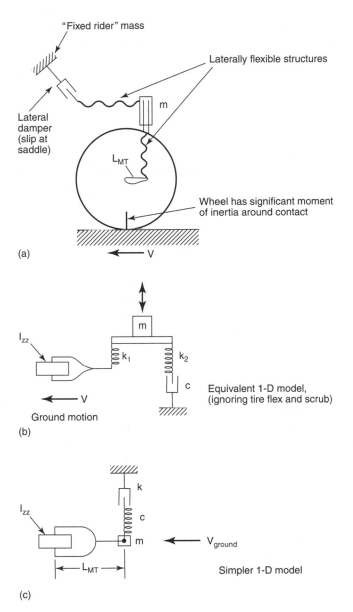

(a)

(b)

(c)

Figure 8.15
Simple shimmy model (trailer oscillates because of motion of ground). (a) Side
view. (b) Top view. (c) Top view of simpler model.

area mass low, and mechanical trail either much less, or much greater, than $\sqrt{I_{zz}/m}$. The great surprise in the relationship is the apparent value of increasing I_{zz}, which presumably helps by reducing the amplitude of steering excursion. The frequency of onset (in radians/s) is $\sqrt{(k/m)[1 + (mL_{MT}^2/I_{zz})]}$, identical to the oscillation frequency of the system at rest, as long as the energy dissipation is not too great (i.e., as long as c is large—see below).

This model fails to represent some potentially very important shimmy physics; for example, it certainly does not include tire sideslip, nor such known problems as flexibly mounted rear luggage. But it agrees with the following qualitative observations by Papadopoulos on his personal bicycle.

- If the bicycle saddle is restrained by pressing it against a door jamb, the head tube has a clearly defined lateral resonance at several cycles per second.
- Shimmy while riding occurs at essentially this same frequency.
- Shimmy can be sustained at widely different speeds depending on the firmness of the rider's connection to the saddle (denoted by c above). By reducing saddle pressure, shimmy onset speed can be raised above 17 m/s. Conversely, by increasing pressure and lateral firmness (by exerting upward pedal forces and contracting the muscles of the buttocks) it was possible to sustain shimmy at 5.5 m/s.

In no-hands shimmy experiments, slightly exceeding the speed at which the onset of shimmy occurred brought about sustained oscillations at a medium amplitude. Higher speeds clearly increased the steady amplitude, but not dramatically. This is suggestive of the "pitchfork bifurcation" of nonlinear vibration theory.

Unfortunately, it is not possible to use this model to predict the speed at which the onset of shimmy will occur, because c is an unknown quantity. The best that can be done is use the onset observations given above to determine the range of c. Taking m as 1 kg and k as 1,000 N/m to give a reasonable static frequency, guessing L_{MT} was 40 mm for the bicycle ridden, and taking I_{zz} for the wheel as half of mass times radius squared, or $(1 \text{ kg})(.35 \text{ m})^2/2$, suggests c values from 260 kg/s (sitting firmly) to 70 kg/s (sitting very relaxedly).

To summarize, bicycle shimmy vibrations apparently depend on the combination of elastic flexibility with inadequate damping. With foreknowledge, the rider can generally learn to provide adequate damping to prevent shimmy or arrest it when it occurs.

Lateral properties of tires

The key to an airplane's behavior is its wings. They must be shaped and sized to provide adequate sideforce (i.e., lift), with minimum resistive drag and minimum redirection of the flow field (induced drag). The aerodynamic forces and moments on a wing shape or airfoil are carefully studied as it is held at various angles in a wind tunnel, and the resulting data are heavily relied upon in airplane design to determine the weight that can be lifted and to achieve stability and maneuverability.

In the mid-twentieth century, it was recognized that the airfoil perspective also applies to the lateral-force behavior of tires. The analog of a wind tunnel in the case of tires is a moving belt. Effort was devoted to predicting and measuring tires' so-called cornering properties. But the amount of research on tires has been far less than that on airfoils, which was substantially underwritten by national defense establishments. For a sound and approachable account of car-tire properties and the kind of testing used to measure them, see Milliken and Milliken 1995. Another excellent reference relevant primarily to automobile tires is Clark 1981, particularly the chapter by Pacejka.

Tires are generally considered to be somewhat flexible vertically for obstruction swallowing, but rigid otherwise. For many purposes this approximation is good enough. But in actuality, the possibility of lateral flex of a tire means that when tires are supporting a side load, they do not travel *exactly* in the direction they are pointed. Also, the finite length of tires' contact with the surface on which they are traveling gives rise to unexpected moments: of greatest interest are those that tend to reorient the steering, namely, a rearward shift of any sideforce (pneumatic trail) and a resistance to twisting motions (scrub torque). In some circumstances, these phenomena can significantly affect vehicle handling and even add to the drag acting on the vehicle.

Pneumatic tires differ from rigid disks in that the primary load creates a finite-length contact patch, which obviously resists yawing (i.e., twisting around a vertical axis). But other unexpected behavior arises when a tire is rolling forward simultaneously with a slight amount of yawing, crabbing, or leaned motion. An element of tread that is laid down at the front edge of the contact patch experiences essentially no lateral force or torque from the road, but as the tire's rim moves forward, this element, which is effectively locked to the ground, ends up laterally offset and possibly twisted relative to the rim before being picked up again. Therefore there is a buildup of sideforce or twisting moment toward the rear of the contact patch or both.

The lateral properties of tires have largely been ignored in discussions of bicycle handling. The main research to include them is the CALSPAN report by Roland (1973). For those attempting to dig into the tire literature, it is important to realize that bicycle tires are fairly similar in construction to motorcycle and aircraft tires, which of are of bias-ply construction and quite different from modern car tires, which typically have radial cords and a circumferential belt. Allen writes:

A major advantage of radial plies is that they eliminate the "Chinese finger puzzle" effect or "squirm" which occurs with bias plies, so that the tread tends to narrow under increased load, resulting in faster wear and greater rolling resistance. National/Panasonic of Japan introduced radial-cord bicycle tires at some time in the 1980s, but some users stated that they had a strange "feel"— probably due to their larger amount of sideslip—and were withdrawn within a couple of years. One compromise, once common for motor vehicles, would be a bias-ply tire with a circumferential belt.

Sideslip

To understand so-called sideslip, imagine a wheel being rolled forward, while at the same time being forced to crab rightward 10 mm for every meter it moved forward (1 percent rightward lateral drift). The contact patch in such a case might be 150 mm long. A little bit of tread that is laid down "on center" at the leading edge of the contact patch will, by the time it comes to the rear of the contact patch, be 1.5 mm "off center." The tire is thus increasingly deformed to the left from front to back of the contact patch. As each bit of tire is picked back up off the ground, it slides back to center, in some cases with an audible squeaking. The result is a net side-force from the tire that opposes the wheel's lateral drift (figure 8.16). In the case of a leaning bicycle, sideforce refers to a force component perpendicular to the plane of the wheel, which is not the same as a force component parallel to the pavement on which the bicycle is traveling.

The ratio of the sideforce to the drift slope (or angle in radians) of an upright wheel is called "cornering stiffness" (K_{CS}). Cornering stiffness is not much affected by ordinary variations in the coefficient of friction, since most of the contact patch is not actually slipping. Rather, the primary factor is the lateral stiffness of the tire cross-section, defined largely by pressure and somewhat by whether the wheel's rim is narrow or wide; a secondary factor is the contact-patch length, determined by wheel radius, vertical load, and inflation pressure.

Since the lateral-force intensity builds up toward the rear of the contact patch, the net force should be considered to "act" at a point somewhat behind the lowest point of the wheel. Theoretically, this point is about one-sixth the contact-patch length behind the midpoint of the contact patch. That distance is known as "pneumatic trail," since it acts just like trail (or caster) in tending to align a steerable wheel with the direction of

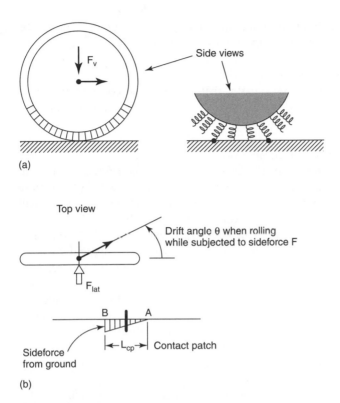

Figure 8.16
Tire cornering-stiffness model: (a) Approximate pneumatic tire as a sequence of radially and laterally springy "fingers" or "spokes" (brush model). (b) A finger touches down at *A*, and because of the drift angle, builds up sideforce until picked up at *B*. Sum of spring forces equals *F*. (c) Top view of simpler model.

travel. For example, when subjected to a sideways push *while rolling forward*, a wheel with a vertical steering axis and no fork offset (and so no trail) will still experience a steering moment tending to reorient the wheel toward the direction in which it is drifting. Any contact-force component perpendicular to the wheel plane is shifted rearward in this way.

When does tire sideforce occur in normal riding? The answer to this question depends partly on a given rider's maneuvering style. In a steady turn a bicycle's wheels virtually line up with the balance plane, so whereas *horizontal road forces* are large, *wheel sideforces* can be small. In ordinary riding, the main origin of wheel sideforces is sudden maneuvers in which the rider's body inertia resists rolling or yawing.[26] Therefore, a point-mass rider generates virtually no wheel sideforces. Skilled body English (i.e., allowing the frame to tilt without enforcing an equal amount of body lean) can be used to reduce these maneuvering sideforces significantly.[27]

Examples of flexible-tire behavior
Tire lateral flexibility has several surprising theoretical consequences that lead to sideforces or yawing moments in response to drift angles, lean angles, or yawing rates.

- A nonsteered cart or car on a side slope has a sideforce acting at each wheel and will drift down the slope. It must be pointed slightly upslope to follow a horizontal line as it drifts. However, if its rear wheels drift more than the front ones when it is traveling straight across a side slope, it will increasingly aim upslope and will eventually travel uphill.
- Whenever a wheel experiences sideslip, extra propulsive power is required. Given a forward velocity V and sideforce F, the dissipated power is VF^2/K_{CS}. This formula is most applicable to vertical-wheel tricycles.
- A bicycle ridden vertically on a side slope experiences no sideforce on the tires. Therefore its sideways drift should be minimal. The relevant measurable property is "camber thrust," the force developed by a leaned wheel carrying a vertical load and rolled while preventing sideways motion. In motorcycle tires, camber thrust has been found (Sakai, Kanaya, and Iijima 1979) to be a little higher than the value theoretically needed to support an in-plane force. The only bicycle measurements (by Roland [1973]) gave a much lower value but are considered to be less reliable (Milliken 2000).
- A bicycle in a steady turn approximates the previous case, because the contact force is nearly in the plane of the wheel. Therefore its lateral drift should be minimal, an important consideration, because lateral drift absorbs power in the amount of drift velocity times lateral force.
- A bicycle ridden straight forward with the rider leaning out of the frame plane engenders purely vertical forces on the ground, which can be divided into components parallel and perpendicular to the wheel plane. The bicycle should creep in the direction of the side toward which the wheels lean. Riding with such a tilt should cause an increase in rolling drag and also a wheel scrub torque tending to steer toward the side of leaning.
- A cart with its wheels on either side tilted inward at their tops (like the letter A) acts differently. The wheels can't creep together, so they build up sideforce (camber thrust) until the ground force is roughly in the plane of the wheels. In a sense they are trying to move together and "squeeze" the cart. They should be toed out for minimum rolling resistance, an important finding as it applies to racing wheelchairs.

• A vertical wheel traveling slowly in a counterclockwise turn twists its rim counterclockwise above the contact patch, that is, scrubbing occurs in the contact patch, leading to a clockwise torque (scrub torque) tending to straighten the path. (It is probably hard to demonstrate this phenomenon on a state-of-the-art metal-belt tire tester, since the tires would run off the edge of the belt. Possibly some kind of tricycle test bed could be used.) On the other hand, a wheel traveling forward in a straight line but leaned left-ward at 45° (from a rear view) appears to be scrubbing the ground in a clockwise sense when viewed from above, leading to a counterclockwise torque tending to turn the steering to the left. Only when the wheel forms the base of an imaginary cone, lying on the ground with its vertex at the center of the turn, is there no scrub torque.[28]

• In general, drag resulting from bicycle-wheel alignment should not become an issue, unless the wheels are substantially canted, leading to scrubbing. But for a tricycle, even vertical wheels can fight each other, especially when the tricycle's frame is deformed by load. Confirming proper alignment of tricycle wheels is difficult. Tricycles cannot benefit from camber thrust when cornering and therefore should show considerable energy loss in hard turns, and even some on cambered roads.

Feel and control: the human-factors domain

Some bicycle characteristics (for example, strength or mechanical efficiency) can be defined by numbers and have relatively well-defined consequences for performance or capability. Others, however, are evaluated primarily by "feel." Foremost among these are aspects of comfort and handling. Coming to grips with such subjective matters requires an especially critical mode of thinking. Among the manifold difficulties in evaluating subjective factors relating to bicycle characteristics are these.

• Human observers are notoriously suggestible. When told that a given bicycle is special for some reason (carbon forks, selected by a world champion, designed for hard cornering), they easily convince themselves that it is.

• Each observer is different in body shape, skills, and expectations. In addition, humans adapt over time and become inured to differences that once were obvious. Soreness and fatigue from a hard ride may also color our judgment about a particular bicycle.

• Most riders have little opportunity to learn exactly what others mean by any particular term used for descriptive purposes.

• Issues similar to those involved in the subjective aspects of bicycling also arise in other fields, such as automobile and aircraft handling or sensory evaluation of foods and audio systems. Some useful techniques that have been developed to minimize bias include the following.

•   Blind testing (the identity of the test specimen is hidden from the subject) or double-blind testing (the test specimen identity is also hidden from the person administering the test).

•   Tester-consistency evaluation (presenting the same specimen, disguised in a different way, to the same or to a different tester).

•   Sensory threshold evaluation (measuring the lowest level at which a factor can be detected).

•   Tester training with standard machines (to develop a consistent vocabulary and rating scale).

•   In Papadopoulos's limited experience of blind testing (varied tire pressure $\pm$ 1 bar, varied bicycle mass $\pm$ 2 kg) or double-blind testing (varied frame stiffness and material) of bicycle characteristics, riders could not demonstrate anywhere near the powers of discrimination among alternatives that they claim to possess. We speculate that many "performance" sensations are imagined; demonstrating otherwise is a promising field for determining what really matters.

Beyond the question of rider feelings and perceptions, considerable work in human factors (Sheridan and Ferrell 1974) teaches us that the human control system has properties and limitations just like those of automatic controllers, and that outside a certain "envelope" of vehicle behavior, stable and accurate control will be compromised, or at least the controller will experience a sense of increased difficulty. A little of this thinking has been applied to bicycles (Roland 1973) and motorcycles (Weir and Zellner 1978). But as far as we know, this early work was never followed up.

In contrast to drivers of cars and airplane pilots, riders of bicycles can use other than manual inputs (body English) to affect the vehicle's steering. In addition, just as for other vehicles, the bicycle responds to these inputs not only in its motion, but also in the force feedback through the control.

A bicycle's response to control actions depends dramatically on its speed. The centrifugal force that can be generated by a given steering angle, and hence the required angle of lean, are proportional to the square of the speed. But we have little or no conscious awareness of these dramatic changes. Adapting our reactions to a changing system is relatively automatic and unconscious.

Our speculation is that we do not normally control balance by controlling steering angles. As standard bicycles have evolved to recover from most disturbances automatically, all we must do in response to sudden disturbances is to allow the bicycle's steering to move however it needs to.

We suspect that preventing road forces from affecting the steering (for example, by adding friction to the steering bearings, adding too much front-assembly inertia, or using power steering to make the fork turn the same as some dummy handlebars) would make a bicycle far harder to ride.

Personal closing note by Dave Wilson

My appreciation for this chapter by Jim Papadopoulos comes from several factors. One is that the chapter on the topic that I wrote for the second edition of this book was the least satisfactory in the book. I had no confidence that I could do better for this edition. I have been editing Human Power since 1984 and have had several contributions on steering and stability. I soon found, through sending the drafts out to experts for review, that there seemed to be no agreement among experts on the topic. Jim states his beliefs clearly and his agreements and disagreements with other authors on specified points. I found his approach refreshingly different from theirs in many respects.

I want to add, however, that I have used two personal experiences as tests on various approaches to steering and stability. One concerned a short-wheelbase recumbent tandem that I bought as a wedding present for my spouse, Ellen, and me. We gave up riding it because it had an unnerving habit of suddenly swerving, particularly when we were riding slowly up a hill, without any obvious input. I blamed Ellen for swaying, but then had a call from John Allen, who had found exactly the same phenomenon when testing a similar configuration of tandem with a very experienced "stoker." No theory has given us any guidance as to why this bike behaved so strangely, nor as to what we should have changed to cure the problem.

The second experience was with my Avatar 2000 long-wheelbase re-cumbent single, which after fifteen years or so of heavy use developed an occasional devastating shimmy. I renewed or tightened all the bearings I could, made an inertial damper clamped near the head tube, and changed the tires and the tire pressures, and this often seemed to cure the problem, only to have it reappear later. Then I found that the shimmy occurred only when I removed the normally ever-present luggage bag on the rest behind my seat. This always contained my tools and heavy lock, spare tires and tubes, and rain gear and usually had a heavy briefcase full of books, etc. But shimmy would be completely eliminated even without the briefcase, so long as the bag was present. This finding still defies explanation for me. The amplitude of the vibration behind the seat was very small, and the mass of the bag was far less than that of me and of the bike, yet somehow the damping or changed the tuning of the natural frequency of the whole machine was brought into a critical range by the removal of the bag.

These are just warnings that confirm Jim Papadopoulos's views: that the theory of bicycle handling and of the dynamic behavior is still a long way from explaining all the attendant phenomena.

Notes

1.   This sensation is deceptive, of course. It is easy to crash a fast-moving bicycle, but we obey an unconscious compulsion not to do so.

2.   An extreme example of such an adjustment is found in a sociable two-person bicycle with one rider heavier than the other. Rather than the heavy side's sinking down, balance is attained by raising that side comically higher!

3.   Of course, matters are different if the body is *not* rigid: for example, a tightwire artist with a balance pole to provide reaction torque.

4.   In bicycling, this relation breaks down when y_{CM} is small enough to approach the radius of gyration of the rider.

5.   Actually, a bicycle nominally at rest can also be balanced in what is called a "track stand," by setting the front wheel at an angle, and rolling the bicycle forward/backward to achieve some lateral motion at the front contact. It is best to put extra body weight over the front wheel. On a bicycle with a free-wheel, a track stand is possible only when the bicycle is facing uphill. In principle, balance with fixed contact positions should also be possible purely by steering the handlebars left and right, although we have not seen it done. But this approach to balancing would not satisfy the definition of a rigid body.

6.   The bicycle must be well aligned in order to travel straight, and the steering bearings must turn freely enough not to impede the small automatic steering corrections.

7.   The turning radius of the front wheel is slightly greater than that of the rear (r) by the amount $L_w^2/2r$.

8.   This is a cavalier treatment of the actual equations of motion, one that essentially ignores the mass of the steering assembly.

9.   Rear steering would make the rear contact the more controllable one, but a rear-steering bicycle is nearly impossible to ride fast. A major reason for this is that although *leftward* acceleration of the rear can be achieved rapidly by steering the rear wheel to the left, the resulting steady turn is *rightward*, that is, with the opposite sense of acceleration. (In the technical discipline of control theory, this is a property that describes a "non-minimum-phase" system.)

10.   These concepts are embodied in ice cycles, skibobs, and hydrofoil boats such as Flying Fish.

11.   As shown in figure 8.3, the front assembly of a bicycle is the front wheel, the fork, the handlebars and any front basket or luggage. The rear assembly is the frame, the rider, and the rear wheel. The two assemblies join at the steering axis.

12.   Mechanical trail can be calculated from other geometrical quantities, using $L_{MT} + L_{FO} = R_W \sin \lambda$. Unfortunately, both L_{FO} and λ can be hard to measure on an assembled bicycle, especially if the bicycle's forks have been bent in a crash.

13.   In principle, whenever a sideforce at the front contact tends to turn the steering, the mechanical-trail lever arm used to calculate steering torque should be augmented by *pneumatic trail* (L_{PT}) (approximately one-sixth of the contact-patch length, as described in "Lateral properties of tires") times $\cos \lambda$.

14.   To be precise, the proper test is to turn the handlebars *while maintaining balance, that is, while keeping the center of mass over the line between contact points.* (And at higher speeds, "over" should be taken in the sense of "maintaining balance.")

15.   This scrub torque arises as a result of turning, because an element of the tire at the front of the contact patch is moving inward relative to its path of travel, and an element at the rear of the patch is moving outward.

16.   John Allen has pointed out that while straight-line steering torque can be eliminated using various methods, it is also desirable to align a bicycle's front assembly so that the rearward and upward (weight transfer) force of front braking has no effect on steering torque. The net force is along a line from the front contact to a point at center-of-mass height over the rear contact; extended forward, that line must intersect the steering axis. When all is said and done, it's probably simplest to make every component of the bicycle symmetric.

17.   An uncontrolled bicycle that is not perfectly symmetrical will generally travel in a curve rather than a straight line (for example, turning and leaning leftward). This tendency can be rectified in any of a number of ways as described above.

18.   Interestingly, conservation of energy predicts that a perturbed bicycle will slightly alter its speed in reverting to steady upright rolling. The energy of the perturbation is transferred into forward motion. A corollary of this is that "pumping" the handlebars from side to side should actually propel the bicycle—and indeed, with hard tires (to minimize sideslip) and adequate trail, substantial handlebar excursions at 1.5 Hz were found to propel a bicycle easily at 3 m/s. Conservation of energy can be simulated properly only with nonlinear equations.

19.   Because a bicycle's steering axis is not vertical, some of the "tipping" torque needed to overcome the front wheel's gyroscopic resistance to continually changing its heading must be supplied through the handlebars. At low speeds, the required steering torque is minuscule, but when, the bicycle is traveling faster, it significantly reduces the geometry-based tendency of the steering to sharpen the turn.

20.   In practice, nonideal tire properties could modify this conclusion. For testing purposes, ideal tire behavior is approached at low loads and high pressure (i.e., short contact patches).

21.   Interestingly, this turn-sharpening criterion would also be satisfied if the inequality signs are all reversed (negative tilt of steer axis and negative trail). But some other stability criterion would in that case very likely be violated.

22.   When the steering axis aims approximately at the rider's shoulders, push and pull forces exert no steering torque.

23.   By this we are not referring to *shimmy*, a high-frequency steering oscillation related to frame and wheel flexibility (discussed below).

24.   Actually a ridden bicycle is never "held" at a given lean angle, and furthermore the manner of holding changes the response when assessing stability. Jones's analysis actually reveals the steering torque arising when one rides in a straight line with one's torso leaning to one side of the frame.

25.   Unfortunately, the formula given for u in the second edition of this book was incorrect.

26.   Measuring wheel sideforce is not simple, but there is a crude way to detect it: if one front-brake block is set close to the front rim (which should be patterned for best results), then even modest sideforces can result in a rubbing sound. The sound from a sudden countersteering to the right is similar to that from leaning the frame to the left while moving straight forward.

27.   It is generally a good idea to limit wheel sideforces, since bicycle wheels are laterally weak. Sideforces are potentially capable of hastening spoke fatigue and even causing wheel collapse.

28.   Scrubbing issues are significant for bearings as well as tires. They are best understood by viewing rotation as an angular velocity vector directed along the spin axis and determining whether this vector has a component perpendicular to the contact patch. A leaned wheel traveling straight forward has its vector along the wheel axle, which is not parallel to the ground. A leaned wheel traveling in a circle has two parts to its angular velocity: rotation about the axle and rotation about the center of the turn. If the rates of these two rotations are properly proportioned (that is, when the axle points at the center point of the turn), the net spin vector is horizontal, and there is no scrubbing.

References

Allen, John. (2001). Personal communication, June 6.

Brandt, Jobst. (2000). Communication to the Bicycle Science e-mail list moderated by Sheldon Brown.

Carvallo, Emmanuel. (1900, 1901). "Theorie du mouvement du monocycle." *Journal de L'Ecole Polytechnique*, Series 2 (Part 1, Volume 5, *Cerceau et Monocycle*; and Part 2, Volume 6, *Theorie de la Bicyclette*).

Clark, S. K., ed. (1981). *Mechanics of Pneumatic Tires*. U.S. Department of Transportation HS 805 952, Washington, D.C.

Cossalter, V., A. Doria, and R. Lot. (1999). "Steady turning of two-wheeled vehicles." *Vehicle System Dynamics* 31:157–181.

Den Hartog, Jacob P. (1985). *Mechanical Vibrations*. New York: Dover Publications. Original work published 1934.

Doehring, E. (1955). "Stability of single-track vehicles." *Forschung Ing.-Wes* 21, no. 2:50–62. English translation by J. Lots available from CALSPAN Library, Buffalo, N.Y.

Forester, John. (1989). "Report on the stability of the Dahon bicycle." Available online at ⟨www.johnforester.com/Articles/BicycleEng/dahon.htm⟩.

Forester, John. (1993). *Effective Cycling*. Cambridge: MIT Press.

Franke, G., W. Suhr, and F. Riess. (1990). "An advanced model of bicycle dynamics." *European Journal of Physics* 11:116–121.

Fuchs, Andreas. (1998). "Trim of aerodynamically faired single track vehicles in crosswinds." In *Proceedings, Third European Seminar on Velomobiles, August 5, Roskilde Denmark*.

Hand, Richard Scott. (1988). "Comparisons and stability analysis of linearized equations of motion for a basic bicycle model." M.S. thesis, Cornell University; portions available online at Ruina Web site, ⟨http://www.tam.cornell.edu/~ruina/hplab/bicycles.html⟩.

Harter, Jim. (1984). *Transportation: A Pictorial Archive from Nineteenth-Century Sources. 525 Copyright-Free Illustrations Selected by Jim Harter*. New York: Dover Publications.

Jones, D. E. H. (1970). "The stability of the bicycle." *Physics Today* (April):34–40.

Kane, T. R. (1977). "Steady turning of single track vehicles." Paper no. 770057, Society of Automotive Engineers, Warrendale, Penn.

Klein, F., and A. Sommerfeld. (1910). *Ueber die Theorie des Kreisels, Vol. 3 Technical Applications*. Leipzig: Teubner.

Man, G. K., and T. R. Kane. (1979). "Steady turning of two-wheeled vehicles." Paper no. 790187 in special publication SP-443, Society of Automotive Engineers, Warrendale, Penn.

Military Standard—Flying Qualities of Piloted Vehicles (MIL-STD-1797). (1990). NASA Scientific and Technical Information Program, Washington, D.C.

Milliken, Douglas L. (1989). "Stability or control?" *Human Power* 7, no. 3 (Spring):9, 14.

Milliken, Douglas L. (2000). Personal communication.

Milliken, William F., and Douglas L. Milliken. (1995). *Race Car Vehicle Dynamics*. Warrendale, Penn.: Society of Automotive Engineers.

Neimark, J. I., and N. A. Fufaev. (1972). *Dynamics of Nonholonomic Systems*. Vol. 33 in American Mathematical Society Translations of Mathematical Monographs, 330–374. Original work published 1967.

Olsen, John, and Jim Papadopoulos. (1988). "Bicycle dynamics—The meaning behind the math." *Bike Tech* (December).

Papadopoulos, Jim. (1987). "Bicycle steering dynamics and self-stability: A summary report of work in progress." Cornell Bicycle Project report, December 15; available online at Ruina Web site, ⟨http://www.tam.cornell.edu/~ruina/hplab/bicycles.html⟩.

Roland, R. D. (1973). "Computer simulation of bicycle dynamics." *ASME Conference on Mechanics and Sports*, 35–83. New York: American Society of Mechanical Engineers.

Ruina, Andy. (1987). Personal communication.

Sakai, H., O. Kanaya, and H. Iijima. (1979). "Effect of main factors on dynamic properties of motorcycle tires." Paper no. 790259, Society of Automotive Engineers, Warrendale, Penn.

Sharp, Archibald. (1896). *Bicycles and Tricycles*. London: Longmans, Green; reprint, Cambridge: MIT Press, 1977.

Sharp, R. S. (1971). "The stability and control of motorcycles." *Journal of Mechanical Engineering Science* 13, no. 5 (August):316–329.

Sheridan, T. B., and W. R. Ferrell. (1974). *Man-Machine Systems: Information, Control, and Decision Models of Human Performance*. Cambridge: MIT Press.

Society of Automotive Engineers, International Automotive Engineering Congress and Exposition. (1978). "Motorcycle dynamics and rider control." Paper no. SP-428, Society of Automotive Engineers, Warrendale, Penn.

Society of Automotive Engineers, International Automotive Engineering Congress and Exposition. (1979). "Dynamics of wheeled recreational vehicles." Paper no. SP-443, Society of Automotive Engineers, Warrendale, Penn.

Weir, D. H. (1972). "Motorcycle handling dynamics and rider control and the effect of design configuration on response and performance." Ph.D. diss., Department of Engineering, University of California at Los Angeles.

Weir, D. H., and J. W. Zellner. (1978). "Lateral-directional motorcycle dynamics and rider control." Paper no. 780304, Society of Automotive Engineers, Warrendale, Penn.

Weir, D. H., and J. W. Zellner. (1979). "Experimental investigation of the transient behavior of motorcycles." Paper no. 790266, Society of Automotive Engineers, Warrendale, Penn.

Whipple, F. J. W. (1899). "The stability of the motion of a bicycle." *Quarterly Journal of Pure and Applied Mathematics* 30:312–348.

Zellner, J. W., and D. H. Weir. (1978). "Development of handling test procedures for motorcycles." Paper no. 780313, Society of Automotive Engineers, Warrendale, Penn.

9 Mechanics and mechanisms: power transmission

Introduction

A transmission is the connection between a vehicle's power source and the driving wheel(s). Its purpose is to transmit power with as little loss as possible, and (in the case of bicycles) to transmit it in a way that enables the rider's limbs to move in as near optimum a manner as possible. A bicycle's transmission therefore encompasses two functions: to transmit power from the rider's feet or possibly her hands (or both), and to do so in such a way that at one favored speed, at least, and perhaps over a range of speeds, the rider either is developing maximum possible power or is producing a lesser amount of power in maximum comfort.[1] In this chapter we review the principles of the standard form and of alternative means of power transmission in bicycles, examine some examples, and discuss some possible future developments in the area of bicycle power transmission.

One starting point for this examination is our knowledge of the generation of human power, which is limited to the circular or linear foot and hand motions used in existing bicycles, rowing sculls, and ergometers. With the exception of the speed variations given by elliptical chainwheels, the foot velocity in rotary pedaling is a constant proportion of the wheel velocity. Therefore, although we may have hunches that there are other foot, hand, or body motions (or combinations of these) that could enable humans to produce higher levels of maximum power (higher than the upper curves of figures 2.4 and 2.10), or equal levels of power at greater comfort, our scientific knowledge confines us to rotary or linear motions as inputs to power transmissions. For this reason we shall be discussing, principally, rotary pedals and cranks and linear sliders.

To start with, then, we shall limit ourselves to discussing transmissions connecting rotary pedal motions to rotary wheel motions, typified by the familiar pedals and cranks. Let us first make a brief review of the historical development of transmissions to indicate how advances resulted from perceived needs.

Transmission history

Power has been transmitted in machinery driven by water, wind, and animal (including human) power since very early times. Human-powered vehicles that preceded the bicycle were mostly various types of pushed or pulled carts or wheelbarrows. McGurn (1987, 13, figure 4) shows an old print of Stephan Farffler, apparently legless, on a hand-cranked three-wheeled chair that he had made to get himself to church near Nuremburg in the 1680s. He also shows (13, figure 5) how some important (self-important?) people circa 1760 had themselves transported in developments of horse-drawn traps that were driven by a servant operating foot cranks on the rear axle. (Was this the first application of foot power to cranks? The treadle grindstone must have preceded it.) This book, however, is about bicycles. Despite claims made by advocates of earlier inventors, claims that historical scholarship has shown to be false, the first bicycle, the Draisienne, was developed in Germany by 1817.

The first bicycle "transmission" was linear: to ride a Draisienne (figure 1.6), one pushed one's foot backward on the ground to propel the vehicle forward. The motion was similar to walking and running. However, in walking, the legs act as spokes of partial wheels, with the body rolling over the feet, being given both support and forward motion. The essence of von Drais' machine was that the legs were relieved of the need to provide support for the body's weight and could just give thrust. Some downward push was still necessary to provide enough friction and possibly to maintain balance.

The next two developments in bicycling transmissions were true transmissions that were approximately linear. Lewis Gompertz in 1821 added a sector-gear hand drive to the front wheel of a Draisienne (figure 1.7) (Ritchie 1975). This hand drive was, no doubt, meant to supplement the foot thrusts, as he provided no footrests. The relatively small amount of power deliverable by the arms, coupled with the need to steer, the evident weight of the vehicle, the solid-rimmed wheels, and the poor road surfaces, must have doomed this design to failure. We have found no reports of its use.

Kirkpatrick Macmillan's velocipede, thought by many to have been developed between 1839 and 1842, also used an approximately linear (actually arcuate) drive, with the feet pushing forward on swinging levers (figure 1.8). This was the first true bicycle transmission, and it enabled the rider to travel long distances with the feet off the ground. Although Macmillan made the rear (driving) wheel of his velocipede larger than the front, it was only about a meter in diameter, and the feet had to move back and forth with each rotation, giving a low gear. However, this arrangement probably suited the road conditions of the day. If one stopped with the cranks aligned with the pull rods (at "dead center"), the machine would have to be moved a little by pushing on the ground before the pedals would provide torque. No thread of development followed from Macmillan's pioneering efforts.

Pierre Michaux or Pierre Lallement, one of whom was the first successful developer of rotary crank drive for bicycles, attached the cranks directly to the front wheel (figure 1.10). This was a somewhat simpler arrangement than Macmillan's and gave the front wheel more freedom to steer, but the wheel diameter was close to that of Macmillan's driving wheel, and so a similar, low gear was the result.

Michaux and Lallement were followed by imitators and developers, as Macmillan was not, and the driving wheel was gradually increased in diameter to provide a better coupling, or speed match, between the human body and the machine (figure 9.1). (This match is sometimes called an "impedance match," but we shall refer to it as a "speed match" or as a "gear ratio.") The high-wheeler (figure 1.13) offered the first combination of a comfortable riding position and an easy rate of pedaling on a two-wheeled vehicle (Sharp 1896).

This high-wheeler gear ratio was preserved when the chain drive was developed to the extent that a step-up drive between the (separate) cranks and the (rear) driving wheel could be used. The resulting safety bicycle was so successful that it is still, in its essentials, the standard bicycle of today.

Thus, by 1885 the principal requirements of a bicycle transmission had been met: to provide a foot motion and a pedaling frequency well suited in average conditions to the capability of the human body to produce power, and to transmit this power from the body (in this case from the feet) to the driving wheel with as little energy loss as possible. The chain drive fulfilled both of these requirements superbly.

Developments to cover nonaverage conditions followed rapidly. A simple approach to low-torque requirements (for downhill travel or level running with a strong tailwind) was to fit a one-way clutch or free-wheel (figure 9.2) to the chain drive, thus permitting coasting with the feet on the pedals. This removed one possibility of braking, that of putting reverse pressure on the pedals, but it also enabled the rider to "bail out" of the vehicle feet first, if necessary (figure 1.18) (Bury and Hillier 1887).

In high-torque conditions, such as starting, hill climbing, or riding in headwinds or on soft ground, riders had to strain at the pedals, often standing on them and pulling up on the handlebars, while pedaling at a very low rate. Scientific testing (see figures 2.16 and 2.27) has confirmed what was perceived intuitively: such pedaling was inefficient. In the twenty years following the introduction of the chain-driven safety bicycle, many different gear-change mechanisms were developed to extend the range of conditions in which a cyclist could pedal effectively and in reasonable comfort. The two most successful types, the multispeed hub gear and the derailleur gear, were developed to cover a wide range of conditions and are the predominant types today. In light of their success, it is perhaps surprising that there seems to be more innovation and development in the area of variable-ratio transmissions than in any of the other aspects of the bicycle.

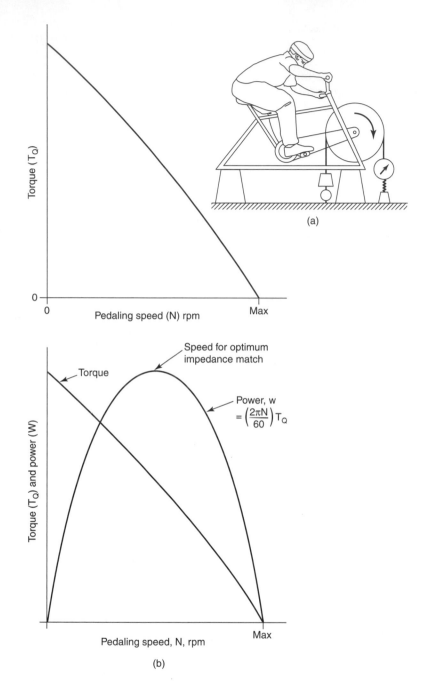

Figure 9.1
Impedance-match curve: (a) An ergometer used to develop a curve of maximum torque versus pedaling speed. (b) Torque and power versus pedaling speed, showing speed for optimum impedance match.

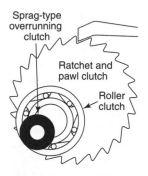

Figure 9.2
One-way clutches.

So much development is occurring in the area of variable-ratio transmissions that to examine more than a few prominent examples of different types would be beyond the bounds of the discussion. Rather, we shall look at some fundamental principles and review alternative possibilities, drawing conclusions where warranted.

Transmission efficiency

Transmission or mechanical efficiency (η_M) is defined in bicycling as the energy output at the coupling to the driving wheel divided by the energy input from the human body, usually via the feet. Both energy quantities in the definition are measured by the product of force and distance. At a wheel or crank, this product can also be expressed as the product of an average torque (T_Q, the tangential force times the radius from the center of rotation at which it acts) and the angle through which it acts (θ, measured in radians). Thus,

$$\eta_M \equiv \frac{T_{Q,\text{wheel}} \times \theta_{\text{wheel}}}{T_{Q,\text{crank}} \times \theta_{\text{crank}}}.$$

The ratio $\theta_{\text{wheel}}/\theta_{\text{crank}}$ is also known as the speed ratio or the gear ratio. A perfect transmission, with an efficiency of 100 percent, has therefore an average torque ratio ($T_{Q,\text{wheel}}/T_{Q,\text{crank}}$) that is the inverse of the gear ratio. In practice, meaning with an efficiency of less than 100 percent, the torque ratio is less than this value:

$$\frac{T_{Q,\text{wheel}}}{T_{Q,\text{crank}}} = \frac{\eta_M}{\theta_{\text{wheel}}/\theta_{\text{crank}}}.$$

Energy loss in a transmission can occur in two ways. One is through friction in bearings and in other components such as the chain. This is the only significant form of loss in "positive-drive" (chain-and-gear) transmissions operating at slow speed. (At high speeds, impact losses become increasingly important.) The other is slip loss, which can occur in transmissions in which the drive is not positive (such as those that use a smooth belt, or some other form of friction or "traction" drive, or an electrical or hydraulic coupling).[2] From this categorization of the forms of energy loss possible within them, we can divide transmissions into two broad types: those with and those without positive drive.

Whether or not the efficiency of bicycle transmission is of great importance is often a matter of debate. Chester Kyle (1995) has analyzed and measured components of aerodynamic drag on racing bicycles and has produced reductions in drag that he believes are significant in allowing riders to win races. From this point of view, it is surprising that there has not been more interest in transmission efficiency, because the losses from transmission seem to be much greater than those resulting from, for instance, the aerodynamic drag from protruding bolt heads and brake cables. As will be mentioned below, it is difficult even to find good information on transmission losses. However, another point of view is that, for the utility or recreational bicyclist who is not putting out maximum exertion to go at top speed, the effect of a 1 percent difference in transmission efficiency is almost always negligible in either the time taken for a trip or the degree of fatigue felt at the end of it.

An example may highlight the relevant issues: an increase in efficiency from 98 to 99 percent would be expected to increase level-road speed by about 0.3 percent, for a gain of twelve seconds in an hour. Clearly, such an improvement would be worth a lot to someone who missed a race win by nine seconds after an hour of solo riding, but not much to a recreational rider.

Positive drives

Chains and toothed belts

The steel roller chain (Kyle 1982), in which a freely rotating lubricated roller surrounds each bushing (figure 9.3), invented by Hans Renold in 1880 and subsequently used in safety bicycles, can, together with a front chainwheel and a rear-wheel "cog" or sprocket, constitute a complete transmission for bicycles that are used principally on race tracks (so-called track bikes on velodromes). Or the rear sprocket may be attached to a one-way clutch or free-wheel (figure 9.2), or to multiratio gears (usually enclosed in the rear-wheel hub and incorporating a one-way clutch, as shown in figure 9.4). Or an overlong chain can be used with guiding-plus-

Figure 9.3
Roller-chain components.

tensioner sprockets or "pulleys" that can force the chain to run on one of many in a nest or cluster of sprockets on the wheel and on the chainwheel (the derailleur gear, figure 9.5).

Despite the importance of drive chain to bicycles and especially to industrial equipment, published research on it has been spotty. Some of the best work is secreted in manufacturers' vaults or stored in the minds of retired engineers, because there is no present commercial value to applying or disseminating it. A relatively complete compilation of the available literature on efficiency and on the effects of tooth form has been made by Matthew Kidd (2000; Kidd, Loch, and Reuben 1999). An idea of the level of sophistication that has been reached in shaping the teeth of multiple chainwheels and sprockets to provide smooth chain shifting can be gained by studying various U.S. patents in this area.[3]

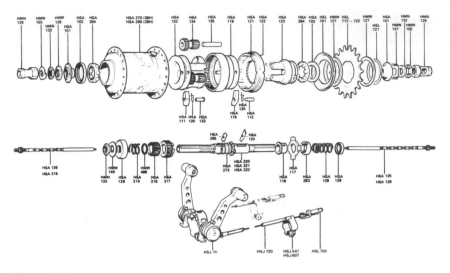

Figure 9.4
Exploded view of Sturmey-Archer five-speed hub gear. (Courtesy of Sturmey-Archer Ltd.)

When new, clean, and well-lubricated, and when sprockets with a minimum of 21 teeth are used, a chain transmission is highly efficient (at a level of maybe 98.5 percent or even higher) and very strong (capable of taking the high tension force from a strong, heavy rider exerting maximum force on the pedals, when the torque may be fifteen times the normal operating torque). Most bicycles in most parts of the world outside North America and Britain have enclosed chains (so-called gear cases), and their transmissions stay in good condition, often for many years of hard use. (A roadster bicycle with enclosed chain drive is shown in figure 9.6.) There is a clear trade-off between the small increase in bicycle weight and cost incurred in using an enclosure and obtaining higher efficiency, lower maintenance, and longer life. Unfortunately (in the opinion of many), chain enclosures have developed an effeminate image, at least in the United States, and are no longer available on standard bicycles. The result is that chains, which tend to be in the path of water thrown up by the front tire and of that carried over by the rear tire, often operate in a mixture of old grease, sand and grit, and salt water. Under these conditions, wear is rapid and is seen as "stretch": the chain becomes longer as the effective pitch of the pin links (that is, the spacing between pins) increases, and the chain therefore rides up the teeth at a larger-than-normal radius (figure 9.7). A remarkable feature of chain drives is that, even in these very poor conditions, they continue to operate, usually reliably, although their efficiency

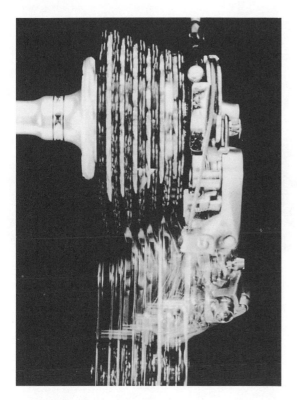

Figure 9.5
Multiexposure photograph of a rear derailleur during the changing sequence.
(Courtesy Shimano Corp.)

decreases. (Surprisingly, at least to the author, it is not known by how much, the efficiency of chain drives decreases in adverse conditions, although some information is given in the tables in this chapter.)

Chains used in multispeed derailleur transmissions wear even more rapidly, for reasons given below. Operation of such chairs may become unreliable as the teeth develop hollows, forming hooks that can prevent the entering chain from seating, periodically carrying links over and producing a slipping effect. (An excellent reference on derailleur mechanisms is Berto, Shepherd, and Henry 2000).

Let us look at ways in which existing chain drives could be improved. A lightweight enclosure of stiff but resilient plastic, such as high-density polyethylene, polypropylene, or Kevlar-reinforced polyester, could be produced to protect the chain and any derailleur mechanism from dirt, water, snow, and sand. (In 2003 some initial components to protect derailleurs

Figure 9.6
Roadster bicycle (Raleigh) with a gear case.

Chain pitch greater than sprocket pitch

Load carried by one or two teeth

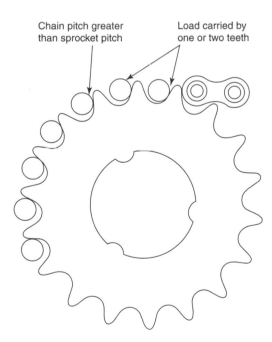

Figure 9.7
Tooth wear from stretched chain.

were coming on the market). But Clemens Bucher (1998) has gone further: he has encapsulated, within the main frame tube of his recumbent bicycle, the chains, a hub gear on a countershaft, and a derailleur, all protected from the weather and safeguarding the clothing of the rider from oil and dirt.

A smaller chain pitch seems desirable to give a wider choice of gear/ratios and to reduce weight. In 1909 the Coventry Cycle Chain Company introduced the Chainette, a small-pitch (8 mm) chain weighing 177 g/m (1.9 oz/ft), which when tested by *Cycling* was found to run "more like a silken cord" than a chain over sprockets. The British racing cyclist F. H. Grubb broke road records on a bicycle fitted with this chain.

Mechanical wear is proportional to the product of force and relative movement between two components in contact (McClintock and Argon 1966). Two links in a chain have to undergo relative movement only when they "articulate" onto and off a sprocket. The angle of articulation is equal to $360°/N$, where N is the number of teeth on the sprocket (Shigley 1972). To obtain the gear ratios that the data in chapter 2 confirm as being most desirable, we do not have much freedom of choice for the diameters of the chainwheel and sprocket. The most severe angle of articulation on a bicycle chain occurs when the chain goes over the rear sprocket. Kyle and Berto (2001) found unusually low efficiencies with twelve-tooth sprockets on derailleurs (meaning, presumably, that smaller sprockets would have even lower efficiencies). A small-pitch chain on sprockets of normal diameter would have a reduced angle of articulation because the number of teeth would be greater. However, with the increased number of joints in the chain, the overall stretch would remain the same.

Whether or not chains of smaller pitch than those currently employed were used, friction and wear would be slightly reduced in derailleur gears if "jockey pulleys" of larger diameter than presently fitted were used. Not only would this decrease the angle of articulation as the chain left each jockey pulley, but it would reduce the rotation rate of the pulleys, which nowadays usually incorporate plain plastic-to-steel (rather than ball) bearings, in an inverse ratio to the increase in diameter. Of much greater import is idler sprockets or rollers in the taut section of chain drives in some recumbent bicycles to avoid front-wheel interference. No matter how slight the change in direction, this could double the frictional loss in the chain drive, because both entering and leaving articulations occur under full tension. Such idlers, when unavoidable, should be as large as the front sprocket if space is available.

There have been many designs of geared transmissions for bicycles in which the chain maintains its alignment in one plane. An early example was the TriVelox gear, in which a cluster of three sprockets was moved in and out of a large-diameter rear-wheel hub while the chain was kept in

Figure 9.8
Whitt's expanding oval chainwheel.

its normal alignment. There have been many designs in which the size or number of teeth in the chainwheel or the rear sprocket or both has been changed during operation. Nothing of this type is presently available on the market so far as is we know. In honor of the senior author of the first two editions of this book, Frank Rowland Whitt, we show in figure 9.8 his expanding-chainwheel gear, which is circular in the lowest gear and increases in ovality as one shifts to higher gears. At the time Whitt developed his gear, oval chainwheels were generally believed to allow a rider to produce either greater power, or the same level of power in greater comfort. This and other similar transmissions allow gear changes under full load. The transmission efficiency of such gears should be slightly higher than that of derailleurs, because they eliminate the small effects of chain misalignment. They promise more certain gear changing than with derailleurs

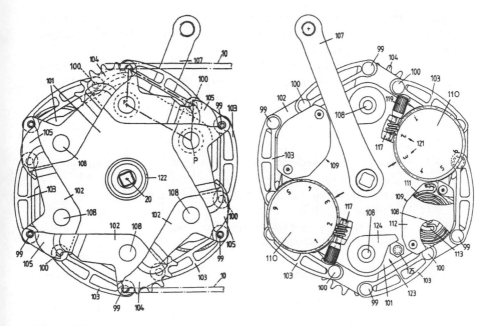

Figure 9.9
Deal drive. (Drawings from Deal's U.S. patent no. 4,618,331.)

(although modern "index-shifting" systems with sprockets designed under the patents discussed above are very effective). The range of chainwheel diameters is less, in general, than can be obtained with derailleur gears. Husted (1985) patented an expanding chainwheel with a wider range. An automatically self-changing expanding-chainwheel gear made by Michel Deal (1986) (the Deal drive) won widespread acclaim but no commercial success (figure 9.9). Automatic transmissions seem appropriate for inanimate and possibly computer-controlled internal-combustion engines, but human beings generally like to choose their point of operation and the exact timing of gear shifts based on upcoming hills or the intention to sprint.

Innovative bicycle transmissions and the patent system

The U.S. and European patent systems have enormous research value for studies in any field, including that of innovative bicycle transmissions. To encourage innovation, patents (which are limited-term monopolies) are granted to help inventors profit from their creations, in exchange for making complete descriptions available to the public. To make this information most accessible, patent offices attempt to categorize inventions into classes and subclasses and to call attention to related work. Although they inevitably fall short in this task, exploration of a relevant subclass is likely to expose a cataract of useful illustrated ideas.

The Internet has dramatically reduced the effort required to search patents, at least for recent patents (e.g., since 1976 for U.S. patents).[4] Some classes and subclasses of patents related to bicycle transmissions are the following.

Class 280 (land vehicles)
280/216   pumps combined with propelling systems (i.e., hydraulic bicycle transmissions)
280/260   vehicles with a train of gearing between crank and wheel
280/261   vehicles with belt or chain gearing
Class 474 (endless-belt power-transmission systems)
474/80   including belt-shifter mechanisms.

A search of an online database on bicycle* and transmission* is also very effective.

Flexing toothed belts

The Berg Company ⟨www.wmberg.com⟩ manufactures a lightweight chainlike belt with flexing articulation, the Speed E flexible drive, for use in vehicle transmissions. Stranded steel cables are used to take the chain tension, and polyurethane "buttons" take the place of the rollers in a steel chain (figure 9.10). This drive achieved brilliant successes and considerable weight savings in the Gossamer Condor, Gossamer Albatross, and Chrysalis human-powered airplanes. In these applications several meters separated the centers of the driving and driven sprockets, which had rather small step-up ratios and were at right angles to each other. As yet there has been no successful application to bicycle transmissions because the small diameters of the rear-wheel sprocket and derailleur pulleys in bicycles have led to fatigue failures of the metal cables and especially of the cable joints (Berg 1981). Bicycling occasionally creates much higher pedaling torques (e.g., when accelerating from rest) than human-powered airplanes, and unlike such airplanes, it virtually demands a ratio-changing gear system to produce these high torques.

The cable-reinforced toothed belts being used to such a large extent as automobile camshaft drives and some motorcycle drives would seem to be good candidates for bicycles, at least for those with hub gears. A beautifully engineered example, by Izzy Urieli, a professor at Ohio University in Athens, and Don Sodomsky, is shown in figure 9.11, installed on a front-wheel-drive bicycle of Urieli's design; the Gates Polychain toothed belt,

Figure 9.10
SpeedE flexible drive. (Courtesy of Winfred M. Berg, Inc.)

10 mm wide, aluminum-alloy chainwheel, and sprocket are presented in figure 9.12; and the sprocket mounted on a Sturmey-Archer hub gear, with a two-speed Mountain Drive bottom-bracket gear in the cranks, is depicted in figure 9.13. Urieli, who enjoys giving no-hands demonstrations of his bicycle's stability and performance, reports total satisfaction with the transmission.

An alternative to a composite toothed belt could be a perforated steel band and associated drive pulleys (figure 9.14) (Belt Technologies 1998). A belt of 301 stainless steel, 12.5 mm (0.5 inches) wide, 0.25 mm (0.010 inches) thick, on pulleys of diameters similar to those of the sprockets used for the chain drive to a hub gear, should be able to take 1,500 N peak loads (about one-third the capacity of an ordinary bicycle chain) and would be extremely light and of high efficiency. It would need protection from grit and other foreign bodies and would require accurate alignment.

Figure 9.11
Uriele front-wheel-drive bicycle. (Courtesy Izzy Urieli.)

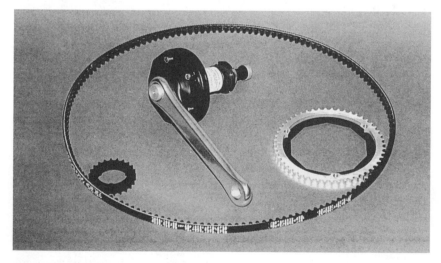

Figure 9.12
Components of Polychain drive. (Courtesy Izzy Urieli.)

Figure 9.13
Three-speed hub and Mountain Drive bracket gear equipped for Polychain drive. (Courtesy Izzy Urieli.)

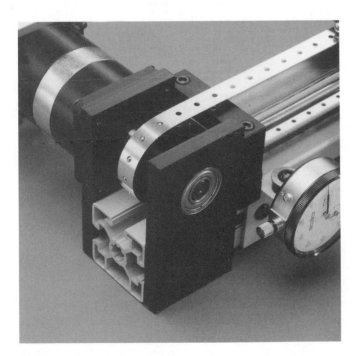

Figure 9.14
Steel-belt transmission. (Courtesy Belt Technologies.)

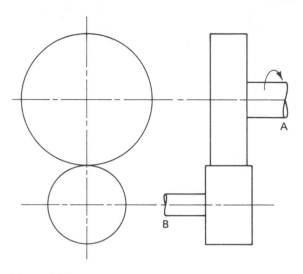

Figure 9.15
Fixed-axis gears. If the gear on shaft A has T_A teeth and that on shaft B has T_B teeth, then one turn clockwise ($+1$) of shaft A will turn shaft B counterclockwise by the quantity $-T_A/T_B$.

Spur-gear systems

Although the word "gear" is used in several different ways in connection with bicycling, in mechanical engineering it refers to toothed spur gears that mesh directly with one another rather than via a chain or toothed belt. When a set of gears is to be designed to yield a particular speed (or torque) ratio between input and output shafts, two alternative approaches to the design are possible. In one, all the axes around which the individual gears rotate are fixed relative to the gear box/casing (figure 9.15); in the other, some of the gear axes themselves rotate around a center (figure 9.16). The latter are called epicyclic gears. Virtually all bicycle spur-gear systems used at present are epicyclic, principally because of the compact arrangement that is possible with epicyclic gears. Though at different times gear-change systems have been developed to fit the bottom-bracket or crank position, these have tended to be large, because they must withstand the full cranking torque. In the rear-wheel hub, connected to the chainwheel by a conventional chain, the torque is reduced by the chainwheel-to-sprocket ratio (usually about 3:1), so that a hub gear can be designed to one-third the peak torque of a bottom-bracket gear. However, the two-speed Schlumpf Mountain Drive gear, shown in an external view in figure 9.13, has been used successfully in the bottom-bracket location.

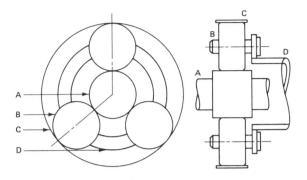

Figure 9.16
Moving-axis (epicyclic) gears. Inputs and/or outputs can be connected to *A*, *C*, and *D*. In a three-speed bicycle hub gear, *A* is on a stationary shaft. In the lowest gear, the chain-sprocket input is connected to *C* and the output (*D*) is connected to the wheel hub. In the highest gear, these connections are reversed. The gear set is bypassed in the middle gear, with the sprocket connected via the free-wheel to the wheel hub.

The calculation of speed ratios in epicyclic gears is illustrated by table 9.1. The design of any spur-gear transmission is highly specialized. Standard mechanical-design texts provide excellent guidance, but they are usually written for industrial applications, for which machines may sometimes be expected to operate for 100,000 hours. A bicycle (or an automobile) has a relatively short operating life (1,000–2,000 hours), and transmissions for bicycles (and automobiles) were developed to their present compact sizes and configurations through early intense efforts to reduce weight, volume, and cost.

At the time of publication, hub gears for bicycles are undergoing a renaissance. For many decades in the twentieth century, the Sturmey-Archer three-speed hub gear and a few similar hub gears were almost universal in many parts of the world, at least on utilitarian bicycles. Around midcentury Sturmey-Archer introduced a model incorporating two epicyclic gear sets, giving four- and five-speed hubs. Although the company advertised that British bicycling champions trained on bikes equipped with its four-speed hubs, the advent of reasonably low-cost ten-speed derailleurs was perceived by the bicycling public as giving more choices. ("Ten speeds" more usually produced six or seven actually useful ratios.) The derailleur "ten speeds" given by two chain-wheels and five rear sprockets grew by increments year by year to three chain-wheels and nine rear sprockets, thus being nominally twenty-seven speeds. These require a widely splayed set of rear forks, strongly "dished" (asymmetric) rear wheels, and narrow,

Table 9.1
Calculation of ratios in an epicyclic gear set (see figure 9.16)

Step	A (shaft and pinion)	B (gear)	C (ring)	D (cage)
Stop rotation of D; turn shaft A −1	−1	$+(T_A/T_B)$	$+(T_A/T_C)$	0
Fix all gears relative to each other and rotate whole gear set +1	+1	+1	+1	+1
Resulting ratio (add)	0	$1 + (T_A/T_B)$	$1 + (T_A/T_C)$	+1

If $T_A = T_B$, then the highest-gear ratio is 1.333 and the lowest-gear ratio is 0.75 (because $T_C = T_A + 2T_R = 3T_A$).

expensive chains. Established manufacturers of hub gears have added more internal epicyclic gear trains to give seven- and eleven-speed versions. A new hub-gear manufacturer in Germany, Rohloff, has produced a fourteen-speed hub gear, the Rohloff Speedhub (figure 9.17) ("Is the Derailleur Dead?" 2000). The spacing of the gears on the hub is almost uniform, at about 13.5 percent difference between adjacent gears. The overall speed-ratio range of over 500 percent is equivalent to that of all but the most extreme derailleurs on all-terrain bicycles. The title of the article in *Mountain Bike Action* that tested the Rohloff hub gear, "Is the derailleur dead?" is appropriate.

Despite the increasing effectiveness of derailleur shifting mechanisms, the shifting of a hub gear can be easier and faster, and the shift can be made when the bicycle is stationary. (The derailleur requires that the bicycle be moving for a gear shift to be made.) A good hub gear and its chain, if enclosed, last indefinitely, in contrast to derailleur gears, which generally need new cogs, chainwheel sprockets, and chains every year if they are in daily use. With all these advantages, it would seem that the derailleur gear is, or should be, "dead."

However, there are three factors that determine the choice between derailleur and hub gear, and they do not at present point in the hub gear's direction: cost, weight, and efficiency. At the time of publication, the price of a Rohloff Speedhub alone is more than that of ten complete department-store eighteen-speed derailleur-equipped bicycles in the United States. One would expect that increased production and competition would bring the price of the Rohloff down sharply in the future. With regard to weight, hub

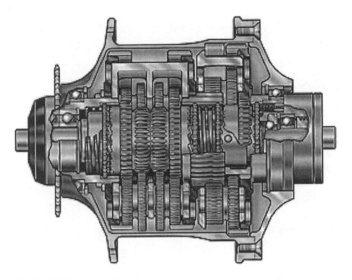

Figure 9.17
Rohloff fourteen-speed Speedhub gear. (Courtesy of Rohloff A. G.)

gears in their wheels and with their shifting mechanisms are presently marginally heavier than are the equivalent derailleur gears. Here again, one would expect that competition could achieve considerable reductions in the weight of what are generally first-generation high-gear-count hub gears. With regard to efficiency of gears, considerable uncertainty exists. This is the subject of a later section, "Transmission efficiencies."

Direct-drive bicycles
Thomas Kretschmer (2000) believes that in some ways bicycles took a retrograde step when the early Michaux-Lallement type of bicycle that developed into the high-wheeler, all versions of which had pedals and cranks fixed to the front wheel, gave way to the safety bicycle, which has a step-up chain drive to the rear wheel. Kretschmer regards the external chain drive as the weak point of utility bicycles and is developing what he terms "direct-drive bicycles" (e.g., figure 9.18) with multispeed hub gears in the front (driving) wheel (figure 9.19).

Shaft drives
Some early safety bicycles used a shaft drive in place of a chain, with one right-angle bevel gear set at the crank spindle and another at the rear wheel (figure 9.20). These drives had a neat, compact appearance but were heavier, less efficient, and much more expensive than a chain drive. Their

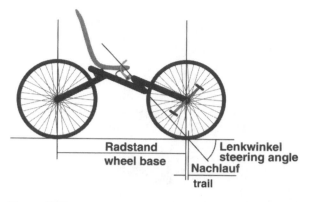

Figure 9.18
Configuration of direct-drive bicycle. (From Kretschmer 2000.)

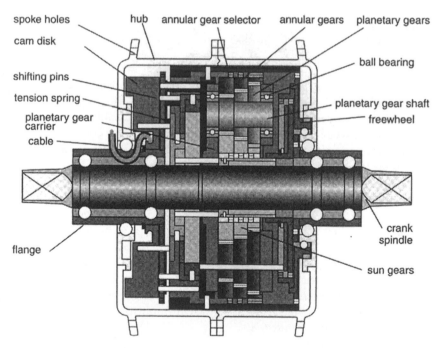

Figure 9.19
Hub for direct-drive bicycle. (From Kretschmer 2000.)

Figure 9.20
Pierce shaft-drive bicycle, 1900. (Courtesy of the Smithsonian Institution.)

fundamental disadvantage is plain: drive torque is transmitted with a moment arm (shaft or gear radius) of about 12 mm, which leads to far higher forces and distortions than the 50- to 100-mm radius of chain drives.

In the period around 1897, most American manufacturers produced at least one shaft-drive model. Tests showed that the power losses were in the range of 3–8 percent (Carpenter 1898), probably because the machining of bevel gears was not very precise. The Waltham Orient pattern using roller pins instead of machined teeth performed well, however, and Major Taylor broke many records using this transmission (Ritchie 1996). Some data on the efficiency of an unspecified shaft drive are given in table 9.8.

Other forms of positive drive

Linear and oscillating transmissions   An early form of linear transmission, an oscillating drive, is illustrated in figure 9.21. When the foot pedal of the transmission is pushed, there is no resistance until the pedal velocity has caught up with the wheel velocity at the setting of the gear-ratio adjustment on the radius arm. For torque to be transmitted smoothly at this point, it is essential that a one-way clutch without backlash or overshoot be used. Such a device is the sprag clutch, shown in figure 9.22. A very sophisticated nineteenth-century version of this oscillating drive is shown in figure 2.24, taken from the cover of *The Bicycle* (Dodge 1996). An oscillating

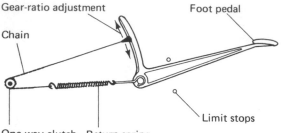

Gear-ratio adjustment   Foot pedal

Chain

One-way clutch   Return spring
(free-wheel)

Limit stops

Figure 9.21
Oscillating drive.

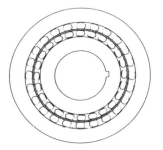

Figure 9.22
Sprag clutch.

drive was also used on the highly advanced (for its time) American Star high-wheeler (see chapter 1).

The linear or oscillating drive has attracted many because it can provide a transmission with a continuously variable ratio, which is apparently well matched to a natural ladder-climbing action of the legs. However, above very low "pedaling" speeds the energy required to bring the legs and feet up to the speed of the wheel is considerable, and this energy will be lost (with the additional metabolic costs of eccentric contraction) if muscles are also used to decelerate the legs and feet at the end of the stroke. Such losses of energy can be reduced by increasing the gear ratio at the end of the power stroke to decelerate the limb. One method for doing this is to incorporate a "fusee," a grooved cone used in some spring-wound time-pieces, which was employed in a modified form in the tricycle Dragonfly II, designed by Steve Ball.[5]

Snek cable drive and the Rowbike Bert and Derk Thijs patented the Snek cable drive in 1998–1999 as a clever extension of the fusee principle just mentioned (see Thijs 1998–1999). The length of the path on the fusee is substantially longer than the stroke of the steel cable that wraps and unwraps around it (figure 9.23a). The Thijses devised a derailleur shifting mechanism to allow the cable to start its stroke on different parts of the fusee, thus providing several effective gears. The fusee itself is mounted on the rear wheel of a vehicle, connected to it by a sprag or other one-way clutch, and contains a rotary spring that rewinds the fusee as the cable is released at the end of a stroke. The cable is pulled by a combination of a sliding carriage to which the pedals are attached and pivoting handlebars, which are pulled toward the chest as the feet are pushed forward. It would be good to know the power-duration curves, similar to those that have been made for conventional drives, for the form of power production involved in this design. The motion required is, however, similar to rowing, as the name of the bicycle that employs the Snek cable drive, the Rowbike, implies, and we know from the data in chapter 2 that top athletes produce amounts of power in rowing that are similar to those produced in rotary pedaling. We might hazard a guess that the contribution of the arms, which can add 20 percent more power to rotary pedaling, at least for short durations, compensates for the losses to be expected from a rowing motion that does not conserve the kinetic energy of the limbs and mechanism at the end of the stroke. We might also expect that the motion of the modification produced by the fusee increases the power output to some extent over that produced in constant-velocity rowing.

Enthusiasts for human-powered vehicles of any description have in any case preferred the judgment of race results to laboratory data. Derk Thijs has won several races and has broken several records on his Rowbike, which has now become popular enough in The Netherlands for one-design races to be held (figure 9.23b).

Hydrostatic drives

Heavy earth-moving equipment (for example) often uses a type of transmission in which the engine drives a positive-displacement hydraulic pump and high-pressure oil is piped to hydraulic motors in the wheels. A major advantage is that a type of variable-angle-swash-plate axial-piston pump permits the output to be varied over a wide range from positive to negative flow, giving a continuous variation in speed ratio. There have been many attempts to apply this type of transmission to passenger automobiles and to bicycles. An apparently insurmountable problem is that the peak efficiency of a hydraulic pump (and of a hydraulic motor) is about 90 percent, so the overall transmission efficiency cannot be much over

(a)

(b)

Figure 9.23

(a) "Snek" cable transmission. (b) A Rowingbike race in Zeeland, the Netherlands, September 15, 2002. Derk Thijs, followed by Nico Rienks (Courtesy of
Thijs Roeifietsen.)

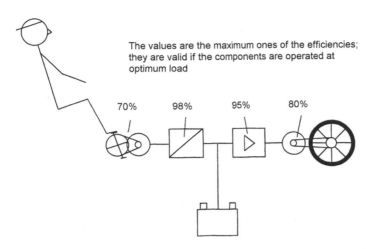

The values are the maximum ones of the efficiencies; they are valid if the components are operated at optimum load

70%   98%   95%   80%

Figure 9.24
Electrical-transmission efficiencies. (From Blatter and Fuchs 1998.)

80 percent: a low figure for the maximum efficiency of a human-power application.

Nonpositive drives

It would be easy to dismiss forms of transmission involving nonpositive drives from further consideration, because the additional slip losses incurred in such drives beyond those involved in positive drives are a considerable penalty for bicycle application. The chain drive of the first safety bicycles had to compete with the direct wheel-mounted cranks of the high-wheelers, which had an efficiency of almost 100 percent, and any great loss in the chain drive would not have been tolerated. The best roller-chain drives appear to have an efficiency of about 98.5 percent, and the losses resulting from the use of such drives would, except in very close races, be imperceptible. We wrote in the second edition of this book that "the slip loss of a V-belt, or a hydraulic coupling, or some form of electrical coupling characterized by a generator driving a motor, would multiply these losses by between 5 and 10 unless very large, oversized transmissions were used. The weight and volume of such transmissions would make them unattractive" (Whitt and Wilson 1982, p. 281). However, since then, there have been major improvements in both the efficiency and size of electrical generators and motors. Juerg Blatter and Andreas Fuchs (1998) in Bern have been producing increasingly promising results.

A simple version of Blatter and Fuchs's system is shown in figure 9.24, and table 9.2 presents component efficiencies. One line of the table

Table 9.2
Efficiency (η) of the Blatter and Fuchs system

Components of the drive train		Generator	Maximum efficiency tracker	Motor control	Motor	Total η	Typical η
Low-cost components (working prototype)	η max.	0.70	0.98	0.95	0.80	0.52	~50%
Optimized components (brushless)	η max.	0.95	0.95	0.95	0.95	0.81	~80%

Source: Blatter and Fuchs 1998.

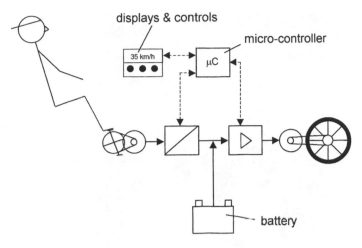

Figure 9.25
Complete system of a fully electric hybrid. (From Blatter and Fuchs 1998.)

shows data obtained in 1998 with low-cost components, and these give a typical overall efficiency of the system as an unattractive 50 percent and the projected future efficiencies, with optimum components, an overall efficiency of 80 percent or more. Blatter and Fuchs point out that although this overall efficiency would be low for a racing bicycle, compared to an everyday utility machine with a chain and other components that are worn and dirty, it could be high. The many advantages of this system could be seen as justifying the electrical transmission: virtually nothing to degrade or get dirty or worn out; a built-in, continuously variable transmission that could be scheduled so that the rider is always pedaling close to her/his optimum cadence and never has to pause to change gears; some battery assistance for hill climbs and quick starts; and much braking energy going back into the battery. It would offer advantages for folding bicycles, in which the oily chain of the typical chain drive is a significant problem; for tandems, on which each rider could pedal at her/his optimm speed; and for recumbents, which normally must necessarily have a long chain (or long chains) that must be routed around pulleys to pass under the rider to reach the rear wheel, because for all of these only electrical wires need connect the generator at the pedals with the motor at the wheel (or wheels). A block diagram of the circuit for a complete system is shown in figure 9.25, and the pancake generator of the present low-cost system is presented in figure 9.26. Fuchs (2000) predicts that energy storage in the future will probably be greatly improved in efficiency and in specific weight by the use of the rapidly improving supercapacitors.

Figure 9.26
Transmission to pancake generator. (From Blatter and Fuchs 1998.)

Another possible exception to the almost total elimination of all except positive-drive transmissions is traction drives, or continuously variable–ratio transmissions (Loewenthal, Rohn, and Anderson 1983). Some well-known types are shown in figure 9.27. The reason these might deserve examination after they have been tried repeatedly and rejected over many years by the major automobile manufacturers is the discovery of lubricants that, under high-pressure contact between two hard surfaces, undergo a reversible change in viscosity such that they can transmit a high shear force (Green and Langenfeld 1974). It seems probable that these lubricants will extend the range of usefulness of traction drives outside the very specialized areas to which they are presently confined. However, it is unlikely that any traction drives will penetrate the human-powered-vehicle field, because bicycles require transmissions that can withstand relatively high torque at low speed in providing a step-up speed ratio. Almost all industrial and commercial transmissions have the opposite characteristics in all three respects: low (relative) torque at high speed giving a step-down ratio. In fact, the torque capability of a standard bicycle chain drive would enable it to transmit 10–15 kW in an industrial drive (although not with an acceptably long life). Traction drives are already much heavier than

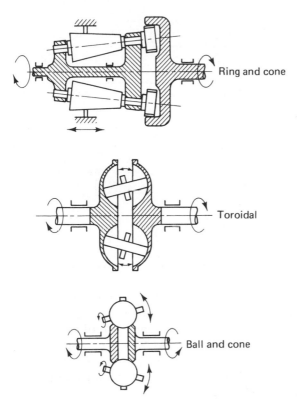

Figure 9.27
Traction drives.

their industrial competitors. Therefore, even the use of the new lubricants seems unlikely to overcome their inherent disadvantages for application to bicycles.

Other transmissions

In the above review of alternative transmissions, we have probably omitted more transmission types than we have included. Among those omitted are several interesting types shown in earlier editions of this book. A fascinating but far-from-comprehensive summary of bicycle patents by Herzog (1984) gives many more that are worthy of examination. It is not clear that the best designs are those that have succeeded in the marketplace. However, space considerations prevent us from showing all potentially interesting transmissions, and we must therefore be content with stating principles and reporting data on transmissions in general use.

Table 9.3

Transmission efficiencies for chain drives

Driver rpm	70	60	50	60	60
Power (W)	100	100	100	150	175
Calculated chain tension (N)	130	151	182	227	265
Number of teeth, driver-driven					
52–11	88.7	91.1	92.5	94.6	95.5
52–15	90.4	92.3	94.7	96.2	97.5
52–21	92.0	93.8	95.2	97.4	98.2

Source: Data from Spicer et al. 1999.

Transmission efficiencies

A full knowledge of the losses occurring in present bicycle transmissions would focus our attention on whether there are problems and, if so, on how to correct them. Unfortunately, there is presently no consensus. We will therefore report data with cautions and comments.

Ron Shepherd (1990) has stated that the efficiency of chain transmissions, including derailleurs, is normally over 99 percent. Spicer et al. (1999), on the other hand, measured efficiencies down to 88 percent on a clean, as-new derailleur system. Their most-significant findings are given in table 9.3. They point out the following.

1. Transmission efficiency decreases as the size of the rear sprocket is reduced.

2. Efficiency diminishes as the amount of torque transferred (or chain tension) is decreased.

3. The maximum efficiency attained is at relatively high power (175 W) and low pedal rpm (60) and in the lowest gear (meaning the largest-diameter rear sprocket, with twenty-one teeth) and is just over 98 percent.

4. The additional losses because of chain offset (the two sprockets not being in line) are negligible.

5. The type of lubrication, or even whether there is lubricant present, has almost no effect on efficiency (see tables 9.4 and 9.5).

These results, which in many cases contradict popular wisdom, were reviewed by Kyle (2000), who had recently himself supervised a similar proprietary study. He confirmed the general findings and accuracy. His own study in collaboration with Berto (Kyle and Berto 2001) has extended the data given here and is recommended for more detailed study.

Table 9.4
Efficiencies of new, clean, lubricated chain drives

Power (W)	Single-speed	Three-speed hub			Six-speed derailleur		
		Low	Direct	High	24T COG	19T COG	13T COG
50	96.0	90.6	93.4	87.3	94.2	94.1	92.1
100	97.3	92.8	95.7	90.9	96.2	96.4	94.9
200	98.1	94.0	96.9	92.9	97.4	97.6	96.9
400	99.0	95.0	97.9	93.9	98.1	98.4	97.8

Source: From Keller 1983 and Fichtel & Sachs A.G. 1987.

Table 9.5
Efficiencies of shaft drive and of clean and rusty chains

Power (W)	Shaft drive + three-speed hub gear			Used chain	
	Low	Direct	Normal	8,000 km, no rust, lubricated	7,000 km, rusty, dry
50	79.2	82.2	77.3		
100	84.8	88.3	83.3	94–96	88
200	86.6	90.0	85.1	97–98	93

Source: From Keller 1983, 71–75.

Other data from various sources (Wilson 1999) are given in tables 9.6–9.9. The highest efficiencies for the hub gears examined are lower than the highest efficiencies for the derailleur gears considered. However, most of these data were taken in static force tests as described by Cameron (1998), which are not as satisfying as power-input and -output data. Nevertheless, the results confirmed many of the trends of Spicer et al. 1999. Whitt also used static measurements to produce graphical data given in figure 11.16 of the second edition of this book and quoted Thom, Lund, and Todd 1956, again following similar trends.

Measurements of transmission efficiency are difficult to make accurately, because they involve subtracting two imperfectly known large quantities (input and output power) to find a much smaller quantity. Average torques must be known to within 0.1 percent to determine losses with accuracy. The conventional "back-to-back," "recirculating," or "four-square" method, involving two loops of chain each driving the other, is dubious when the sprocket ratio differs from unity, because the efficiencies

Table 9.6
Transmission efficiency of a Shimano Nexus seven-speed hub

Gear	Ratio	First test	Second test
1	0.632	0.91	0.91
2	0.741	0.94	0.93
3	0.843	0.87	0.87
4	0.989	0.86	0.89
5	1.145	0.86	0.87
6	1.335	0.92	0.93
7	1.545	0.91	0.91

Source: Data gathered by Angus Cameron, reported in Puckett 1999.

Table 9.7
Transmission efficiency of a Shimano seven-speed hub gear

Gear	Relative distance per revolution of crank	Transmission efficiency at 100 W power	Transmission efficiency at 200 W power
1	2.9	0.87	0.92
2	3.3	0.90	0.915
3	3.8	0.76	uncertain
4	4.4	0.865	0.87
5	5.2	0.82	0.83
6	6.0	0.92	0.92
7	7.0	0.91	0.91

Source: Data gathered by Jan Verhoeven, reported in Puckett 1999.

of step-up and step-down transmissions are different, but in this method the efficiencies of these transmissions are combined. Quasi-static tests raise the concern of oil-film thicknesses being reduced, and in any event, careful averaging is needed, because chordal action alters mechanical advantages by several percent during the passage of a single tooth. In dynamic testing, such averaging is automatically provided by inertial effects.

What is the optimum number of gear ratios?

To some people, the question that heads this section is a strange one: there are enthusiasts who believe that the only authentic bicyclists are those who ride fixed-gear, single-speed machines, even over mountain passes. At the

Table 9.8
Transmission efficiency of a Sachs Elan twelve-speed hub gear

Gear	Meters per revolution of crank (28-inch wheels)	Transmission efficiency at 100 W power	Transmission efficiency at 200 W power
1	2.2	0.87	0.92
2	2.7	0.91	0.95
3	3.2	0.925	0.965
4	3.8	0.90	0.91
5	4.3	0.90	0.91
6	4.8	0.905	0.905
7	5.1	0.88	0.88
8	5.7	0.88	0.88
9	6.1	0.88	0.88
10	6.6	0.855	0.86
11	7.1	0.865	0.87
12	8.5	0.86	0.88

Source: Data gathered by Jan Verhoeven, reported in Puckett 1999.

Table 9.9
Efficiencies of a Shimano Deore LX derailleur drive

Number of teeth on chainwheel	Number of teeth on rear sprocket	Meters per revolution of crank (28-inch wheels)	Transmission efficiency at 100 W power	Transmission efficiency at 200 W power
22	28	1.7	99	98.5
22	24	2.0	98	98
22	21	2.2	96	98.5
22	18	2.7	96	96.5
32	21	3.1	93.5	95
32	18	3.8	93.5	94.5
32	16	4.2	94	94
32	14	5.0	94.5	93.5
42	16	5.7	93	93
42	14	6.6	91.5	91.5
42	12	7.6	89.5	91.5
42	11	8.3	88	91.5

Source: Data gathered by Jan Verhoeven, reported in Puckett 1999.

other extreme are enthusiasts at the opposite pole who install, for instance, two seven-speed gear systems in series, thus having the nominal choice of forty-nine gears. Can we use analysis to determine whether either is right, or whether there is a happy mean between the extremes?

The starting point for such an analysis is the qualitative power-torque-speed graph (figure 9.1). Every person capable of pedaling can be measured on an ergometer to produce her/his maximum steady-state torque (something that would need precise definition if we were being quantitative) at all possible pedaling speeds, from zero (at which the torque will be at a maximum) to the maximum possible spinning speed (at which the torque will be zero). Very roughly, the "curve" that results when these data are plotted is a straight line. Another curve can then be drawn as the product of torque and speed, which is power output. This curve is roughly a parabola, having a maximum at close to half the maximum possible pedaling speed. Each person's graph would be different. Some people produce maximum power at 120 rpm and can spin up to 180 rpm, whereas others can manage only half these speeds. A person's graph will change depending upon the level of training or of fatigue. Let us look at the case of a person who is rested and well-trained and has the output shown by the graph of figure 9.1, which will remain essentially unchanged.

Let us further suppose that this person is starting from rest on a fixed-gear bicycle, putting in maximum effort. She will start at the origin of the graph, the 0-0 point, and, as the bike speeds up, will "climb up" the power curve, decreasing in torque, increasing in speed, and increasing power output. If the road on which she is riding is uphill, the rider may never reach her pedal rpm for maximum power output: she will never be at the optimum gear ratio. On the other hand, if the road is level or downhill, the rider will quickly attain the maximum-power condition, and if she continues at maximum effort, will continue to increase speed but will decrease power output until the power requirements of the bicycle match the reduced output of the rider. If the event is not a race, the rider is more likely to reach the optimum speed for the bicycle's gear ratio and then simply to drop to a lower effort so as to stay at that speed.

Now let's consider someone who has a two-speed gear. Let us suppose that the low gear is 50 percent of the high (measured as *la developpement* or in gear inches, for instance). The rider starts in low gear and quickly reaches the optimum speed for that gear ratio and develops maximum power. Should she change to the high gear at that point (figure 9.28)? No! She would then be pedaling at half the optimum rpm and would develop, at maximum effort, about 15 percent less than maximum power. The rider should continue to speed up in the low gear until her power output has dropped to the same level in low gear as that which she could produce if in high gear. Switching at that point is optimum, because continued accel-

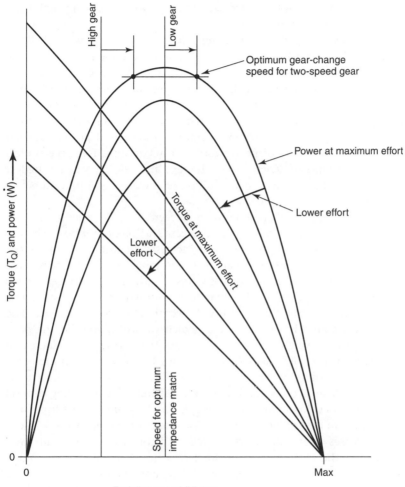

Figure 9.28
Power loss at gear shifts.

eration in low gear would produce decreasing power, whereas in high gear, it would produce increasing power.

This leads us directly to our being able to determine (if we had all the data we need) the optimum number of gears (presumably having equal ratios). We can estimate from figure 9.28 that having two gears, one 50 percent of the other, results in the rider's power output dropping by perhaps 4 percent between gear changes. Increasing the number of evenly spaced gear ratios, perhaps to seven, might lead to steps of, say, 20 percent between changes, and a loss of power output of perhaps 1 percent at each change. The average loss of power during a maximum-effort acceleration would be less than this. Then, when the terminal velocity was reached, it might be either at maximum power output, or at worst just 1 percent below maximum.

Doubling the number of ratios and decreasing the speed ratio between steps to, say, 13 percent might decrease the maximum power loss to a fraction of 1 percent. (Normally as one increases the number of ratios one simultaneously increases the overall gear-ratio range, so that there is not a direct correlation between number of ratios and the range between steps.) The gain in average power output by increasing the number of steps can be easily determined. Adding ratios is not, however, a low-cost procedure. The increased cost comes in money terms (which we will ignore for the present), in increased weight, in increased gear losses (because of increased chain offset, or multiple gear sets in series, for instance), and in increased time devoted to changing gears.

These increased costs are much more difficult to assess. They depend on the state of development of multigear technology and on the ingenuity of a designer in seeing ways to increase the number of ratios without increasing losses or overall weight, which itself constitutes a loss in power. In most cases the costs must be estimated rather than calculated. That is an accepted role for the designer, the athlete, and the manager. Here, we will be content simply to assert that there is no case to be made for having forty-nine gear ratios on a bicycle. We may even have gone beyond the useful range with twenty-seven, except that they are far from evenly spaced, and that the number of useful ratios with such a gear is probably twelve.

Range of variable gears

As mentioned above, there are still some people who believe that "genuine" bicyclists ride one-speed, fixed-gear bicycles, partly because riders of high-wheelers perforce had this system. But even fixed-gear riders competing in road races would often have sprockets on either side of their rear wheels, perhaps one having thirteen teeth and the other sixteen. Before a long hill they could dismount, remove the rear wheel, turn it around and replace it. The gear range, if defined as the high gear divided by the low gear, so obtained would be 16/13, or 1.23. Two-speed hub or bracket gears often had a range of 1.33 (*Variable Gears* 1909). Three- and four-speed hubs were produced in wide, normal, and close ratios, a typical three-speed hub having a range of 1.74.

Conclusions

The efficiencies of present transmissions using chains and derailleurs or hub gears are in the 80–98 percent range, the higher efficiencies being produced at high power (torque) levels. At a power level of 100 W (typical for everyday commuters), the efficiency of the best derailleur gears and the best hub gears is about 95 percent, although in general, hub gears are a little lower in efficiency than clean, new derailleur gears. There is scope for future improvements in efficiency and in alternative designs that are longer lasting, more convenient to use, or lower in cost. There is also scope for weight reduction, for protection against deterioration, and for some input motions of the feet alone or of the feet plus the hands that would allow a higher maximum power to be delivered than with existing circular constant-velocity pedaling. It is presumed that such alternative input motions could also give greater comfort at less than maximum effort.

Notes

1.   Important secondary aims include a reasonably stiff connection, no undue jerking, the down-moving leg's being able to lift the up-moving leg, no large variations in the kinetic energy of the legs, and the ability for the bicycle to be wheeled backward.

2.   To forestall frictional slip, nonpositive transmissions often resort to elevated contact pressures, which lead to unnecessarily large losses when the amount of torque is small. To get around this, some such drives employ "pressure increasers" that operate only when high amounts of torque are applied.

An insidious form of slip loss arises from belt stretch, even when no obvious slippage is occurring. A stretched belt shrinks as it travels from the taut side to the looser side, and the driving sheave seems to rotate faster than expected, by the ratio (taut length)/(slack length).

3.   Of particular interest are patents no. 4889521 (1989), 5632699 (1997), 5569107 (1995), 5545096 (1995), 5514042 (1995), 5192249 (1991), 5188569 (1993), 5162022 (1992), and 5133695 (1992). All are U.S. patents, although all of the inventors are from Japan or Taiwan.

4.   The U.S. Patent and Trademark Office search page has the following URL: 〈http://uspto.gov/patft/index.html〉 (see "Bibliographic Advanced Search").

5.   Two photographs and a short description of the Ball tricycle are given in the second edition of this book (Whitt and Wilson 1982, 298–299).

References

Belt, Technologies. (1998). *Design Guide and Engineers' Reference for Metal Belts*. Agawam, Mass.: Belt Technologies.

Berg, Winfred M. (1981). Personal communication with D. G. Wilson.

Berto, Frank, Ron Shepherd, and Raymond Henry. (2000). *The Dancing Chain —History and Development of the Derailleur Bicycle*. Self-published. Available through Van der Plas Publications, San Francisco.

Blatter, Juerg, and Andreas Fuchs. (1998). "Fully electrical human-power transmission for ultra-lightweight vehicles." In *Proceedings of the Extra-Energy Symposium, IFMass, Koln, Germany*.

Bucher, Clemens. (1998). "Recumbent with encapsulated drive train." Paper presented at Third European Seminar on Velomobile Design, Rosskilde, Denmark.

Bury, The Viscount, and G. L. Hillier, eds. (1887). *Cycling*. London: Longmans, Green.

Cameron, Angus. (1998). "Measuring drive-train efficiency." *Human Power*, no. 46 (Winter):5–7.

Carpenter, R. C. (1898). "The efficiency of bicycles." *Engineering* (August 19–26).

Deal, Michel. (1986). "Variable ratio transmission." U.S. patent no. 4,618,331.

Dodge, Pryor. (1996). *The Bicycle*. Paris and New York: Flammarion.

Fuchs, Andreas. (2000). Personal communication with D. G. Wilson, June 12.

Fichtel & Sachs A. G. (1987). "Versuchs Bereit (Research Report) no. 5162, Fichtel & Sachs, Schweinfurt, Germany, February 23.

Green, R. L., and F. L. Langenfeld. (1974). "Lubricants and traction drives." *Machine Design* (May 2):108–113.

Herzog, Ulrich. (1984). *Fahrradpatente* (Bicycle patents). Kiel, Germany: Moby Dick Verlag.

Husted, Royce H. (1985). "Expandible sprocket." U.S. patent no. 4493678.

"Is the derailleur dead? Trail-testing the Rohloff Speedhub 14." (2000). *Mountain Bike Action* (March).

Keller, J. (1983). "Comparative efficiencies of shaft drive and clean and used chains" (in German). *Radmarkt* (Germany), no. 2:71–75.

Kidd, M. D. (2000). "Bicycle-chain efficiency." Ph.D. diss., Department of Mechanical Engineering, Heriot-Watt University, Edinburgh, Scotland.

Kidd, M. D., N. E. Loch, and R. L. Reuben. (1999). "Experimental examination of bicycle chain forces." *Experimental Mechanics* 39, no. 4 (December).

Kretschmer, Thomas. (2000). "Direct-drive (chainless) recumbent bicycles." *Human Power*, no. 49:11–14.

Kyle, Chester R. (1982). *Chains for Power Transmission and Materials Handling*. New York: American Chain Association and Basel, Switzerland: Marcel Dekker.

Kyle, Chester R. (1995). "Bicycle aerodynamics. In *Human-Powered Vehicles*, ed. Allan Abbott and David Wilson, 141–156. Champaign, Ill.: Human Kinetics.

Kyle, Chester R. (2000). Personal communication with D. G. Wilson concerning the paper of Spicer et al.

Kyle, Chester R., and Frank Berto. (2001). "The mechanical efficiency of bicycle derailleur and hub-gear transmissions." *Human Power*, no. 52:3–11.

Loewenthal, S. H., D. A. Rohn, and N. E. Anderson. (1983). "Advances in traction-drive technology." Technical paper no. 831304, Society of Automotive Engineers, Warrendale, Penn.

McClintock, F. A., and A. S. Argon. (1966). *Mechanical Behavior of Materials*. Reading, Mass.: Addison-Wesley.

McGurn, James. (1987). *On Your Bicycle: An Illustrated History of Bicycling*. London: John Murray.

Puckett, Giles. (1999). "Hub gear efficiencies." Message posted to mailing list "bikelist.org," January 28, from Australia.

Ritchie, Andrew. (1975). *King of the Road*. Berkeley, Calif.: Ten Speed Press.

Ritchie, Andrew. (1996). *Major Taylor: The Extraordinary Career of a Champion Bicycle Racer*. Baltimore: Johns Hopkins University Press.

Sharp, Archibald. (1896). *Bicycles and Tricycles*. London: Longmans, Green; reprint, Cambridge: MIT Press, 1977.

Shepherd, Ron. (1990). "Derailleur pulley resistance." In *Proceedings of the seventh international cycle-history conference*. San Francisco: Bicycle Books.

Shigley, J. E. (1972). *Mechanical Engineering Design*, 2d ed. New York: McGraw-Hill.

Spicer, James B., Christopher J. Richardson, Michael J. Ehrlich, Johanna R. Berstein, Masahiko Fukuda, and Masao Terada. (1999). *On the Efficiency of Bicycle Chain Drives*. Baltimore: Johns Hopkins University.

Thijs, Derk. (1998–1999). "Patent no. WO 00/12378, application no. 99941870.0-2425. Snek transmission."

Thom, A., P. G. Lund, and J. D. Todd. (1956). "Efficiency of three-speed bicycle gears." *Engineering* 180 (July 2):78–79.

Variable Gears. (1909). No. 2 in *Cycling Penny Handbooks*. Birmingham, U.K.: Pinkerton Press. Reprint.

Whitt, Frank Rowland, and David Gordon Wilson. (1982). *Bicycling Science*, 2d ed. Cambridge, Mass.: MIT Press.

Wilson, Dave. (1999). "Transmission efficiencies." *Human Power*, no. 48 (Summer):20–22.

Recommended reading

Whitt, F. R. (1979). "Variable gears: Some basic ergonomics and mechanics." In *Developing Pedal Power*. Milton Keynes, U.K.: New Towns Study Unit, Open University.

10 Materials and stresses

Introduction

After reviewing a little of the history of the use of various materials in early bicycles, we examine the properties of old and new materials. We show that the most probable cause of unexpected structural failures in bicycles is "low-cycle" fatigue, and that "notch" and "fracture" toughness are other lesser-known but important properties. "Young's modulus" should be high enough to impart adequate stiffness to a bicycle's frame, but a designer can also compensate for using lower-modulus materials by increasing the size of cross-sections (e.g., tube diameters). We recommend good design practices in various materials and make introductory references to the voluminous materials engineering literature.

A brief history of bicycle materials

The makers of early bicycles employed "traditional" materials, principally woods reinforced with metals, as used in the earliest vehicles. Until the Bessemer and open-hearth processes were developed in 1855–1865 to produce low-cost steel, the metals used in bicycles and other vehicles were mainly low-strength wrought and cast iron, including "malleable" cast iron, along with some bronze and brass. The greater availability of steel and the development of other materials suitable for use in vehicles in the period from 1869 to 1880 allowed inventors and designers to produce tension wheels, rubber (solid) tires, and tubular-steel construction with rolling (instead of rubbing) bearings (Wilson and Saleh 1993). Continued refinement of materials and improved design have resulted in a reduction in current bicycle weight to about one-third of that common for early machines. Aluminum and titanium alloys have been prominent among the successful alternatives to steel for bicycle frames. The stiffness and strength of resins reinforced with carbon fibers, as well as the toughness imparted by Kevlar fibers, have resulted in an outburst of almost-free-form configurations at the end of the twentieth century. New fibers such as Zylon could bring about further remarkable design variations in bicycles.

However, for most of the past century, the principal materials used for the frames of bicycles have been steels (i.e., iron plus carbon): low-carbon for inexpensive machines, medium-carbon for the middle-range models, and chrome- (or manganese-)molybdenum steel alloys with medium carbon content for the best competition cycles. Inexpensive frames are made of straight-gauge tubes formed from steel strip, rolled and electrically welded along the seams, and later welded to the other frame components.

The best steel frames are made from seamless tubes, drawn over a shaped mandrel to be thinner in the middle than at the ends.[1] One method of joining these tubes to form the bicycle frame results in connections that are stronger than the tubes themselves: low-temperature silver brazing of the tubes into externally tapered end-sockets, called lugs. However, it seems that for the kinds of overloads to which a bicycle is typically subjected, even a high-temperature unreinforced joint made by welding the tubes together can yield a frame that is almost as good as one made by the above method, if the welding is carried out skilfully.

Strength of materials

Factors of safety
All structures are designed to be stronger than is strictly necessary in "normal service" (something that has to be defined closely). The ratio between the actual load that would cause a structure to fail and the service load the structure is expected to carry is called the "factor of safety" (it could also be called a factor of ignorance). Bicycles are built with much greater strength than the normal loads that result from smooth-road riding call for; that is, the factor of safety in bicycles is high. Fairly standard bicycles are used in circus acts to carry five or ten people or even more. A well-known advertisement of a few years ago showed fifteen men carried by a commercially available bicycle. In China the overwhelming proportion of freight is still carried by bicycle, often in large unit loads (Lowe 1989).

Yet bicycle frames and components quite often break in service from carrying one person alone. Spoke and axle breakage are perhaps the most common failures in which bicycles are involved. Aluminum-alloy handlebars all too frequently break off (three have done so for the author), as do handlebar stems and even cranks. In some makes and models of bicycles there have been a series of failures of the front forks, with often terrible consequences for the riders.[2] Steel-wire cables are often incorporated into bicycle-brake and gear-change mechanisms with appalling ignorance of manufacturers' specifications aimed at avoiding failures attributable to fatigue.

How can these single-rider in-service failures be reconciled with the large factors of safety built into the design of bicycles? There is ignorance among some designers about several vital matters: of the actual loads to which a bicycle is subjected; of the particular strength properties that govern eventual bicycle failures; and of the effects of notches, small-radius bends, or other defects in bicycle tubes and components.

Loading
Some real-world loads to which bicycles are subjected are far greater than the forces of steady pedaling along a smooth road. The greatest of these

loads arise from jarring bumps, with other large forces resulting from strenuous braking or pedaling. When a bicycle is subjected to such large service forces, its factor of safety is considerably smaller. The use of front- and rear-wheel sprung-and-damped suspensions in a bicycle greatly reduces the forces resulting from bumps it encounters. At the time of publication (2004), however, the overwhelming proportion of bicycles worldwide have unsprung suspensions, and most of the following remarks apply to them.

To define the actual service loads a bicycle endures is a difficult task. Some users will ride up and down curbs, possibly with friends sitting on the handlebars, the crossbar, or the carrier. Others will be unable to avoid riding over deep potholes in the road or may traverse jarring terrain at high speed. Still others will bolt heavy toolboxes to the frame or to the carrier. The degree to which such practices constitute use or abuse is different for different types of bicycle.

The principal information needed for a more analytical approach is a collection of loading data gathered from actual bicycle use by a wide variety of people over at least a year. (It will be argued below that the loads that typically bring about failures in bicycle components are the unusual stresses from, for instance, occasional impacts with potholes, rather than a buildup of material damage from smooth-road bicycling.) Later in this chapter, methods of measuring and analyzing the real-life loads and stresses that bicycles encounter are discussed. But even without such information, it is still possible to perform useful analysis: a good approach to evaluating the use of new materials and new, different configurations is to compare them with the strength and other characteristics of present successful designs. (We give later an example of this, contrasting high-strength aluminum and steel tubes.)

The tubes in a bicycle's frame usually experience, during riding, a combination of bending, shear, torsion, and tension or compression. Appropriate sizes for the frame's components have been arrived at by experience, not by analysis and prediction.[3] And even with advanced engineering software it would be difficult and expensive to analyze all the combined stresses that act on a bicycle frame and thereby to improve its design more than marginally. Given these circumstances, it is important to recognize that with regard to bicycles, at least, empirical development (namely, try it out, strengthen the parts that break, and lighten those that don't) is a valid practical approach to optimization. But hand in hand with that approach must go an awareness that innovation will inevitably bring failures.

Relevant material strength

Another reason for the seemingly large factors of safety built into bicycles is that the standards that are used to assess bicycle safety are frequently

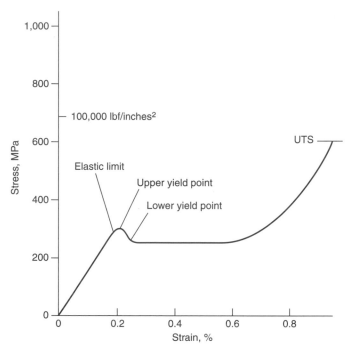

Figure 10.1
Yielding of steels.

inapplicable to or inappropriate for bicycles. For instance, the factor of safety for a bicycle is often based on the "ultimate tensile stress" (UTS) of the material used in constructing the bicycle, that is, the force that must be reached to pull a straight rod or tube made from that material into two pieces. Yet bicycles would collapse (probably without actual fracture) at a load considerably less than that which might cause their tubes to tear apart. Even the "yield stress" (YS) of a particular metal, that is, the level of stress above which the metal is permanently deformed, is considerably greater than the typical stress levels causing failures in bicycles (figure 10.1).

Most bicycle failures actually occur because of fatigue of the materials of which the bicycle is constructed: repeated stress variations creating small cracks, which then grow a little larger with each succeeding stress cycle. Such stresses are always less than UTS.

Fatigue

Metals and other materials "tire." We all know that we can break the soft steel wire in a paper clip by bending it so that it "takes a set" (i.e., exhibits plastic deformation) and then reversing the bend repeatedly. A failure brought about by this type of loading and involving anywhere from 2 to perhaps 50,000 cycles is called low-cycle fatigue (LCF). (Sometimes the plastic deformation is essentially invisible, occurring only at the root of a notch.) Materials also fail by being stressed up to millions or even hundreds of millions of times through somewhat lower stresses (incurring no plastic deformation) in so-called high-cycle fatigue (HCF).

By subjecting many specimens of a material to fully reversed cyclic stresses on special fatigue-testing machines, starting at a high stress level and lowering the stress level for each successive specimen, curves similar to those in figure 10.2 are obtained. The left-hand scale in the figure is that of the cyclic stress amplitude, and the horizontal scale measures the number of times the stress is applied. The stress level at the left of the graph is for one single application of load, and the value producing failure is the UTS. If a lower level of stress is applied, the material specimens can withstand many more stress reversals before failure occurs.

In some materials (most steels and some titanium alloys), a stress magnitude can usually be found below which most test specimens do not fail even after 10^8 cycles. We call this stress the "endurance" or "fatigue" limit. But this special "everlasting" behavior may be counteracted if the material is subjected to occasional overloads. In any event, endurance limit seems to have little bearing on practical bicycle durability.

Unfortunately, the seemingly simple concept of fatigue ramifies into a bewildering variety of special cases. (See Fuchs and Stephens 1980 for an extensive introduction. Other valuable references are the *SAE Fatigue Design Handbook* (SAE Fatigue Design and Evaluation Committee 1997), volume 19 of the *ASM Handbook* (*Fatigue and Fracture*) (ASM International 1996), and Collins 1981.) Here we merely touch on some of the most relevant concepts.

Fatigue behavior largely defines bicycle durability, but it would be wrong to assume that a satisfactory design must keep stresses below the endurance limit. Large loads (from impact, start-up, or swerving), although relatively infrequent, are *far more severe* than the normal loads encountered in general smooth pedaling, which therefore are usually less than the endurance limit and irrelevant to failure. The main concern in developing a design that is satisfactory in its ability to withstand stresses is to achieve a long-enough life, that is, for the bicycle to survive enough hard bumps that it lasts for, say, a decade or two of use. For most users this would entail at most a few hundred or a thousand really hard bumps. Redesigning the same bicycle to last through millions of such bumps would more than double its weight without serving much useful purpose.

Fatigue is best understood with respect to the simple scenario of fully reversed fixed-amplitude loading cycles, possibly superposed on a fixed

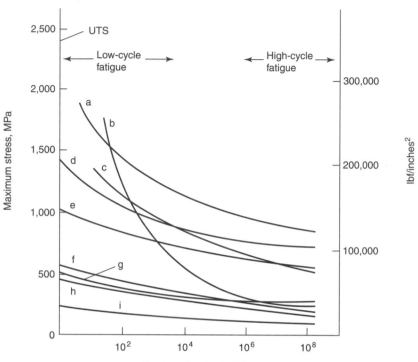

Figure 10.2
Endurance limits of various materials: (a) Kevlar + epoxy (curve for boron + epoxy is similar); (b) "S" glass and epoxy; (c) graphite and epoxy; (d) 4130 chromium-molybdenum-alloy steel; (e) titanium IMI 318 alloy; (f) 7075-T6 aluminum alloy; (g) medium-carbon steel; (h) 2024-T6 aluminum alloy; (i) magnesium. The HCF ends of curves (d) and (g) show the endurance limits for these two steels.

"mean" stress level. However, real-world loadings include the occasional crash, the infinitude of almost imperceptible road vibrations, and everything in between. (The range of loadings a material experiences is sometimes described as a "loading spectrum," in which the amplitude of the various loads that the material encounters is plotted against frequency of occurrence.) Whenever failure occurs, it is invaluable to ascertain whether the cause was a few (up to a thousand) very high forces (LCF) or many (even millions) of medium forces (HCF). A skilled failure analyst can often distinguish these cases by microscopic inspection of the fracture surface (ASM International 2001, 2002). Making the distinction is important, because HCF and LCF failures are best ameliorated by somewhat different strategies.

Bicycle failure through low-cycle fatigue

The available evidence very clearly suggests that bicycle failures are predominantly due to LCF. For example, a spoke in a bicycle wheel is subjected, in every revolution of the wheel, to a brief, moderate reduction in stress by an amount based on rider weight. If spoke failure occurred after a bicycle was ridden on a perfectly smooth road for 20,000 km (giving about ten million cycles), it would be a classic case of HCF, and a 10 percent reduction in rider weight would assure infinite life. But in actuality, spokes also suffer occasional more drastic *stress reductions* (including total loosening at a bump), plus even rarer substantial *stress increases* from stand-up pedaling, off-center bump strikes, swerves, and the like. These cycles of three times or greater magnitude clearly raise the possibility of LCF. But how can responsibility be assigned to either LCF or HCF?

The answer is found in the relative magnitude of the stress variation of smooth riding, versus the stress variation of the occasional large load. For many fatigue curves for bicycle spokes, if the very largest stress variation to which the spokes are subjected does not exceed the UTS and cause instant breakage, it follows immediately that a stress variation of less than half this value will allow a life exceeding 100 million stress cycles and may essentially be discounted. In our case, the factor of $3\times$ suggests that the typical smooth-road load is essentially irrelevant to spoke life.[4] Breakage at ten million wheel revolutions would not exemplify a smooth-rolling-based stress variation slightly above the endurance limit, but rather some number (hundreds, or perhaps thousands) of much rarer severe loadings. Then a 10 percent decrease in rider weight wouldn't eliminate failure: it might just extend life several-fold.[5] (According to Jobst Brandt's [2000] personal experience, it is possible for spokes to last for many more than ten million revolutions. This observation implies a reduction, through a gentle riding style, of the number of high loads encountered but doesn't in itself rule out the slow occurrence of HCF or LCF.)

The foregoing picture is muddied by the presence of stress concentrations (notches, cracks, or pressure points; see below) that may lower a material's endurance limit to as little as 15 percent of the UTS, or even less if extremely sharp notches are present in superstrong (= brittle) materials. But the same principle remains widely applicable. For example, a peak pedaling force of two times body weight[6] can easily be ten or fifteen times the typical level-road pedaling forces of 0.1–0.2 times body weight. Since the occasional peak load doesn't break or even bend the pedal or crank, the level-road pedaling forces are below any possible endurance limit and may thus be discounted in the calculation of fatigue. Landing from a jump could well impose forces as high as five times body weight (or even more) on a single pedal.

In summary, apart from catastrophic collisions or crashes, most bicycle failures can be attributed to fatigue. And except for cases of very smooth riding style and conditions, most bicycle fatigue failures fall into the classification of LCF. They are caused by just a few hundred or thousand bumps, sprints, etc.: loading events that normally take place just a few times per day.

One consequence of this is that aggressively used bicycles can be expected to have a finite life. This could be a few months for one person or a few decades for another. Many riders would prefer a lightweight bicycle meant to last only a few years to pushing around something double the weight and meant to endure for millennia. This perspective has largely remained unacknowledged by manufacturers, which sidestep the recognition of finite life in the frequently satisfied hope that a rider will lose interest in a given bicycle before it has accumulated much damage.

Another consequence is that the choice of optimum material for bicycle construction (and of design that can best use a particular material) is a subtler matter than merely choosing an adequate endurance limit. If a tube or component is designed to be stressed uniformly, the material of greatest UTS will probably support a given varying load with the least weight. On the other hand, if stresses are elevated in a small region such as a notch (see below), then the highest loads will cause considerable straining of the material in which the notch occurs. In this case the greater ductility of a medium-strength metal may offer a considerable life advantage compared with the highest-strength materials (Kern and Suess 1975) (see also Collins's (1981, 387) comment on "fracture ductility."

A useful choice for the maximum stress in a particular material is that stress that causes failure after 3,000–6,000 reversals.

Stress raisers

A final reason that bicycles seem to require huge factors of safety is "stress raisers" or "stress concentrations." Holes, cracks, and sharp notches or "inside corners" in a bicycle's frame all raise the stress that the bicycle undergoes to considerably above the levels calculated for the adjacent material (by factors of three, six, or more). In a brittle material these stress raisers are usually responsible for sudden, dangerous failure. Even in a tougher material they can sometimes induce brittle failure when there is an overload. More to the point, they usually reduce a material's fatigue strength considerably.

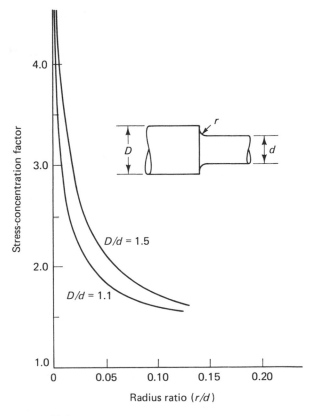

Figure 10.3
Stress-concentration factors. (From Peterson 1974.)

As an example, figure 10.3 shows the stress-concentration factors (taken from Peterson 1974) for the bending of a solid rod that has a sharp change in diameter. (The figure is also applicable to the case of a solid-titanium bottom-bracket spindle, which fatigued at the sharp shoulders provided to locate the high-hardness steel-bearing races.) The theoretical maximum stresses that the rod can sustain are related primarily to the inverse square root of the radius of curvature at the junction of the two diameters (See McClintock and Argon 1966, equation 11.35), and secondarily to the ratio of the diameters of the change section incross. A series of disastrous failures in a certain make of front fork just above the fork crown was traced to incorrect heat treatment coupled with a sharper-than-specified radius of curvature, which increased local stresses.

Changes of section should therefore be gradual. High-quality frame lugs are tapered, filed, and cut with decorative patterns to transfer stress

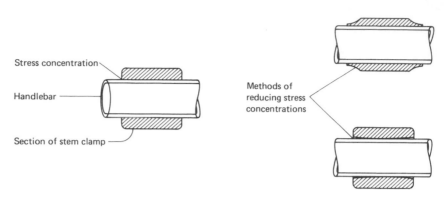

Figure 10.4
Stresses at joints.

gradually from tube to lug and then to a connecting tube. Likewise, the walls of high-quality frame tubes are gradually tapered to provide greater thickness at the ends, where the stresses from bending are highest.

Inexpensive handlebars sometimes break because their manufacturers have not taken into account the serious weakening effect of the clamp that joins them to the bicycle's stem. At the point where the handlebars are clamped into the stem, the bending moment from the forces exerted by the rider's arms at the bar ends will be at a maximum. If the clamp on the handlebar stem fits well, it acts something like a sudden change in wall thickness. However, instead of drastically increasing stresses, the clamp mostly induces microscopic slip at its edge. This leads to so-called fretting damage, which significantly lowers the handlebars' fatigue strength. Two alternative methods of reducing this problem are shown in figure 10.4. (An external sleeve is common on handlebars for competition.)

Toughness versus brittle behavior

LCF strength, elastic modulus, and density are probably the most important physical properties for bicycle materials. But only a little less significant are properties of *toughness* and *ductility*.

Bicycles, used dynamically in all kinds of situations, will always be susceptible to unforeseen high stresses. Whether these stresses arise from unexpected bumps, pedal strikes, overtightened clamps, scratches, hidden stress raisers, errant hammer blows, or even fatigue cracks, it is prudent to demand of a bicycle's design that (1) overload should never result in the rider's being thrown to the ground (i.e., the bicycle should not lose all its strength) and (2) any damage sustained because of exposure to stresses should show up as a visible deformation, giving some warning of incipient failure. A glass bicycle, because it fulfills neither of these requirements, would be extremely dangerous.

There are several important aspects to "forgiving" behavior, and we will merely touch upon them here. The first is a material's ductility (i.e., its ability to "yield," to be stretched permanently, without weakening it). The ductility of a material is measured by the permanent "set" a rod made of the material can accept when it is pulled in a tensile-testing machine until it nearly breaks. A brittle material like glass has no ductility, whereas a soft material like lead will stretch considerably before failing. To put things in perspective, strong materials usually approach their breaking stress when stretched elastically between 0.3 and 0.7 percent. The ability to yield generally permits a material to endure an additional ten to one hundred times this amount of stretch, without much stress increase, and without actually breaking.

Most lower-strength metals exhibit considerable ability to yield. This has permitted the vast majority of metal structures to succeed, even in the face of impact, earthquake, imprecise hole drilling such that only one of a set of bolts initially carries a load, or welding that creates high shrinkage stresses. The main result of a material's ability to yield is that stresses do not immediately build up to exceed the UTS: rather, the stressed material "gives" and permits nearby material to shoulder some of the burden.

For one example, it is common experience that some plastic films and some synthetic-rubber inner tubes will tear easily once a small cut has been made in them, whereas other films of approximately equal strength and thickness will "blunt" a cut and greatly resist tearing. For another, glass can never be firmly bolted to another component, nor is it feasible to make practical bolts out of glass. In fact, glass is cut by making a shallow scratch, which creates an extremely high-stress "breaking line" (a stress raiser or notch). (A ductile steel plate, on the other hand, is essentially unaffected by such scratches.) It happens that a few brittle materials, like grey cast iron, are fortunately not sensitive to scratches or notches. However, this behavior remains the exception rather than the rule, and they still will snap rather than bend when overloaded.

The ductile yielding of a component is often visible, especially in bending. For instance, a front fork that has been stressed beyond the yield point may show cracking or flaking of the brittle enamel.

Not only materials, but structures themselves, can be "brittle," if poorly shaped. For example, narrowing a very short section in an aluminum bar will weaken it sufficiently that any attempt to bend the bar will simply concentrate deformation on the slender section. With very little overall bending, material in the slender section will stretch and then fail, and the bar itself will seem brittle. (This is the principle behind perforations in toilet paper.) If a bar or chain must have a particular "weak spot," it is better to weaken the rest similarly, in order to prevent such brittle behavior.

If a structure must have a deep notch, or if it risks someday developing a crack, additional properties of the material of which it is constructed come into play. Even though most of any deformation in the structure will be concentrated in the notch region, some materials have such excellent cohesion and ability to harden that they are able to force nearby regions to deform also, before finally breaking. This property is termed "notch toughness." It is measured by striking a notched bar with a pendulum hammer to measure the energy absorbed.

In summary, ductile materials appear to overcome the dangers of brittleness. But their effectiveness is very much shape- and circumstance-dependent. In the case of a single overload to failure, combinations of notch depth and sharpness, speed of loading and possibly low temperature can suppress ductility where it is most needed, thus giving a brittle-appearing fracture. Fortunately, some materials can be designated notch tough, as determined by energy absorbed in a notch impact test, and they undergo dramatically more plastic flow near a notch, that is, with considerable bending before breakage.

Notch sensitivity

As stated above, glass and most other brittle materials are considerably weakened by shallow scratches, but as a rule ductile materials are not. However, this immunity to being weakened by scratches does not apply in fatigue. Fortunately, although a sharp scratch in a material may theoretically elevate the stress by a concentration factor of, say, 5.0, it may be observed that the apparent fatigue strength of the material is not necessarily reduced by the same factor of 5.0. The discrepancy is described by means of a factor defining notch-sensitivity in fatigue, given by Farag (1989) as the following ratio:

$$\frac{\left[\dfrac{\textit{fatigue strength without a stress concentration}}{\textit{fatigue strength with a stress concentration}} - 1\right]}{\textit{theoretical stress-concentration factor} - 1}.$$

Fortunately, it has been found that a ductile material is not very notch-sensitive in fatigue to scratches that are shallower than the material's grain size. Furthermore, the sensitivity of ductile materials in LCF is somewhat less than in HCF.

The distinctions among these various definitions of toughness are probably too fine for most people concerned with bicycle design and manufacture. Should the use of a new material in a bicycle frame or components be contemplated, its notch and/or fracture toughness should be compared with that of bicycle-frame steel before development work is undertaken.

Testing

It is our understanding that expert application of engineering methods has played very little part in bicycle design.[7] It is far more common in bicycle design to establish an empirical test that prior successful bicycles barely pass and require new designs to pass it as well. Some low-volume innovators do not test even in this way, and some in-service failures are the likely consequence of their neglect of testing. When bicycles begin to be used in new ways (e.g., for down-mountain racing), the dangers involved in riding them increase substantially, because there is no experience on which to build.

Relatively low-level pass-fail tests have been developed over the years to ensure that a bicycle will be safe at least for ordinary, relatively gentle use. These are embodied in U.S. regulations (16 CFR 1512 [Code of Federal Regulations on Hazardous Substances, administered by the Consumer Product Safety Commission]; search ⟨www.access.gpo.gov/nara/cfr/index. html⟩), in Japanese (JIS D9301) and other foreign national standards (DIN 79100; ⟨www.tbnet.org.tw/stander/JIS.HTM⟩ is a Taiwanese compilation), and in ISO standards (ISO 4210; see also ⟨www.iso.ch⟩). The requirements imposed by such regulations form the basis for product recalls and for assigning culpability when a nonconforming product causes injury. Because of the investment required to build the proper fixtures for testing products, testing companies (e.g., SGS U.S. Testing in Fairfield, New Jersey; ⟨www.ustesting.sgsna.com/page_b.html⟩) have found some customers willing to pay to have tests carried out. In addition, proprietary tests that are much more strenuous than those required by the regulations have been developed by companies gambling that a need for them will be recognized (for example, the EFBe testing company ⟨www.efbe.de/e1servic.htm⟩ and the VELOTECH testing company ⟨www.velotech.de/englische_seiten/ about_velotech.htm⟩).

The advent of off-road riding and racing has made stronger bicycle frames and components necessary. Somewhat more stringent standards for bicycle construction are under development, but the process is slow. And not every participant in the writing of the new standards agrees that more demanding tests than those currently employed are desirable (in effect, they are concerned that the results may confuse the casual user or could harm the perception of their own products).

Although it may take a long time to win acceptance, we strongly favor more informative and more stringent testing, firmly grounded in accepted engineering practice and an understanding of service requirements.

· Tests should provide *quantitative information about a bicycle's strength or toughness* (not just a pass-fail assessment) or at least should locate the result within an ascending series of "duty levels."

· Both full-bicycle and individual-component tests are needed. Inasmuch as an inappropriate handlebar clamp can be just as responsible for handlebar failure as the handlebar itself, thought must be given to damage caused by a mating component. A handlebar should be shown to have adequate strength in a "standard stem clamp." Similarly, a stem should be shown not to lower the strength of a "standard handlebar." As an exception, special handlebar–stem combinations could be tested together *only*, but then each component would have to be labeled as unsuited for other mates.

· Fatigue tests should focus on LCF so that everyone will know that bicycle life is expected to be finite. Experience would teach sellers and users what duty level is usually appropriate for a given weight of rider and type and amount of riding. Any service failures experienced would clearly indicate that the rider must choose equipment rated for a higher duty level than the equipment that failed.

· Static failure tests should determine what load can be supported with no permanent set. Static yield strength is not an accurate indicator of durability, but it is simple to measure and provides a useful design target.

· Some kind of "energy" or "retained strength" test should be performed on parts that are commonly bent in crashes: pedals, cranks, handlebars, and forks. Either the terminal strength or dissipated energy after imposing a typical deformation of (say) 50 mm can be determined; or for a pass-fail evaluation, an energy of (say) 50 mm required strength can be imposed via a drop weight.

Much like tire mileage ratings, the information obtained from such testing would allow riders to make more rational choices about the durability of what they buy. Such a testing scheme as the one outlined above need not interfere with existing requirements, which would simply fall within one of the lower duty levels.

Setting design goals

In setting goals for an improved bicycle, there is a temptation to say "the stronger the better ... and the lighter the better!" Here we offer a brief critique of those obvious-sounding goals.

1. The goals of strength and stiffness are somewhat in competition with that of lightness. If tube diameters may not be increased, then all three properties are proportional to the thickness of the tube wall. Only by increasing tube diameter substantially can stiffness and strength increase substantially while weight decreases. There's a practical limit to this, based on what fits between a rider's legs and on danger of a thinned tube's denting or crumpling as does an aluminum beverage can (which is a low-energy failure mode).

2.   The goal of increased strength or stiffness is somewhat in conflict with elastic compliance for comfort over bumps. Actually the main arena for this competition is the handlebars, and to a lesser extent the seat, cranks, and fork; the rest of the frame plays no role in softening bumps. There is probably value in trying to increase the handlebars' torsional stiffness while maintaining or decreasing vertical stiffness.

3.   There is a potential danger in unbalanced strengthening: overall bicycle ductility may degrade. For example, in a frontal impact both the frame and fork may deform a total of 80 mm. If the same frame is strengthened somewhat, when tested alone, it may also absorb more energy than in its unstrengthened form. But when the two parts are tested together, the weaker fork alone will have to take up the entire deformation (no energy into the frame), which may cause it to break in two. It is therefore important to test the energy absorption of the entire assembly.

4.   Interest in lower bicycle weight is never-ending, which makes bicycle weight a hugely effective marketing tool for sellers and a satisfying bragging point for consumers. However, the value of weight reduction in increasing speed seems to be overstated. Even on the steepest mountain roads, adding 1 kg of mass to a bicycle (say, a 1–2 percent increase in its overall mass) will make only a slight difference to the climbing time: say, thirty seconds or so out of an hour-long climb. This is rarely enough of a difference to catapult a typical contestant onto the winner's podium. And the expected speed difference on the level or lag developed in a sprint is just about unmeasurable. From this perspective, only those who are already good enough to place in races have justification for weight shaving. The vast majority of us never even attempt long uphill rides, and a conventional 12-kg machine should serve well, even in most competition.

5.   There is also a widespread conviction that bicycle strength or stiffness enhances either power production or propulsion. However, there is no theoretical reason why this should be so, at least for stiffness increases beyond the current level. Furthermore, we are aware of no experimental demonstration of this point. (To the contrary, top riders seem to succeed on relatively flexible, lightweight bicycles.)

David Malicky (1987) performed a double-blind test of stiffness perception, using bicycles with frame tubing of relatively high and relatively low wall thickness (with mass added so the bicycles would weigh the same). The test population of racers could not perceive any difference in stiffness among the bicycles.

Much of the bicycle flex in high-force pedaling occurs not in the frame, but in the handlebars and crankset. The most significant potential increases in stiffness involve bracing those components more effectively.

The upshot of this discussion is that, compared to current (2003) sport bicycles, reduced weight or enhanced stiffness theoretically should offer virtually no performance advantages and may not even be detectable by the rider.

Other material properties and criteria for choice

Being strong enough to endure in-service stresses and tough enough to absorb damage without disintegration and to shrug off effects of stress raisers are necessary but not sufficient conditions for considering a material suitable for bicycle construction. Some other requirements are the following.

- The density of the material must be such that the resulting structure is light (but not necessarily "ultralight").
- The resulting structure should not be unduly flexible. (The property defining material flexibility in table 10.1 is the elastic or Young's modulus (E). The usual engineering metals have virtually identical ratios of modulus to density.)
- The cost of both tube fabrication and tube joining must be reasonable.
- Joining one piece to another should be possible with minimum loss of strength in the parent material(s) or in the joint.
- The material should intrinsically resist, or should be easily protected from, corrosion.

Joining properties

The ends of a frame member, where the bending moments are normally highest, are also the points at which the member must be joined to other members. The means for joining one member to another must therefore both preserve the tube's strength and even surpass it in the joint itself. As mentioned above, high-quality, lightweight steel frames use tubular angled sockets, called "lugs," into which the frame tubes are brazed. Low-alloy steel of medium strength (800 MPa) can be brazed with regular brass or bronze brazing alloy at temperatures up to 950°C, but some higher-strength steels (1300 MPa) require the use of silver-alloy brazing solders (no hotter than 650°C) to minimize thermal degradation of tube strength.

Welding, in contrast to brazing, melts a small quantity of the material into the joint, both from the parent metal and, when one is used, from a filler rod. All heat treatment in the weld region is disturbed, and in addition, the constituents of the alloy that results from the weld are somewhat uncertain. The shrinkage of the solidifying and cooling metal will introduce thermal stresses. Furthermore, improper welding technique may lead to significant invisible flaws in the finished weld. Welded aluminum frames must be heat treated, first to relieve these stresses and then to restore most of the original properties to the metal. The endurance limits (measured at 6,000 cycles) for alloys 2024 (Duralumin) and 6061, popular for bicycle frames and components, are reduced by over 35 percent when the properties enhanced by heat treatment are diminished.

Components of aircraft wings and parts of fuselages have been glued together with high-strength adhesives in highly controlled circumstances. With suitable close-fitting lugs, this procedure is also satisfactory for aluminum bicycle frames, with no degradation of properties, no thermal stresses, and considerable savings of time, costs, and energy. Some high-quality frames are made in this way; the tubes and the lugs are sometimes threaded for added reliability.

A combination of adhesives and lugs, sometimes involving compression, has also been used for the fiber-composite frames tried out recently by some frame makers (for example, the experimental frame shown in figure 10.5).

The frames of several human-powered aircraft have been constructed of carbon-fiber composite tubing. The joints have been made by wrapping adjacent butting tubes with "prepreg" (resin-impregnated) tape and then thermally curing it. The wheel shown in figure 10.6 use spokes of Zylon polybenzoxazole (PBO) fiber, which has the outstanding properties shown in table 10.1.

Corrosion resistance

Nonferrous metals and plastics are more resistant than steel to atmospheric corrosion. The surface treatments necessary to ensure satisfactory service of these materials are minor operations compared with the plating or enameling processes required for steels. On this account, the use of these materials for the less stressed parts of bicycles has been generally satisfactory and will no doubt be extended in various ways. Some organic fibers (e.g., Zylon) degrade under ultraviolet light and therefore need a protective coating or covering when they will be exposed to sunlight.

Cost

At present, low-cost steel (easily shaped and joined) is the least expensive material for making a bicycle. It is possible that high-strength fiber-reinforced plastics may eventually win a place among the materials for

Table 10.1
Properties of some materials used in bicycle construction

Material	Modulus of elasticity, E (GPa)	Ultimate tensile strength, UTS (MPa)	Elongation at failure (%)	Fracture toughness, K_{IC} (MPa-m$^{0.5}$)	Fatigue limit/UTS (5×10^8 cycles)	Density (Mg/m^3) (specific gravity)
Steels						
Medium-carbon	200	520	26	51–54	0.5	7.85
AISI 4130 (CrMo)	200	1,425	12	?	0.5	7.85
Columbus Record, Super Vitus	200	835	?	?	0.5?	7.85
Reynolds 525 (CrMo)	200	700–900	>10	?	0.5?	7.85
Reynolds 531	200	700–850	>10	?	0.5?	7.85
Reynolds 631/853	200	1,250–1,450	>10	?	0.5?	7.85
Reynolds 725	200	1,080–1,280	>8	?	0.5?	7.85
Reynolds 753	200	1,160	>10	?	0.5?	7.85
Tange Champion Pro 1,2,3	200	893	>10	?	0.5?	7.85
301 stainless steel, annealed	193–214	515–758	40–60	100	0.47	8.03
301 stainless steel, "full hard"	174	1,200–1,275	8–9	?	0.43	8.03
Aluminum alloys						
2024-T4 (Duralumin)	73.1	470	20	24–44	0.29	2.80
6061-T6	68.9	310	12	23–45	0.31	2.80
7075-T6	71.7	570	11	?	0.28	2.80
Magnesium	44	248	5–8	?	0.37	1.79
Titanium alloys						
IMI 125 (pure)	105–120	390–540	20–29	55	0.5	4.51
IMI 318	105–120	1,000 (approx.)	8	55–115	0.55	4.42

Composites (fiber direction)						
"S"-glass-epoxy	90	1,500	3.5	42–60	0.16	2.63
HT-graphite-epoxy	200–300	3,600?	1.25	32–45	0.25?	1.75
Boron-aluminum	262?	1,025	0.65	?	0.7	2.60
Boron-epoxy	207?	1,200		46	0.80?	1.99
Kevlar-49-resin	70–80	1,380	2.75	10?	0.70	1.37
DuPont Zytel FE 8018 NC-10 (glass-nylon)	2.3	59.9	14	?	?	1.18
Fibers						
Carbon-graphite	230	1,700	1.5			
Toyobo Zylon (PBO)	280	2,700	2.5			
Kevlar-49	109	1,400	2.4			
Nylon-46	2.2	66.8–100	65	3	?	1.2
Polyester	3.2	85	20–25	?	?	1.4
Woods (e.g., ash, beech, oak)	9–16	35–55	11–13			0.67

Sources: ⟨www.matls.com⟩; manufacturers' data; fracture toughness mostly from Ashby and Jones 1980. The properties of composites are lower, sometimes much lower, in compression, and in tension in directions across the fiber alignment. Properties vary greatly in dependence on the materials used and on the mode of failure. ? indicates unknown, uncertain, or scattered data.

Figure 10.5
Frame of composite tubes glued in metal lugs.

Figure 10.6
Wheels with Zylon (PBO) fiber. (Courtesy of Spinergy, Inc.)

bicycle construction because of the automated production that these materials allow in the form of fiber-filled sheet molding compound (SMC), commonly employed to make car-body panels. An alternative automated approach, though one requiring far larger presses, is to produce the frame as two sheet-metal halves and resistance-weld them together, as for a light motorcycle.

Example: calculation of use of aluminum alloys versus steel

Useful charts comparing one material with another have been produced, especially, perhaps, those by Ashby (1992). For instance, Ashby shows how classes of engineering materials can be ranked when their properties have been plotted on a graph of fracture toughness versus density, two properties of importance to bicycle-frame performance, or modulus versus strength. Ashby warns readers that when materials are being compared on the basis of only two properties, the problem the comparison is intended to solve has usually been oversimplified. Nevertheless, one can often determine that two materials are satisfactory in all other relevant respects and that one can therefore legitimately restrict the focus of one's comparison to two properties. As an exercise, we shall compare the characteristics of bicycle frames in respect of weight and stiffness when they are designed to the same LCF life.

We are fortunate to have examples of successful components such as steel-tube diamond-pattern frames before us, and we can simply compare with them the size and mass of the same components produced in alternative materials, such as aluminum alloy, to give the same performance.

Let us suppose that we wish to compare the weight of an aluminum-alloy frame designed to have the same strength and stiffness as a steel frame. Although we know that the loading of a bicycle frame can be complex, we choose simple bending as the loading method used for comparison because it will serve well when we are just comparing one material with another. Both frames will be constructed from circular tubes. (We will ignore for the moment the question of joining the tubes.)

Any standard engineering reference book, such as *Marks' Handbook* (Baumeister 1978), will give the stiffness (force per unit deflection) of a cantilevered beam as

$$\frac{3EI}{L^3},$$

where E is the modulus of elasticity, I is the section moment of inertia, and L is the length of the beam. For a circular tube,

$$I = \frac{\pi D^4}{64} [1 - (d/D)^4],$$

where D is the outside diameter of the tubing and d is the inside diameter.

For steel and aluminum tubes of the same length to have the same stiffness, the product ED^4 must be the same for both materials if the diameter ratio d/D is, at least for the moment, specified as identical for the two materials. Table 10.1 shows that the modulus of elasticity of aluminum alloy is about one-third that of steel. Thus, the tube diameter of the aluminum-alloy frame must be $3^{0.25} = 1.316$ times that of the steel frame.

Now we must ask this question: if the aluminum-alloy frame is as stiff as the steel frame, will it be safe from fatigue failure? The maximum stress in a circular tube for a specified load and tube length is given by the relation

$$\frac{maximum\ stress}{load} = \frac{32L}{\pi D^3 [1 - (d/D)^4]}.$$

Therefore, the maximum stress in an aluminum frame of equal stiffness to a steel frame is $1.316^{-3} = 0.439$ of the peak stress in a steel frame (again, for the same ratio of inside to outside diameter).

The fatigue-limit stress in the strongest of the three aluminum alloys listed in table 10.1, 7075-T6, is $0.278 \times 570 = 159$ MPa. The fatigue-limit stress in the steel-alloy frame is $0.5 \times 1{,}425 = 712.5$ MPa. Therefore, the ratio of the fatigue-limit stresses is 0.223, which is much less than the ratio of the peak stresses (0.439), and the aluminum-alloy frame will be much more highly stressed (perhaps dangerously so). However, we have used the fatigue-limit stresses as "surrogates" for the stresses that could be accepted for far lower stress cycles, in LCF. Such use presumes that the shapes of the S-N fatigue-failure curves for steel and alloy 7075-T6 are similar. They are not! The 7075 curve rises much faster than that of the steel (figure 10.2), so that the degree to which the aluminum-alloy frame will be more highly stressed relative to the LCF limit will be less than the above numbers would indicate. Our lack of knowledge of typical LCF loadings of different bicycle types in different uses makes this an area of great uncertainty. We will proceed with the conservative approach of using HCF, or fatigue-limit, data.

The weight of the two frames would be proportional to ρD^2, where ρ is the density:

$$\frac{weight\ of\ aluminum\text{-}alloy\ frame}{weight\ of\ steel\text{-}alloy\ frame} = \frac{(\rho D^2)_{aluminum}}{(\rho D^2)_{steel}} = \frac{2.80}{7.85}(3^{0.25})^2 = 0.618,$$

giving a substantial advantage to the aluminum-alloy frame.

An alternative method is to design the aluminum-alloy frame to have peak stresses that are the same proportion of the fatigue-limit stresses as for the steel-alloy frame and then compare the results. Then,

$$\frac{D_{\text{aluminum}}}{D_{\text{steel}}} = \left(\frac{712.5}{159}\right)^{1/3} = 1.649$$

The ratio of the weights of frames made from the two materials would then be

$$\frac{weight\ of\ aluminum\text{-}alloy\ frame}{weight\ of\ steel\text{-}alloy\ frame} = \frac{2.80}{7.85}(1.649)^2 = 0.97.$$

Therefore, fortuitously, the weights of the two frames have turned out to be virtually identical when the frames are designed for the same proportional fatigue-limit stresses. The aluminum-alloy frame would, however, be much stiffer. (An aluminum-alloy frame of a road-racing bicycle is shown in figure 10.7.) Bicycle designers therefore have some freedom to trade off among stiffness, stress, and weight by changing not only the diameter of the tubes used in constructing the bicycle's frame but also the ratio of inside to outside diameter or tube thickness. In the comparison above, the aluminum tubes would be 65 percent larger in diameter than those in the steel frame, and the thickness would be larger by a similar amount. A track bicycle,

Figure 10.7
Aluminum-alloy road-racing-bicycle frame. (Courtesy Trek.)

which will be exposed to few bumps and potholes, could well be designed to an equal-stiffness criterion in aluminum rather than steel. It would seem unsafe to do so for touring bicycles, which often are loaded with heavy bags and travel on rough streets, conditions that would build up fatigue damage.

This illustration was meant to provide simply an example of how to use material-property data in bicycle design. The important principles presented in the illustration are to use fatigue-stress limits (ideally LCF at 3,000–6,000 cycles) rather than UTS, to consider stiffness as well as strength, and to take successful components as models of stiffness and strength because of the great uncertainty in the actual magnitude, type, and frequency of loads the bicycle will carry.

Nonmetallic components

Plastics and composites (fiber-reinforced resins) are now competitive with metal components in such demanding applications as airplane structures, racing and sports cars, and chemical plants, and they have made substantial inroads, rather surprisingly, in opposite poles of bicycle production. On the one hand, the frames and the wheels of the lightest racing bicycles, especially those designed for world records, are now commonly made from carbon-fiber-reinforced polymers or Kevlar-fiber-reinforced polymers; on the other hand, the lowest-priced children's tricycles are often made by blow molding or by injection molding, using unreinforced polymers able to flow into simple molds.

These low-priced bicycles (and tricycles) for children usually employ plastic bearings, which must be made with larger clearances than plain metal bearings; that is, the fit is "sloppier." As discussed in chapter 6, good-quality plastic wheel bearings would not noticeably slow a bicycle. However, manufacturers appear to have realized that adult cyclists will not accept plain bearings of plastic. (There is one exception: most manufacturers produce the lightly loaded "jockey pulleys" in most derailleur gears with plastic bearings. A "cottage industry" of manufacturers of replacement pulleys with sealed ball bearings has developed in response. The author's experience is that the life of these replacement pulleys is still nasty, brutish, and short.)

The use of toothed belts of reinforced rubber together with wide-range multispeed hub gears, which was discussed in the last chapter, has considerable attractions. Nylon derivatives are successfully used for motorcycle rear-wheel sprockets and could perhaps be used for bicycle sprockets.

Glass-reinforced nylon is being used commercially for wheels for BMX off-road bikes (figure 10.8). The higher weight and lower stiffness of these wheels would make them unattractive for road or track racing. How-

Figure 10.8
Glass-reinforced-nylon (Zytel) wheels. (Courtesy E. I. dePont de Nemours & Co.)

ever, a rather astonishing range of composite wheels are used for such races (e.g., figure 10.6), with the wheels providing the advantages of lighter weight and lower aerodynamic drag than metal-spoked wheels to such an extent that racing regulations prohibiting their use are continually being formulated, applied, changed, and sometimes abandoned.

Alternative frame materials

Wood
Bicycles with wooden frames (see figure 10.9) have been made and ridden with satisfaction at regular intervals since the earliest "hobbyhorse" days around 1817. In the 1870s metal construction became dominant, but there were regular revivals of wood frames (including some of bamboo) until the end of the nineteenth century. The Stanley shows in the United Kingdom of this period included bicycles with completely wooden wheels fitted with pneumatic tires; an early Columbia with such wheels is on display at the science museum in London. Various wooden-framed bicycles dating back to the 1890s are still ridden by proud owners in veteran-cycle rallies. Although wood was used regularly up until the 1930s for wheel rims (for

Hand-Crafted Wooden Italian Bicycle.
Like the legendary mahogany Chris Craft motorboats and the wood-paneled station wagons of the 1940s, the Gianni Hand-Crafted Wooden Italian Bicycle—a Hammacher Schlemmer exclusive—was designed and built to apply wood's superior structural qualities to an entirely new mode of prestige transportation. In fact, this touring bicycle, with its main frame constructed of sturdy oak, has been hailed by competitors in the Tour de France for its unusual design and solid ride. The seat is made of cherry wood, and the elliptically-curved spokes and wheel rims have been formed from laminated marine-quality okoume (a tropical softwood), giving the wheels extra flexibility, aiding in water resistance, and reducing aerodynamic resistance. The steel and chrome gears and disk brakes are made in Japan by Shimano, a name noted for excellence in the cycling world. The 7-speed bicycle has pneumatic tires, a front reflector and a battery-operated tail-light.

Figure 10.9
"Hand-crafted wooden Italian bicycle" in Hammacher-Schlemmer catalog, 2001.

both sew-up and clincher tires), and wooden mudguards and seat pillars were not unknown, the wooden frame did not appear again until the 1940s, when metal needed to be conserved for use in World War II. However, in the United Kingdom at least, wood became scarcer during the war than steel. Wood is structurally similar to a fiber-reinforced plastic with all the fibers in one direction (e.g., a "pultruded" fiberglass rod). In that "strong" direction, such desirable woods as dry Douglas fir exhibit as good a strength-to-weight ratio and modulus-to-weight ratio as a high-strength steel.

This justifies the use of wood in the highly loaded wing spars of many airplanes. Indeed, whenever an I-beam or tube construction is selected to carry tension and bending only, wood is a fine choice.[8] Unfortunately, the tubes that make up bicycle frames are also subjected to torsion, and with no helical fibers, a wood rod or tube would be absolutely unacceptable as regards strength or stiffness. And as a final disadvantage, wood damaged in a crash can present dangerously sharp fractured ends.

Molded plastics
Since the recent advent of relatively large moldings in plastics (sometimes reinforced), there have been several attempts to market molded bicycle frames. These bicycles have generally been bulkier in appearance and more flexible than steel-framed bicycles. As new polymers and polymer-fiber combinations and improved manufacturing methods are developed, mass-produced composite frames will become less bulky, lighter, and stiffer. There are certainly advantages for general everyday use to a frame made of an inexpensive material that is completely resistant to corrosion.

Composite tubes

While early fiber-reinforced composites earned reputations as dangerously brittle and highly susceptible to fatigue, continued development has propelled them into the forefront of engineering materials (see ASM International 2002). (Their main downfall is cost, inasmuch as their production requires skilled hand labor and long curing times.) They have superb properties, but only if fiber surface cleanliness and chemistry are carefully controlled, fiber density is maximized and voids are minimized, and curing reactions take place consistently. If these things can be done, carbon-fiber-reinforced plastics demonstrate endurance limits close to the UTS. In addition, composites have fracture toughness similar to that of steel (see Ashby 1992), can give warning of overload, and even hang together after failure.

The bulky shape of molded plastic frames can be avoided if the frame is constructed along conventional lines, using fiber-reinforced tubes fitted into joints. These, with fibers aligned with the principal stresses, can have higher tensile strengths and Young's moduluses than strong steels. However, such alignment is not often possible in frame tubes subjected both to bending and twisting. In addition, the fibers do not exhibit one of the desirable properties of metals: they do not stretch appreciably before breaking. Also, the composite fiber structures have much poorer properties across the grain than with the grain, giving a composite of varying properties, most much less attractive than those of the fiber. Although the properties of isolated carbon and Kevlar fibers are well known, the properties of usable forms, such as tubes, made from such fibers are not. For the second edition of this book, I went to a distinguished colleague, James H. Williams Jr., to ask what advice could be given to readers on the fatigue strength of composites. He gave me a paper on the topic, which stated that it is difficult to quote values on the strength of composites, and that the failure of composites in fatigue is like "sudden death." I was inclined to believe this to be unduly pessimistic until an airliner crashed in New York City in late 2001, allegedly because of the failure of the carbon-fiber-reinforced composite tail fin. This has given a wake-up call to structural engineers that composites can conceal defects deep in the material.

Aluminum alloys

The example worked in a previous section confirms manufacturers' claims that it is possible to produce aluminum-alloy frames that are stiffer, lighter, and stronger than steel frames (although the LCF "strength" used in the example does not imply the superiority of such frames in situations with either destructive overload or gentle, HCF loading.) This possibility arises because low-density aluminum can make use of the structural advantages of large tube diameter, without suffering from a too-thin, wrinkle-prone tube wall.

A significant improvement in aluminum alloys is metal-matrix composite (MMC), a mixture of aluminum with a powder of ceramic such as silicon carbide or aluminum oxide. Mixing aluminum with ceramic materials increases both strength and modulus, though with some decrease in ductility. The leading commercial supplier of MMC is DURALCAN. Aluminum MMC enjoys wide use on at least one brand of bicycle.

Nickel

The use of nickel tubing for bicycle frames followed the use of aluminum in the 1890s, no doubt in an attempt to produce a rustless frame. The firm manufacturing the frames, however, existed for only a short while during the bicycle-boom period, when cost was of less importance than it later came to have. Nickel was and is more expensive than steel, but it is strong and rigid and can be welded satisfactorily. It is seldom used in its pure form but is a major component, with chromium, of stainless and high-strength steels.

Titanium

Titanium in various alloy compositions (see table 10.1) is now used for corrosion-resistant heavy engineering equipment, for the spars and skin of high-speed aircraft, and for the disks and blades of jet-engine compressors. Satisfactory methods for welding titanium using inert-gas shielding to avoid weld deterioration have been developed. For bicycle use, titanium is corrosion-proof. Titanium frames are usually left in their as-welded state rather than being painted.

Titanium has a density and an elastic modulus just over half those of steel, and its fatigue-limit stress is 70 percent that of steel. It is therefore possible to arrive at tube sizes (in a manner similar to the method used above to compare aluminum-alloy with steel) that produce frames as stiff as those of steel while being lighter and having the same or greater fatigue life. The cost of the raw material has also dropped substantially, partly because of supplies from Russia released after the collapse of the Soviet Union, so that the production of high-quality titanium frames, though still limited, has increased considerably.

Magnesium and beryllium

The only other metal likely to be considered for bicycle components and frames is magnesium and its alloys. The density of magnesium is considerably lower that that of titanium and aluminum, which to some extent compensates for its relatively low tensile strength, and for its very low modulus of elasticity, which is one-fifth of that of steel. A magnesium alloy dubbed "Elecktron" was used fairly satisfactorily for making bicycle rims in the 1930s, and in the early 1990s there were serious plans by the

Norwegian energy producer Norsk Hydro to produce a die-cast magnesium frame, the Kirk Precision (Kirk 1990, see also ⟨www.ntnu.no/gemini/1993-dec/52a.html⟩ and ⟨www.firstflightbikes.com/KirkPrecision.html⟩). Some frame members had I-sections, and some had U-sections closed off by an added piece. Evidently production of the Kirk Precision has been suspended.

A few samples of a beryllium bicycle frame were made by American Bicycle Corporation, presumably achieving an astounding stiffness-to-weight ratio, but at a cost exceeding $20,000. Beryllium is not only expensive, but also toxic, and these two factors together render it essentially useless for most bicycle construction.

Frame design

The classical theory of truss structures was developed to apply to assemblies of bars with pinned (i.e., in-plane-rotation-allowed) joints. If insufficient bars are present, such structures may collapse (for example, a rectangular frame may collapse into a parallelogram). The simplest way to make sure such assemblies do not collapse is through a process known as "triangulation," in which each truss is constructed such that the open spaces between the bars are bounded on three sides.

Because slender tubes are relatively easy to bend, they act almost as if they were pin-jointed even when they are welded into structures. Therefore, triangulation remains generally desirable when such tubes are in use. A conventional bicycle frame may appear triangulated, but in fact this is not the case. The head tube, top tube, seat tube, and down tube actually form a quadrilateral rather than a triangle. (This quadrilateral has sometimes been braced by a diagonal tube, which does create triangulation.) Triangulation is apparently achieved in a small-sized frame, in which the head tube has virtually no length. However, in such a frame, the front forks act as a long lever arm to "twist" that joint. If the joint were truly pinned, the bicycle would collapse; therefore, in the welded or brazed cases of such a frame, stresses and deflections will be high.

The greatest loads on a bicycle arise from hard, near-vertical impacts, and in that direction the frame structure can be quite strong. But the front forks are not triangulated (they are like a diving board: a cantilevered beam). In addition the frame is not braced against torsional loads. Greater triangulation is possible: for example, the forks could be braced fore and aft like those in Pedersen's bicycles as described by Evans (1978). However, such bracing proves not to be very worthwhile: the forks also need lateral bracing (or must aim to intersect at the wheel's contact with the ground) and lacking this must retain rather stout proportions. Pedersen's supposedly triangulated bicycle had virtually no bracing against out-of-plane loads (and also eliminated the rider's normally firm lateral connection at the saddle, both factors militating against precise steering).

Sensible frame structures

A desirable structure for a bicycle frame generally must be economical to manufacture. For instance, adding a triangulating tube to a standard diamond frame may improve structural efficiency, but at the potential disadvantage of requiring the manufacturing and joining of several nonstandard parts. The greatest torsion is transmitted between a bicycle's handlebars and bottom bracket, so a stout direct tube is one natural step. In a bicycle equipped with such a tube, the chain tension tries to pull the rear wheel toward the bottom bracket, so relatively stout chainstays is another natural step. The seat tube must also be able to resist torsion from the head tube, so it should taper from a stout bottom to a slender top. Fore and aft compressive struts from the top of the seat tube would reduce vertical bending requirements in the down tube and chainstays.

Apart from maintaining integrity under loads, there are other structural functions that a bicycle's frame must perform. For example, it must prevent the wheels from rubbing the brakes in normal operation.[9] The frame should not deform under the action of steering forces (if it does, it will feel unpleasantly imprecise to maneuver). The stem and handlebars, which form an essential part of the vertical compliance affording some vibration comfort, should offer vertical resilience without compromising the stiff reactions offered to pedaling torsion.

When developing a frame structure for a bicycle, it is natural to ask whether the loads to be borne might be affected by its design (i.e., its stiffness or strength). Clearly, those forces that arise from nondamaging impacts will be mitigated if the bicycle's compliance (including that of the suspension) exceeds that of both the rider and the tire. And in damaging impacts, frame yielding will certainly limit the peak forces. But it is unlikely that other nonimpact forces would be affected by the stiffness (e.g., it is unlikely that a rider would either pedal or brake significantly less forcefully if the stiffness were lower).

Estimating loads

What load amplitudes are really experienced by bicycles? The best way to determine this is to instrument the bicycle (see below) and to ride in a way calculated to create high forces. But there are other ways to estimate loads approximately. The force on a bicycle's handlebars, and for that matter on the front wheel, is really governed by the resisting hand and arm strength of the rider, potentially a brief 850–1,700 N per arm. Bump forces are sometimes large enough (in excess of 3,000 N) to damage wheels. When a person jumps off a chair and lands fairly stiff-legged, which is comparable to landing a bicycle with cranks horizontal, forces can briefly exceed 4,000 N per foot.

Apart from impact against a curb, the maximum rearward bending of a bicycle's front fork occurs under peak braking when the rear wheel starts to lift and the front contact force points at the rider center of mass. (This can be simulated by locking the front brake and winching the rider forward with a rope attached to his belt, until the rear wheel just lifts.) The maximum forward bending arises from the vertical force of a hard landing.

The maximum (start-up) crank torque acting on a bicycle is probably based on the relatively low strength of the lifting leg. Assume it can pull up with 450 N, then the forward pedal bears at least body weight plus 450 N, and even more if the pedaler lunges downward. Therefore an occasional pedaling torque of 350 N-m seems entirely possible. But the more common case is simply body weight applied to alternate cranks, for less than half that value.

Usually the torque that acts on a bicycle from crank-axle loading is from the left crank only, so the bicycle's left taper or spline or cotter pin carries both bending and torque simultaneously, whereas the right taper carries the torque from the left, alternating with a torque-free bending of slightly higher magnitude than on the left. But landing from jumps changes this picture entirely: higher loads overall act to bend both sides (the right a little more than the left) at the same time that a large torque is transmitted through the axle.

Laboratory testing is never likely to replicate all the forces experienced by a bicycle in actual riding, so we should look at those suspected of causing most failures. Such testing will not detect unusual weaknesses that may be introduced in other parts of the structure by new designs. A conscientious tester must carefully examine each new design aspect he encounters and apply judgment to decide whether it calls for additional tests.

Instrumentation for stress

The preferred method for measuring frame stresses is the use of bonded resistance strain gauges, which if employed carefully can also determine the loads that act. Just a few millimeters in size, gauges alter their electrical resistance in response to being stretched minutely, and appropriate circuitry produces a corresponding output voltage that can be displayed or logged. In the last decade, compact, portable solid-state data loggers such as Somat ⟨www.somat.com/products/2100.shtml⟩ have made it feasible to acquire data at a high rate from several gauges for later examination. (Such professional tools are rather expensive, but now some adaptations of the

Palm handheld computer offer part of this functionality for a fraction of the price; see ⟨www.imagiworks.com⟩.)

The greatest value of electronic stress measurements is not necessarily to determine stress at the various points on the bicycle where the gauges are placed. This gives very little information for improving the bicycle's design, because that stress is determined not only by the material thickness at the measuring point, but also by the load-transmission properties of the entire structure. Rather, with enough gauges wisely placed plus careful calibration, it is possible to deduce the loads acting on the bicycle at a particular point such as the pedal or seat or the wheel's contact with the ground. It is these loads that must be known for analyzing or testing a new bicycle. Once determined on the road, they can be applied repeatedly in the lab, and rational structural optimization can proceed.

In this way, the locations and directions of the highest stresses acting on the bicycle can be predicted, so that strain gauges can be applied appropriately. Once that has been done, the load can be applied precisely in a laboratory setting, and the stress can be determined with high accuracy. There are just a few ways to learn these "hot spot" locations. One is to determine failure-initiation points, although if the failure being studied is an in-service rather than a lab failure, the loads acting on the bicycle at failure will not be known. Another is to use a computer analysis method known as finite-element analysis (FEA) to calculate the stresses occurring throughout the bicycle and visually emphasize the hot spots. Considerable judgment and validation is needed to do this effectively. A final method is the use of brittle lacquer (one brand is Stress Coat). A translucent brittle lacquer is applied to the bicycle and cured, the frame is loaded and unloaded, and then the lacquer is examined to determine the regions in which the crack density is highest. The main drawback to this method is its sensitivity to temperature variations, which can lead to results that are spurious when applied to conditions for actual riding outside the laboratory. The method can be made somewhat quantitative by simultaneously coating a calibration fixture (since batches and curing conditions may vary). The frame's sensitivity to cracking is then determined by measuring crack spacing at sample points of known strain.

Without access to modern electronics, measuring a time-varying stress or load on a bicycle is difficult, but not impossible.

· A properly attached pointer (similar to that on a torque wrench) can reveal tube twisting or bending, and a pencil attached to the pointer end can reveal the extremes of deformation. For such methods the advanced recording technology of a century ago (a clockwork-driven roll of paper) would still be very useful today.

- For slow enough loadings, an inexpensive dial indicator gives an accurate quantitative measurement.

Despite widespread availability of powerful software, we are disappointed to be required to conclude that bicycle structural analysis seems still to be in its infancy. One of the more sensible studies of such analysis was conducted by Peterson and Londry (1986).

Wheels

Bicycle wheels are special lightweight structures that must bear principally radial loads, but also some amount of lateral and tangential (braking or driving) load. When a disk of steel is used to carry very high loads (for example, railroad wheels) the thickness naturally required to support the weight is automatically sufficient to withstand the other force components that act upon it. However, if the radial strength of a steel disk were lowered to the levels required to support a bicycle, the resulting disk would be paper thin and totally unable to support the lateral forces acting upon it. (In fact, even the radial load alone would induce a different type of failure called "buckling," i.e., bending and collapse.) Part of the solution to this dilemma involves maintaining some lateral stiffness in a bicycle's wheel by joining the slender rim to two separated hub flanges. But this alone would still leave a very thin and buckling-prone structure. The ingenious solution is to create tension in the spoke or "sheet" supports.[10] This permits them to stay straight and stiff under compressive loads, forestalling their own buckling, although still permitting buckling of the wheel rim if tension is too high.

If we were unable to use tension in this way, all the material in a bicycle wheel's spokes would have to be collected into just a few spokes, fabricated as thin-wall tubes or channels so as to resist compression buckling. Rim sections would have to be strengthened to support bending loads over longer spans. The resulting structure would be hard to "true," and we expect it would also be heavier. This approach has been adopted for composite construction, which not surprisingly does not show the weight reduction expected of such superior materials (although it possesses the advantage of never being warped by yielding, so that truing is not necessary).

Conventional tension-spoked bicycle wheels[11] can suffer various kinds of structural damage:

- rear-axle bending and breakage (fatigue from repeated ground impacts when one bearing is far inboard);

- hub-flange breakage (especially when spokes are radial, rather than being virtually joined end to end in pairs, with a tangent-spoking pattern);
- spoke-elbow fatigue due to variation in spoke tension (in concert with loose-fitting hub contact and absence of assembly overload, which improves the head support and reduces the residual stress in the elbow); also spoke fatigue in which the spoke is bent near a misaligned nipple;
- spoke-body failure that arises from a spoke's rubbing against a neighbor;
- nipples unscrewing because of repeated loosening;
- cracks in the rim near spoke holes (cyclic section bending, plus cyclic spoke force);
- braking wear-through of rim sidewalls, giving a potential for explosive separation due to tire pressure;
- lateral buckling of rim (insufficient torsional stiffness, poor spoke bracing angle, high spoke tensions, and sideload, especially in combination with a spoke-loosening radial load);
- rim radial untruth (from radial impact) and lateral untruth (possibly torsional yield due to overload on one flange only); and
- rim denting from local penetration of tire by a sharp obstacle.

Not only are the structural mechanics of wheels intriguing in their own right, but the stresses, deformations, and buckling tendencies of wheels have an impact on bicycle durability and performance. These aspects can be investigated both experimentally and theoretically.

Experimentally, one kind of useful lab test involves holding the wheel's axle or hub body and applying a known force (usually radial or lateral) at any of a number of positions on the rim. Rim motion can be measured by dial indicators or a direct-current differential transformer,[12] preferably mounted on a nonloaded arm secured to the hub, if the wheel-holding fixture itself is not extremely rigid.

Shifting the loading point around the bicycle's wheel has two main effects. When the load is far from the measuring point, out-of-plane rim motion can be reduced or reversed. In addition, nonradial spoke orientation at the load point (clockwise or counterclockwise, to left flange or right flange, or at a symmetric point of the pattern) effectively adds a tangential or lateral load component, which can shift the entire rim in one direction or another. While the load is acting, spoke tensions (or rather changes in tension from the unloaded state) can be determined by frequency (from plucking or soft-hammer tapping, best if spokes are not interwoven) or by a spoke tensiometer.

A similarly interesting study can be carried out by tightening one spoke and noting the resulting change in rim shape and in spoke tensions. (Typical results can be viewed in images from Papadopoulos 1992 on computer-based truing.)

Another potentially useful experiment is to place strain gauges on one or more spokes (preferably in pairs, to cancel any effect of bending). In the laboratory, this is potentially a more accurate method of determining tension change.

But the real need is to log load data while riding. Lacking a recording unit on the rotating wheel, the signals transmitted by instrumentation installed on the spokes must taken off by slip rings or radio telemetry, as described by Gavin (1996), who focused on riding straight and upright. Unfortunately, when more extreme loading is to be studied, an instrumented spoke may not be the best approach. It will rarely be near the point at which the wheel makes contact with the ground when the highest loads occur, and determining wheel orientation, to decode the strain signal into load magnitudes, adds an extra burden. However if the bicycle is ridden long enough, an instrumented spoke should be an excellent way to learn what spokes of similar construction must endure.

Analysis of spoked wheels

A bicycle wheel is essentially a curved beam (a ring) mounted to a rigid hub by many tensile springs. In structural studies it is known as a "curved beam (ring) on elastic foundation." Primarily intended for use in airplanes, much spoked-wheel mathematics was worked out between the two world wars. Foremost were a series of papers in *Philosophical Magazine* by A. J. S. Pippard, consolidated in his 1952 book. (As well as investigating wheels with few spokes, Pippard also pioneered the useful approximate approach of representing a finite number of ordinary spokes as a "sheet," i.e., an infinite number of infinitesimal spokes, also known as an "elastic foundation." Hetenyi 1946 presents a wide variety of elastic-foundation calculations, including some involving rings.) Burgoyne and Dilmaghanian 1993 verified the applicability of Pippard's theory to contemporary bicycle wheels. (See also the follow-up discussion by Papadopoulos [1995].) Biezeno and Grammel (1956) also delved extensively into wheels with few spokes (specifically, four- and six-spoke wheels, although the approach could be adapted to wheels with other numbers of spokes).

It is also reasonable to analyze wheel response by computer, typically with an FEA program (for examples, see Burgoyne and Dilmaghanian [1993], Gavin [1996], and Brandt [1981]). The main disadvantage of this approach is that even when the results follow a simple formula, the method merely gives pages of numbers for each particular case. But FEA offers the compelling advantage of dealing with almost any degree of complexity, without requiring commensurate analytical training on the part of the researcher. In particular it can include important effects such as spoke tension that tends to destabilize the wheel laterally, and nonlinear behavior like rim yielding and spoke "dropout" (when a spoke's initial tension is driven to zero and it becomes unable to resist further loading).

Perhaps the most authoritative writer on the real-world behavior of bicycle wheels is Jobst Brandt (1981). His book includes some early FEA calculation of spoke tensions (due to the radial and tangential loading of a two-dimensional wheel), but its greatest value is in the practical advice it offers for maximizing spoked-wheel durability. (Brandt also takes great pains to debunk popular myths about the working of tension spoking.) Over the years of Internet newsgroups and e-mail forums, Brandt has produced a valuable legacy of amplification, explanation, and advice that is well worth tracking down.

In the arena of beam analysis, curvature leads to some interesting and unexpected phenomena. For example, a loop of a stretched helical tension spring is deformed purely in torsion even though there is no cross-sectional rotation. And a ring being "rolled" slightly inside out deforms purely by bending, although there is no lateral displacement and no slope change of cross sections. The continuous change of direction of a curved beam means that a twisting moment at one point is transmuted into a bending moment at another.

The most important elastic property of spokes is their tensile stiffness, although their built-in tension also becomes important for lateral deformation. The relevant strength is that in fatigue, which will be very sensitive to details of the head fit and positioning in the flange hole (these may be improved by an intentional overload during assembly).

The main properties of bicycle-wheel rims are the following three "beam elastic stiffnesses."

- In-plane (radial) bending stiffness (EI_R), most easily measured by elastically compressing a bare rim across its diameter, with the rim joint at the $45°$ point. (To understand overload flat spots, the in-plane bending moment causing yield can also be determined from diametral compression testing.)
- Out-of-plane (lateral) bending stiffness (EI_L) (see below for a method to measure this). EI_L is large for ordinary rim designs, and it may be that taking its value to be infinite will have little effect on calculated deformations and stresses. Likewise, the lateral bending moment sufficient to cause yielding is very high, suggesting that lateral bending failure of bicycle wheels is virtually impossible.
- Torsional stiffness (GJ). One could determine GJ by cutting the rim and loading it like a single coil of a helical spring. However, there are also

nondestructive ways to measure GJ. One is four-point bending: the rim is supported at 12 and 6 o'clock and pressed down at 9 and 3 o'clock. If both the vertical displacement and the cross-sectional rotation can be measured for a given load, both bending and torsional stiffnesses can be determined algebraically. (In effect, what is needed is the translational stiffness and the location of the instantaneous center.)[13]

Conventional bicycle-wheel rims have adequate lateral bending stiffness, but the torsional stiffness of such rims can be extremely low, especially if the cross section does not include a hollow box. It is informative to perform four-point-bend tests on a rim for tubular tires, either bare or spoked into a wheel, and then to repeat the test after a slit has been cut around the outer circumference with a saw. This dramatically reduces GJ without altering EI_L, with obvious effects on lateral stiffness. Consistent with this result is the observation that lateral deformation of a rim into a "saddle" or "potato chip" shape is very nearly a deformation of pure twist (i.e., the instantaneous center is about halfway between the rim and the hub).

In view of the obvious importance of rim torsion, we speculate that one reason that rims go laterally untrue is torsional yield, perhaps from impacts on one flange only. This could be explored further by cutting a damaged rim and noting how it "springs."

With the foregoing as a preamble, it is now possible to describe typical wheel structural response.

Radial load

In-plane wheel mechanics has been mythologized for decades by people who envisioned (1) a relatively rigid rim or (2) loose spokes. In fact the hub of a bicycle's wheel does not "hang from the upper spokes." It is true that the tension of the upper spokes is needed to counterbalance those of the lower spokes, but a radial load from the ground affects only the lower spokes.

A bicycle-wheel's rim is flexible in bending, and its spokes are relatively stiff as long as they retain some preload. The external load acting on the wheel is transmitted to the hub by just a couple of spokes, because the rim is not stiff enough to transfer the load to more of them. (Modern deepsection "aero" rims are, however, much stiffer than traditional rims.)

With radial-pattern spoking of a symmetric (front) wheel, loaded between spokes, the rim moves inward over a region of two or three spokes, slightly outward just outside that region (this is the well-known overshoot of a beam on elastic foundations), and somewhat less outward around the rest of the wheel (a consequence of a fixed-circumference rim subjected to a flat spot). A radial load acting on a wheel does not significantly raise the tension of any of the spokes. Under such a load, the wheel will deform slightly to the side away from the spoke where the load is applied (lateral deformation is described below).

Furthermore, if the spokes of a wheel are not radial, the load applied to the wheel may be applied to a nearby spoke either clockwise or counterclockwise to point where the load is applied, which will tend to move so as to wrap further around the hub. Then the rim will rotate slightly relative to the hub and also display a slight wavy pattern of radial motion, as the tangential motion interacts with clockwise or counterclockwise restraints.

Tangential load

The two primary analytical resources on tangential loads in wheels are Pippard's (1952) analytical formulae and Brandt's (1981) two-dimensional finite-element analysis. Following the approach of Pippard and of Hetenyi (1946), a simple formula for tangential load can probably be derived, but to our knowledge it has not been. Part of the reason that it has not is that tangential load is not a cause of significant spoke tensions, although this may change with the advent of powerful disk brakes for use on bicycles.

A bicycle wheel's rim is very stiff in the tangential direction, whereas the spoke restraint is very soft. The rim rotates almost as a rigid body, except that the tangential resistance offered by each spoke leads to compression ahead of the point where the load is applied and tension behind it, with a continuous variation around the rim. Where compression is increased, the rim bulges outward (i.e., it stretches the spokes). In addition there is a marked radial sine wave due to interaction of the compression with spokes located clockwise and counterclockwise of the location where compression is applied. The result is that a tangential load does not much affect spoke tensions, even when torque is high, as with a front hub brake. A rim brake leads to two opposed tangential loads, with very little tangential deformation.

Lateral load

This is by far the most complex case (and the one most relevant for wheel strength and life). A bicycle wheel's spokes are laterally quite soft, its rim is laterally soft, especially if its torsional stiffness is low, and high spoke tensions result in a tendency for lateral buckling. It is particularly interesting to reproduce an experiment conducted by Brandt in which a rim is spoked to a single hub flange only (no spoke bracing angle) and shows marked buckling proclivities as soon as the spokes are even moderately tensioned.

Perhaps the only valid quantitative analysis of lateral load is that presented by Pippard (1952), which is densely mathematical. This is one arena in which a simple formula may not be so feasible. (However, a spreadsheet version of Pippard's equations should be easy to create.)

With a lateral load applied to a wheel at 6 o'clock, the rim there moves to the side. At 3 o'clock and 9 o'clock, it moves somewhat less, but in the opposite direction from that at 6 o'clock. At 12 o'clock, it moves in the same direction as at 6 o'clock, only far less. In other words, even with modest tensions, the effect of a lateral load is to produce a tendency to buckle in a shape somewhat similar to a potato chip, the preferred buckling shape.

Also, in a situation that falls between a case of purely radial and purely tangential loads, spoke tensions are quite strongly affected near the load point (the rim is not torsionally stiff enough to spread the load to all spokes). That is to say, some spoke tensions near 6 o'clock increase markedly under such a load, whereas others decrease. Of course, bicycle wheels in service never see lateral loads without an accompanying, and considerably larger, radial load. Interestingly, the zone affected by a lateral load encompasses some ten spokes on a typical bicycle wheel, whereas the zone affected by a radial load involves only two or so. This means that just a few spokes away from the load, the radial component of a load has almost no effect, so spoke tension increases that are due to lateral load are almost unabated. This can be demonstrated easily by sitting on a bike at rest and having an associate repeatedly pluck a single spoke just a little bit away from the bottom of the wheel, as the frame is leaned to either side. (It's also informative to perform this test on an upright bicycle as it is rolled a few inches forward.)

We believe that lateral wheel mechanics deserve much more study than has to date been conducted. Lateral stiffness plays an important role in bicycle-wheel collapse, and lateral loads may often bear responsibility for spoke fatigue.

Wheel buckling

The classical buckled shape of a bicycle wheel is sometimes likened to that of a potato chip: the points at 12 o'clock and 6 o'clock move to one side, and those at 3 o'clock and 9 o'clock move to the other. In essence, by becoming laterally wavy, the rim sits at a smaller radius, thus reducing the amount of energy stored in the stretched spokes. The torsional stiffness of a rim thus deformed is very low, especially in a rim without a hollow "box" cross section, and this wavy shape is primarily an expression of torsional distortion.

Wheel buckling is promoted by tight spokes and hub flanges positioned close together (it is rare on a front wheel) and probably also by low torsional stiffness of the rim. It can be brought on by squeezing some spokes together by hand in the construction process. Otherwise it is seen when a substantial lateral load acts on a wheel (usually in concert with at least an ordinary radial load); examples include the sideforce of falling, the sideforce of a violent swerve, and the sideforce of being struck by another vehicle.

Buckling may or may not involve yielding. There are wheels that can be snapped into a buckled shape and then later snapped back with no ill effect. Some spokes (from the flange away from the rim motion) remain taut after buckling: they are what hold the unyielded rim in its deformed configuration. Tightening these spokes, which seemingly would pull a buckled rim back to planarity, actually increases the wheel's waviness.

In some cases a buckled wheel may be observed after an accident, and a zealous plaintiff's lawyer may argue that spoke tensions were improper and caused the buckling and an ensuing crash. To discredit such an assertion, it suffices to force the rim into a plane again, at which point spoke tensions can be measured.

Wheel evolution

Modern tension-spoked wheels are developing in ways that seem appropriate. These include incorporating a large box area in rim cross sections to increase torsional rigidity and strength and attaching spokes to the far edge of a rim to increase bracing angle. (A large bracing angle, coupled with short "column length" circumferentially, is part of why smaller wheels are so much stronger laterally than larger ones.)

Notes

1.   The trapped mandrel is removed by an ingenious process: the tube is subsequently enlarged by being "rolled" along its whole length, like bread dough, to become a little thinner in the circumferential direction.

2.   Not all structural failures are equally dangerous. The key is "structural redundancy": if one part breaks, is another able to hold everything together, even if imperfectly, at least long enough for the rider to stop himself safely?

3.   Bicycle frames embody constructional sophistication rarely found even in airplanes, including tapered diameters, varying wall thicknesses, nonround shapes, and even some shaped "monocoque" shell structures.

4.   To be more certain of this conclusion, we would need a fatigue curve for pretensioned spokes attached to holes in a hub flange.

5.   From the opposite perspective, if the original failures *were* due to HCF, then the high-cycle flatness of fatigue curves means a rider who is 30 percent heavier than the original rider could experience spoke failure in a single day's riding, which isn't often seen. However, the author and his wife, being neither weighty nor aggressive bicyclists, riding a tandem having apparently standard wheels, often had eight spoke failures per day of riding, presumably contributed to by stress raisers at the wheel flange.

6.   In aggressive start-up accelerations, a pull-up force on the rising pedal equivalent to body weight translates to roughly two times body weight on the descending pedal. In addition the pedaler may exceed this force by "bouncing" a little.

7.   Happily the situation seems to be changing. The new sport of off-road racing has blended higher loads with untried designs, resulting in a rash of bicycle failures. Now many bicycle manufacturers employ skilled engineers, have increased their use of engineering software and modern test equipment in the design process, and have joined a movement to create stricter durability requirements for bicycles.

8.   In fact, when *solid* wood is being compared to *solid* steel *of equal weight*, the wood is far superior. At one sixteenth the density, it is four times as thick, and ends up sixteen times as stiff in bending. However, even apart from the torsion problem this superiority cannot practically be used. For example, a crank made of wood would be far too bulky.

9.   When a leftward side load acts on the ground contact of the bicycle's front wheel, the tendency of the forks to deform leftward as a parallelogram brings the rim of the front wheel near the left brake pad. On the other hand, this force also moves the left fork tip up (and the right one down), which brings the rim near the right pad. Whether by luck or design, these two tendencies have been observed to cancel out on a standard fork, leading to zero relative rim motion in the vicinity of the brake.

10.   A "sheet" of spokes is not like a continuous sheet of metal: it is more like a sheet of metal with a large number of near-radial slits. Spoke beds from each flange are like two sheets of metal with oppositely directed angled slits.

11.   Although early airplanes often borrowed from bicycle technology, the tension-spoked wheel was initially an invention by Cayley for his experimental airplanes early in the nineteenth century.

12.   This is a direct-current-powered version of a linear variable differential transformer. Such economical electronic displacement transducers can easily resolve submicron displacements with virtually no force. A good source is ⟨www.transtekinc.com⟩.

13.   When a cross section of a rim displaces laterally relative to the hub, it also generally rotates in proportion, as if the rim were twisting. Any small motion combining displacement and rotation can always be described as "rotation about an instantaneous center" and is easy to visualize once the instantaneous center is located.

References

Ashby, M. F. (1992). *Materials Selection in Mechanical Design*. Oxford, U.K.: Pergamon Press.

Ashby, M. F., and D. R. H. Jones. (1980, 1986). *Engineering Materials, Parts 1 and 2*. Oxford, U.K.: Pergamon Press.

ASM International. (1996). *ASM Handbook. Volume 19: Fatigue and Fracture*. Materials Park, Ohio: ASM International.

ASM International. (2001). *Engineered Materials Handbook. Volume 21: Composites*. Materials Park, Ohio: ASM International.

ASM International. (2002). *ASM Handbook. Volume 11: Failure Analysis and Prevention*. Materials Park, Ohio: ASM International.

Baumeister, T., ed. (1978). *Marks' Standard Handbook for Mechanical Engineers*, 8th ed. New York: McGraw-Hill.

Biezeno, C. B., and R. Grammel. (1956). *Engineering Dynamics*. London: Blackie. Original work published 1939.

Brandt, Jobst. (1981). *The Bicycle Wheel*. Palo Alto, Calif.: Avocet.

Brandt, Jobst. (2000). "Touring wheel." Message posted to mailing list "rec.bicycles.tech," March 4.

Burgoyne, C. J., and R. Dilmaghanian. (1993). "The bicycle wheel as prestressed structure." *Journal of Engineering Mechanics* (ASCE) 119, no. 3:439–455.

Collins, J. A. (1981). *Failure of Materials in Mechanical Design*. New York: John Wiley & Sons.

Evans, David E. (1978). *The Ingenious Mr. Pedersen*. Dursley, U.K.: Allan Sutton.

Farag, Mahmoud M. (1989). *Selection of Materials and Manufacturing Processes for Engineering Design*. London: Prentice-Hall.

Fuchs, H. O., and R. I. Stephens. (1980). *Metal Fatigue in Engineering*. New York: John Wiley & Sons. (Second edition published 2000.)

Gavin, H. P. (1996). "Bicycle-wheel spoke pattern and spoke fatigue." *Journal of Engineering Mechanics* 122, no. 8:736–742.

Hetenyi, M. (1946). *Beams on Elastic Foundation*. Ann Arbor: University of Michigan Press.

Kern, Roy F., and Manfred E. Suess. (1975). *Steel Selection: A Guide for Improving Performance and Profits*. New York: John Wiley & Sons.

Kirk, Francis G. (1990). "Bicycle and frame therefor." U.S. patent no. 4,921,267.

Lowe, Marcia D. (1989). "The bicycle: Vehicle for a small planet." Paper no. 90, Worldwatch Institute, Washington, D.C.

Malicky, David. (1987). "Rider perception of bicycle frame variables." T&AM 492 Undergraduate Research Project Report, Department of Theoretical and Applied Mechanics, Cornell University, Ithaca, NY.

McClintock, F. A., and A. S. Argon. (1966). *Mechanical Behavior of Materials*. Reading, Mass.: Addison-Wesley.

Papadopoulos, J. (1992). "Method for trueing spoked wheels." U.S. patent no. 5,103,414.

Papadopoulos, J. (1995). "Discussion of 'The bicycle wheel as a prestressed structure.'" *Journal of Engineering Mechanics* (ASCE) 119, no. 3 (July):847–848.

Peterson, Leisha A., and Kelly J. Londry. (1986). "Finite-element structural analysis: A new tool for bicycle-frame design." *Bike Tech* 5, no. 2.

Peterson, R. E. (1974). *Stress-Concentration Factors*. New York: John Wiley & Sons.

Pippard, Alfred John Sutton. (1952). *Studies in Elastic Structure*. London: Edward Arnold.

SAE Fatigue Design and Evaluation Committee. (1997). *SAE Fatigue Design Handbook*, 3d ed. Warrendale, Penn.: Society of Automotive Engineers.

Wilson, David Gordon, and Tarik Saleh. (1993). "The influence of materials developments on the design and construction of early cycles." In *Proceedings of the Fourth International Cycle History Conference, Boston, Mass.*, ed. David Herlihy, 49–56. San Francisco: Bicycle Books.

III HUMAN-POWERED VEHICLES AND MACHINES

11 Unusual human-powered machines

Introduction

In this chapter we aim to expand your experience, and perhaps to make you want to use, or even to design and make, some interesting human-powered devices other than bicycles. This aim has an obvious relationship to bicycling, which is an activity having a transportation component that can usually also be accomplished by the use of a motor vehicle. People in the developed world who choose to bicycle generally do so for reasons connected with their own health and well-being and that of the region in which they live and perhaps out of concern for the earth as a whole. There are rather similar, but far more limited, choices that such people can make for mowing grass and clearing snow, for example, and for recreational boating. The role of human power in the modern high-technology world has, alas, to be restricted. Only a very few enthusiasts bicycle across North America, Russia, Asia, or Australia for pleasure. Although we are engaged in some advocacy for human power in this chapter, we are not recommending that human power should be used for such prodigious feats as bicycling across a continent, or to clear snow from a supermarket parking lot, or to cut the grass of a golf course. However, even in large countries like the United States, over half the daily "person-trips" by automobile are of under 8 km (5 miles), a distance most people can easily cover on a bicycle in most weather conditions. Likewise, most lawns and driveways are of sizes that can easily be handled by human-powered devices. The past enthusiasm for reducing what has been characterized as "back-breaking" labor through the incorporation of gasoline-engine- and electric-motor-powered devices has led to an almost total neglect of efforts to improve human-powered tools. In consequence, there is today an unfair competition between highly developed modern electric hedge clippers, for example, and manual shears that have not been sensibly improved for a hundred years. Perhaps we need a new series of Kremer prizes (see below) for specified achievements in human-powered tools.

We have chosen to give in this chapter a series of examples of human-powered tools and of record-breaking and other interesting vehicles (other than standard bicycles) for use on land, on and under water, and in the air. Each example deserves several pages of description and discussion, but the available space will not permit such an extensive treatment. I have selected some interesting features in each case and hope that readers wanting more information will examine the references cited to find out all they want to know.

Figure 11.1
Commercially sold vertical-axis pedaled lawn mower.

Human-powered lawn mowers and snow removers

The first two editions of this book included illustrations of Michael Shake-spear's pedaled lawn mower. In view of the extremely limited budget and time Shakespear had available, it was beautifully designed and executed. His achievement might have inspired others. A commercially sold pedaled mower came on the market later that employed a vertical-axis, high-speed rotary blade that, because of the power required for this type of cutter, made a slowly advancing cut of only about 300-mm width (figure 11.1). However, it did cut long grass and weed stalks, often missed by reel-type mowers. A compact and stylistic pedal-powered riding mower with a central reel was built by Chris Toen in the late 1990s in the Netherlands (figure 11.2).

Another type of lawn mower that would cut long grass was sold in North America and probably elsewhere for much of the early part of the last century is shown in figure 11.3. In this mower, a so-called sickle bar or row of clippers in front of the wheels of a push mower was driven from a cylinder cam that would seem to have a high degree of friction. This type of cutter has no intrinsic system of removing and collecting the clippings.

Figure 11.2
Chris Toen's riding mower. (From *HPV Nieuws*, The Netherlands.)

The author has discussed pedal-powered riding mowers in *Pedal Power* (McCullagh 1977). I believe that the energy required to pedal a machine across soft ground (a lawn) is so high that the only way in which pedaling would be superior to pushing a mower would be for the pedaler to be either stationary or moving slowly, while the cutter, presumably light in weight, covered a considerable area.

Snow removers

The use of snow shovels at the first snowfall of the winter always seems to produce reports of heart attacks. Shoveling snow is another example of a heavy task involving the use of the muscles of the arms and back and of having the back bent uncomfortably. It would be more efficacious and put less stress on the body to use the big muscles of the legs and to have

Figure 11.3
Sickle-bar push lawnmower. (Photo by Ora E. Smith.)

a more natural posture; presumably this would also be less likely to over-strain the heart. It would be delightful to have a small lightweight device that, from leg operation alone, would scoop up a quantity of snow and project it in a desired direction, as one does with considerable effort using a snow shovel. Nothing like that has been on the market, or even in the patent literature, so far as can be learned from searches carried out by the author and his students. His favorite tools for clearing snow are shown in figures 11.4 and 11.5. The first of these is an old push-plow purchased at a garage sale. He made and installed a fiberglass "blade" with a mild-steel cutting edge. He likes to demonstrate that, on the asphalt surface of his driveway (about 50 m^2) he can clear snow in about half the time it takes his neighbors with similar driveway areas, using their engine-powered snow-blowers. Figure 11.5 shows the Sears, Roebuck version of an old device sometimes called the "Swedish snowblower." One pushes the handle of the snowblower while lifting it, so that the blade cuts along the surface of the ground and the snow fills the "bucket." Then one pushes down on the handle while continuing to push forward, so that the device rides up over the snow on its round underside. Using this device, it is easy to push

Figure 11.4
The author's push snow plow. (Photo by Ellen Wilson.)

Sliding and dumping

Figure 11.5
Sears, Roebuck's "Swedish snowblower" for heavy snow.

Figure 11.6
Varna I tricycle designed by George Georgiev for world ski champion Daniel Wesley. (Courtesy George Georgiev.)

heavy wet snow (for which using the snow-plow is heavy work) a considerable distance over snow, which can be built into a long ramp, before dumping it.

We do, however, need better human-powered snow-removal devices, efficient, fun to use even for older and nonathletic people, and compact when stowed.

Human-powered land vehicles

Speed machines

Figure 11.6 shows a tricycle (Varna I) designed by Georg Georgiev especially for Daniel Wesley, a world ski champion and Olympic gold medalist and a double amputee, and figure 11.7 shows Wesley on a mono-ski, also designed for him by Georgiev.

A great many other records have of course been set in many other HPVs, and we do not have space here to pay tribute to them. However, we would like to make space for a commercial of sorts. In the second edition of this book, the author applied the methods recommended in earlier chapters to produce a forecast that the maximum speed of a streamlined HPV pedaled by someone equivalent to the great Eddy Merckx (five-time winner of the Tour de France), through a 200-m measuring section would be 65.4 mile/h (29.25 m/s) (Whitt and Wilson 1982). The author was subsequently very proud that the Du Pont prize for the maximum speed reached by an

Figure 11.7
Mono-ski designed by George Georgiev for Daniel Wesley. (Courtesy George Georgiev.)

Figure 11.8
Cheetah fully faired recumbent bicycle.

Figure 11.9
Varna Mephisto, in which Sam Whittingham (seated) achieved 80.55 mile/h in 2000. (Courtesy George Georgiev.)

HPV over a flying 200-m course (with restrictions on maximum wind and slope) was won by Freddy Markham in the Easy Racer Gold Rush (designed and built by Gardner Martin) at 65.484 mile-h in May 1986. This apparently accurate forecast held for seven years, when it was proven pessimistic when the Cheetah fully faired recumbent bicycle (figure 11.8), designed and built by a team from the University of California, Berkeley, and ridden by 1989 U.S. pursuit champion Chris Huber achieved 30.7 m/s (68.7 mile/ h) on a high-altitude desert highway in Colorado. Eleven years after the Cheetah set its record, on October 10, 2000, Canadian bicycle racer Sam Whittingham rode the Varna Mephisto recumbent bicycle (figure 11.9) at 36.01 m/s (80.55 mile/h) on a long flat asphalt road near Battle Mountain, Nevada. He improved on this to reach 81 mile/h in October 2002. Second in speed to Whittingham on both record-setting occasions was Matt Weaver in a HPV he designed, built, and rode, the Kyle Edge (figure 11.10a). By lying on his back, looking at a small monitor connected to a TV camera in the vehicle's nose, he was able to design a fairing having a flow that was predominantly laminar, and therefore having a low drag (see chapter 5). His speed and power versus distance are shown in figure 11.10b.

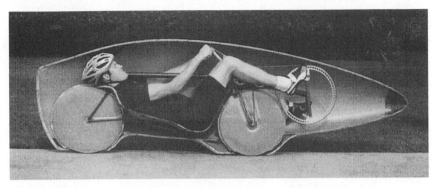

(a)

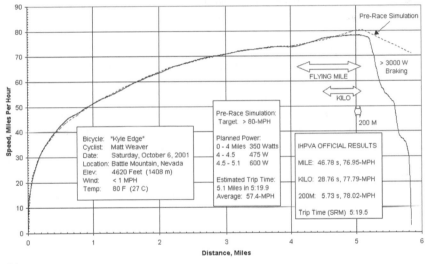

(b)

Figure 11.10
(a) Matt Weaver in half fairing of the Kyle Edge HPV. (b) Matt Weaver's speed
and power versus distance in his Battle Mountain run, October 6, 2001, Nevada.
(Courtesy Matt Weaver.)

Figure 11.11
Flevobike Rug-an-rug (back-to-back) tandem.

Back-to-back tandems
The configuration for recumbent tandems in which the second rider (the "stoker") faces backward has become popular, especially in Europe. Figure 11.11 shows a Flevobike Rug-an-rug tandem. The configuration solves the problem in recumbent tandems in which both riders face forward of great length and consequent torsional flexibility of the frame. Back-to-back tandems have independent drive to each wheel, and the machines can be easily separated in the center for compact folded transport. A further advantage is that luggage can be stored midship between the two seats. The principal disadvantage of such tandems is that the stoker has to become accustomed to riding backward.

Rail cycles
During and after the time when railroads were being built, railcars generally powered by arms were used to take workers along the rails and to inspect the track. The term "rail cycle" or "railbike" is currently used, however, to apply to a new sport: cycling on abandoned railroads (Mellin 1996). We show two examples: figure 11.12 is of Richard Smart's Railcycle for recreational cycling (also used in a slightly modified form by track-maintenance people on the London Underground), and figure 11.13 is of Charles Henry's record-breaking faired machine Snapper and his team. Railroads have been and are still being abandoned in many countries, leaving priceless rights-of-way connecting towns across (usually) picturesque rural routes having very low maximum gradients. These are ideal for bicycling. The rails are removed from many such trails and they are paved over for general vehicular use. In many other cases, however, the rails remain, and

Figure 11.12
Richard Smart's Railcycle, as used by the London Underground. (Courtesy Richard Smart.)

it is on these, principally, that enthusiasts for rail cycling practice their sport (see ⟨rrbike.freeservers.com⟩).

As mentioned above, Charles Henry's Snapper achieved the fastest average speed in speed trials on rails at Interlaken, Switzerland, in 1999. Its builders, Robert Stolz, Bruno Guhl, and Henry himself, based the design on a road machine with a 17-inch front wheel and a 20-inch rear wheel driven by a 2 × 7 transmission. Roller-skate wheels were used on the sides of the rail under the cockpit to keep the vehicle's wheels centered. The layout has a compact long wheelbase (1.5 m) with a single-tube chassis, on which the fairing is mounted. Despite a high cross-wind during the speed trials, the Snapper reached a speed of just over 70 km/h (19.46 m/s).

There is little doubt that streamlined (faired) rail cycles should be the fastest HPVs. However, to achieve top speed they would probably need to have a special narrow-gauge track, perhaps of 200 mm, which, for a record attempt, would probably be in a circular track of, say, 200-m diameter. This would confer the following advantages over pneumatic-tired bicycles on a highway (figure 11.14).

Figure 11.13
Record-breaking Snapper rail bike by Charles Henry. (Courtesy Charles Henry.)

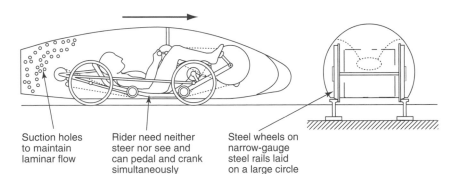

Suction holes
to maintain
laminar flow

Rider need neither
steer nor see and
can pedal and crank
simultaneously

Steel wheels on
narrow-gauge
steel rails laid
on a large circle

Figure 11.14
Proposed HPV record rail vehicle.

1.   Wheels made of steel (or of other hard material) used on steel rails would have very low rolling resistance.

2.   The wheels would not have to be steered, so that they would be mostly inside the fairing and could run in narrow gaps (thus avoiding large "pumping" losses).

3.   Because the wheels would not have to be steered, the rider's arms could be used to add power to that delivered by the legs. An increase of 20 percent in power output could be expected for a short-duration effort. Also, the rider would not have to see ahead, if the track were circular and well protected, so that the air drag associated with a window and a heads-up position could be eliminated.

4.   The streamlined enclosure would be much better aligned with the relative air flow and much steadier than that on a typical road machine. Boundary-layer suction (see figure 5.16) could be used to produce laminar flow and thereby to reduce the aerodynamic drag to perhaps 10 percent of the normal level by means of an expenditure of power that would be only a small proportion of that saved.

5.   The maximum speed would therefore be determined not so much by the human power output matching the aerodynamic and rolling losses, as for road machines, but predominantly by the ability of the rider to deliver the required kinetic energy to the vehicle as the rider's own energy and power output are being depleted.

All-terrain vehicles
ATBs are well known and fall outside the topic of this chapter. However, there are many other types of all-terrain vehicles under development, such as that depicted in figure 11.15, for use by paraplegics.

Utility machines
In China (for example) not only are standard one-speed bicycles responsible for the overwhelming preponderance of person-trips made each day, but they also carry over 90 percent of the ton-miles or tonne-kilometers. However, even in China, bicycles are used only a little in hilly cities such as Chongqing. A human-powered school bus capable of carrying up to twelve youngsters, photographed in Kanpur, India, is shown in figure 11.16. In the Western world, entrepreneurs are continually improving on human-powered freight-carrying vehicles and rickshaws (figure 11.17).

A multihuman-powered land vehicle, the Thuner Trampelwurm
The Thuner Trampelwurm is a unique type of human-powered "road train" (figure 11.18). Although other linked trains of HPVs exist, none is as radical as the Trampelwurm, a brainchild of the Swiss artist Albert Levice. Ten

Figure 11.15
An all-terrain HPV for paraplegics. (Courtesy Mike Augspurger, One-Off-Titanium.)

two-wheeled trailers, each for carrying one person, are hooked up behind a long-wheelbase recumbent tricycle in such a way that they follow the leader almost perfectly—almost as if on rails defined by the path of the leading trike. It was a difficult task for a group of students led by Hansueli Feldmann at the Engineering College of the Kanton of Bern in Biel, Switzerland, to come up with a usable system. They designed a good compromise with almost perfect tracking and enough stability to drive up to about 15 km/h without the train's beginning to snake back and forth. Even so, hydraulic yaw dampers are required at the connecting links. A similar pitch-stability problem was solved by using the trailer units in pairs, with each pair having one pinned and one sliding coupling. This also allows the train to be shortened easily, which comes in very handy if only a few people want to use it. Each unit has a seat and pedals or a linear drive or a rowing mechanism, as well as a roof made of canvas on a tubular frame.

    Four complete Trampelwurms were built by unemployed persons at the city of Thun in Switzerland and extravagantly decorated by local schoolchildren. The city of Thun owns and operates three of the vehicles. A part-time staff of six people runs the project, taking bookings and performing the repairs that are frequently necessary. Another ten people are en-

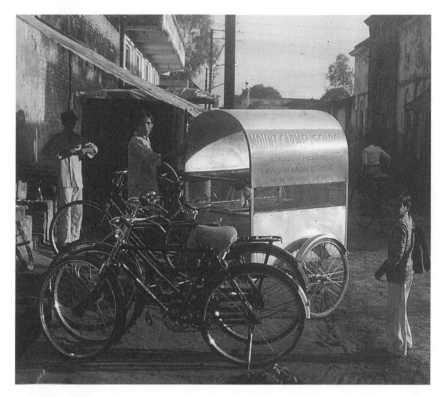

Figure 11.16
Human-powered school bus in Kanpur, India. (Photo by Dave Wilson.)

Figure 11.17
Human-powered freight-carrying vehicle.

Figure 11.18
Thuner Trampelwurm. (Courtesy Theo Schmidt.)

gaged as drivers; the vehicles ply for customers in the pedestrian part of Thun and are available to be booked privately. Although as heavy as an automobile and as long as any legal road vehicle, the Trampelwurm can negotiate the most crowded and narrow pedestrian areas in safety and can also travel on typical roads as long as they are not too steep. Parties enjoy the tricks the drivers perform, like catching up with their train's own tail, forming a temporary human-powered merry-go-round, or diving into a particular steep narrow tunnel in roller-coaster fashion. The trains operate from April to November, and the number of people transported per year is on average 5,700.

Electric-assist bicycles

The Velocity power-assist drive system
The Velocity system, invented by Michael Kutter of Basel, Switzerland, was the first commercially available power assist for electric bicycles to use the "human-power amplifier" approach: a sensor measures the rider's pedaling rate and controls the battery-powered electric motor according to a programmable function determining the feel of the ride, from "economical" to "wow!" Because of the predictable relationship between speed and power, a further torque sensor is not required. This system has since been used in many other designs, but none offers the same dynamic range as the Velocity bicycle (figure 11.19), which allows smooth acceleration while pedaling from a standing start to about 30 km/h without changing gears. This is possible because of Velocity's unique way of mixing human and motor inputs: instead of the usual adding of torques at constant speed, the epicyclic hub-gear differential adds speeds at constant torque, acting as a virtual continuously variable transmission. Pedal gear changing is still required for adapting to gradients.

Figure 11.19
Velocity assisted-human-powered bicycle. (Courtesy Theo Schmidt.)

The bicycle can operate in three modes.

• Pedal only. In this mode, the gear range is automatically quite low, so that it is always possible to reach any particular point the rider wishes to, even uphill with discharged batteries.
• Motor only. In this mode, the speed is automatically limited to less than 20 km/h, which encourages the use of the bicycle's pedals and prevents motorized misuse.
• Combined use. In this mode, the speed is a function of the rider's effort even when the motor is fully on, so that the "power-amplifier feeling" is always active. The speed that can be attained remains a function of the rider's fitness, as with an unmotorized bicycle, but at a higher level.

The Velocity system offers great riding fun and high average speeds for ordinary people and simultaneously prevents excessively unsafe peak speeds.
The first series of about 30 Velocity bicycles was built with Cannondale frames, some with and some without suspension. The 2000 model is the Velocity Dolphin, which uses a custom-welded aluminum frame and, as before, a removable battery case with integrated charger. A few recumbents and also a Leitra tricycle have also been equipped for experimenting and racing.

Human-powered water vehicles

Speed vehicle: the Decavitator hydrofoil

In 1989, Du Pont, through the IHPVA, offered a prize of $25,000 for the first single-person human-powered water vehicle to reach 20 knots (10.29 m/s) before December 31, 1992, or for the vehicle of that type that had reached the highest tested speed (under strict rules) by that date. The competition for this prize stimulated much activity in the area of human-powered water vehicles. The MIT group that successfully mounted the Daedalus human-powered airplane effort (see below) returned from Greece earlier than they had expected, because of their immediate success in accomplishing their goal there, full of pent-up energy, and decided that the water speed challenge would be a suitable outlet. They chose to use the propeller employed in the Daedalus aircraft, thereby giving themselves an advantage unforeseen by the people who set the rules, which allowed records to be set with a higher maximum wind speed than the upper limit on water current. A water vehicle getting its thrust from the air has a bonus when running with the maximum allowable tailwind. Developing the hydrofoil craft that would cope with the high-level thrust from the airscrew and with the waves of the Charles River basin where the craft was tested was, however, nontrivial, as the boat's name, Decavitator (figure 11.20), implies.

Recreational and utility watercraft

A wide range of single-hull and catamaran boats driven by pedals and propellers are now available for purchase, and Circle Mountain Industries produces a pedal-conversion kit, shown in figure 11.21, that can be fitted to one's own favorite boat. Some of the available human-powered boats position the rider high over a narrow hull and provide a front rudder connected to bicycle handlebars, so that the boat is kept upright by the same actions as in bicycling. Philip Thiel (1991) has produced a range of utility boats, including a barge (figure 11.22) for cruising canals, propelled by one or two people, with sleeping berths for two, toilet, kitchen, and space for a folding bicycle on which errands can be run for needed supplies of bread, cheese, fruit and wine (when in France . . .).

Vel'Eau 12

Vel'Eau 12 (figures 11.23 and 11.24) is a human-powered boat with seats for twelve persons, six on each side facing one another, offset to allow for ten pedal drives that are connected to the longitudinal propulsion shaft located under the boat's floor. All drives except the helmsman's have freewheels, so the danger associated with multiple fixed pedals is removed. An

Figure 11.20
Decavitator world-record-setting pedaled hydrofoil. (Courtesy Mark Drela.)

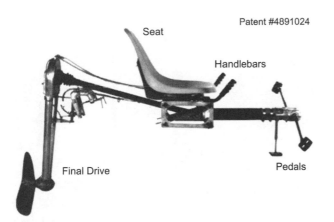

Figure 11.21
Human-powered "outboard" boat drive. (Courtesy Circle Mountain Industries.)

Figure 11.22
Human-powered runabout boat. (Photo by Dave Wilson.)

Figure 11.23
Vel'Eau 12 human-powered boat. (Courtesy Theo Schmidt.)

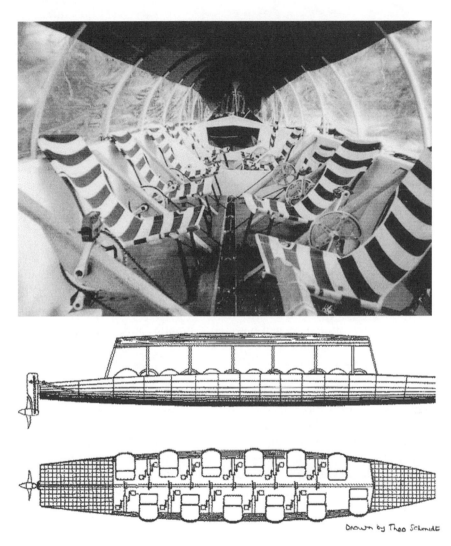

Figure 11.24
Design drawings and inside view of Vel'Eau 12. (Courtesy Theo Schmidt.)

arrangement of universal joints and a telescoping section allow the propulsion shaft, exiting at the uppermost point of the boat's transom, to connect to the propeller/rudder unit in such a way that it can be steered almost 90° to either side and also lift 90°, for example, in shallow water or for clearing the propeller. Internally, the propeller unit contains a simple untwisted chain drive with a step-up ratio of about four. The two-bladed propeller has a diameter of 550 mm and a pitch of 700 mm.

Vel'Eau 12 is 12 m long and 1.3 m wide. The hull is hard-chine and made from 6-mm marine plywood glued and sealed with epoxy. Plastic hoops support a removable canvas roof with clear sides. There is also a leeboard to prevent excessive sideways drift in windy conditions. The complete craft weighs about 250 kg.

Vel'Eau 12 is easily driven by as few as two persons. The all-day cruising speed is about 5 knots with crews of four to ten average persons. Vel'Eau 12 is owned by the French company Eco-Inventions, which rents it out to groups; these groups often take camping equipment along for week-long trips, mainly on the Saone River.

Human-powered submarines

The first international human-powered submarine race was organized by the IHPVA in 1989 and the third in June 1993. The fall 1993 issue of *Human Power*, edited by P. K. Poole, offers guidance on design parameters for human-powered submarines. The issue also contains descriptions of the design and construction of five submarines by teams competing in the 1993 race. Not described in that issue, but nonetheless very successful in submarine competitions from 1993 to 1999, was the Omer series of submarines, designed and built by students at the Ecole de Technologie Superieure (ETS) in Montreal (see also this school's helicopter project, discussed below.) Figure 11.25 offers a view of Omer 3. Its hull is formed of a carbon-Kevlar composite, with a maximum diameter of 610 mm and a length of 2.75 m. The maximum speed reached is 6.98 knots (3.6 m/s, 8 mile/h). At this speed the propeller is rotating at 210 rpm, with the pedaling speed at 70 rpm. The propeller has variable pitch, continuously controlled by an on-board microprocessor (the team credits this system with its successes over other craft, none of which has such a system at the time of publication [2004]).

Human-powered airplanes

The Kremer prizes

Human beings have tried to imitate birds for at least two millennia. Leonardo da Vinci sketched a helicopter, and many experimentalists in the

Figure 11.25
Human-powered submarine Omer 3. (Courtesy Ecole de Technologie Superieure, Montreal.)

nineteenth century dedicated themselves, and sometimes their lives, to
demonstrating human-powered flight, without success. Some short "hops"
were achieved in the 1920–1960 period. In 1959 Henry Kremer, a British
industrialist, was persuaded in a moment of weakness to offer a prize of
£5,000 (then equivalent to about $20,000) for the first human-powered
aircraft to fly a figure-eight course at least ten feet from the ground around
two pylons a half-mile apart. Paul MacCready's Gossamer Condor won the
prize in 1977. Kremer was delighted and agreed to offer a series of prizes,
starting with one for a human-powered crossing of the English Channel (La
Manche). This prize (£100,000, then worth $180,000) was won by another
MacCready plane, the Gossamer Albatross, in June 1979. (The prize money
awarded in this case seems to be a substantial sum. However, if a government commissioned the development of a human-powered plane to cross
the channel, it would cost many times this. Kremer's generous prizes
stimulated an enormous amount of interest and activity, out of proportion
to the amounts involved.)

The Gossamer aircraft have been given a great deal of well-deserved publicity. We will illustrate human-powered flight with two other remarkable aircraft series: the Musculair and the Daedalus. We will also mention some helicopter and dirigible projects. A full description of human-powered aircraft is given by Roper (1995).

The Musculair aircraft

Musculair I was built by the late Gunter Rochelt and his son Holger, a supposedly compromise design to win two Kremer prizes. The younger Rochelt, characterized by Roper (1995) as "not particularly athletic," was designated as the pilot. However, he won the non-U.S. figure-of-eight prize in June 1984, shortly after the Musculair I's first flight. He also took his younger sister Katrina up for a short trip, the first passenger flight of a human-powered aircraft. In August of that year he won a second Kremer prize, for speed around a circuit. The rules governing the Kremer prize allowed energy storage to be used (from the pilot's energy, which had to be stored on board the aircraft and generated during the ten minutes preceding the flight.) The Rochelts decided that they could do better without energy storage and proved themselves right. They also used less than a quarter of the wing area of the Gossamer Condor and no wing bracing. It was a remarkable achievement (especially for someone "not particularly athletic"!).

The Musculair I was unfortunately destroyed in February 1985 in a traffic accident while being towed in its trailer. The Rochelts built a new version, Musculair II (figure 11.26), aimed at winning another speed prize. Successive Kremer prizes could be awarded for a 5 percent improvement in speed over the previous record. Holger Rochelt won a third Kremer prize for what had earlier been regarded as an unbelievable speed of about 48 km/h (over 13 m/s and nearly 30 mile/h) on October 1, 1985. Roper (1995) reports that the Royal Aeronautical Society, the prize administrators, closed the speed prizes shortly afterward because it was considered impossible to exceed Holger Rochelt's speed by another 5 percent.

Daedalus

Students, at the Massachusetts Institute of Technology, along with faculty members and other advisors, designed and built a biplane, Chrysalis, as a preparation for an attempt to win the cross-channel Kremer prize. (The author was one of the many permitted to power and pilot it.) Later they built the monoplanes Monarchs A & B and won a Kremer speed prize. They could not hope to surpass the Rochelts' speed, however, and in 1985 they began considering a remarkable flight that had no monetary prize: a re-creation of the mythic flight of Daedalus, from Crete to Greece. The destination chosen was actually the island of Santorini, 119 km

Figure 11.26
Musculair II at Basel Air Show, 1985, in flight near a DC-3 and a Concorde.
(Courtesy Ernst Schoberl.)

from the launching site on Crete. The aircraft they created for this
endeavor, the Daedalus (figure 11.27), and its Greek pilot Kanellos Kanel-
lopoulos were successful on the first attempt, even though the aircraft
broke up as it was hit by a rather violent crosswind while coming in to the
beach.

Most human-powered aircraft before the Daedalus had used a lightweight steel-cable chain substitute (see chapter 9) to transfer the pilot's
power from the pedaled shaft to the overhead propeller shaft at right
angles. Power was transmitted Daedalus by means of two sets of bevel gears
and a torque tube. The design was a remarkable accomplishment, simultaneously saving weight, achieving greater reliability, and producing a higher
transmission efficiency with a drive that, when applied to bicycles, has
generally been heavier than the chain drive it has replaced, less efficient,
and more prone to failure.

Figure 11.27
Daedalus world-record-distance human-powered airplane. (Courtesy Mark Drela.)

Human-powered helicopters

In 1980 the American Helicopter Society offered a prize (named after the helicopter pioneer Igor I. Sikorsky) for the first human-powered helicopter to hover for one minute and to reach a minimum height of 3 m at least for a moment. The winning helicopter must also remain over a 10-m square, hold at least one member of the crew nonrotating, and obtain lift solely through rotating elements. The prize ⟨http://www.vtol.org/hph/hph/html⟩ has not yet (as of 2004) been won. The following notes owe much to a communication from the supervisor of the ETS project Helios, Doug Furton (2000), who has also given his permission for the illustrations included here to be used. (A spirit of sportsmanship was cheerfully evident when the author was told by competitors that the ETS human-powered helicopter had the highest probability of succeeding in capturing the prize.) Five designs for human-powered helicopters are shown in figure 11.28 with notes by Furton (2000).

Da Vinci III
A group of students at California Polytechnic was the first to fly a human-powered helicopter, the Da Vinci III (figure 11.29), in 1989, hovering for

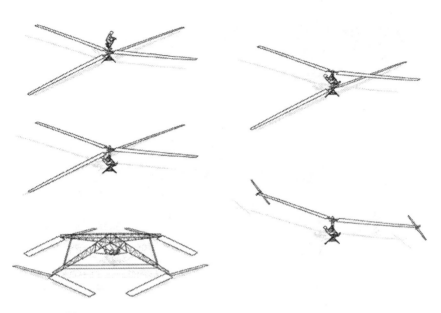

Figure 11.28
Conceptual designs of a number of human-powered helicopters. Those in the left-hand column were developed under the direction of Akira Naito in Japan. The lowermost of these, the Yuri I, holds the world record of 19.46 s at an altitude of 200 mm. The Da Vinci III, on the lower right, was the first and only other human-powered helicopter to fly. Helios, upper right, is most similar to the early two-rotor Naito designs. (Courtesy Doug Furton, Ecole de Technologie Superieure, Montreal.)

seven seconds and reaching a height of 200 mm. The Da Vinci III employed a two-bladed rotor, 30 m in diameter, driven by tip propellers, a system that cancels out net torque on the nonrotating frame carrying the pilot. The entire aircraft weighed 44 kg (96 lbm). The flight was cut short by a structural failure in one rotor blade.

Yuri I
An especially noteworthy human-powered helicopter project was conducted by Akiro Naito, who before retirement was responsible for a long series of successful aircraft designed and built by his students at Nihon University in Japan (Naito 1991). The first four helicopters, A Day Fly and Papillon A, B, and C were developed and constructed with his students. After his retirement he designed and built Yuri I and Yuri II (named for his wife Yuriko), which were almost a one-man effort by a consummate analyst and craftsman (Naito is also the world champion micro-origamist). The

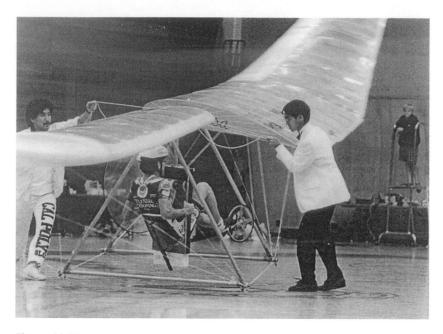

Figure 11.29
"Da Vinci III" leaving the ground. (Courtesy Bill Patterson.)

Yuri I (figure 11.28, lower left) had a large frame of Japanese cypress supporting four rotors, each 10 m in diameter, two rotating clockwise and two counterclockwise by means of a Kevlar wound-rope drive, the whole of the craft weighing 32 kg (71 lbm). It flew in March 1994 for 19.5s and reached a height of 200 mm, an official world record. Later it flew for 24 s and reached 700 mm, but this flight was not officially observed and could not be classified as a world record. Yuri II is shown in figure 11.30.

Helios
The student team at ETS that produced the human-powered submarine Omer turned its attention to the Sikorsky prize in early 1998. The team's efforts to develop a human-powered helicopter to capture the prize are ongoing; what is presented here is merely a progress report. The team reasoned that the two previous human-powered helicopters that had been successfully flown (the Da Vinci III and Yuri I) had used "ground effect": the reflection of the air flow "downwash" upward to provide added lift when an airfoil is close to the ground. However, the first Naito design, with two concentric counterrotating rotors, suffered from severe buffeting as the lower rotor turned through the wake of the upper rotor. The ETS team

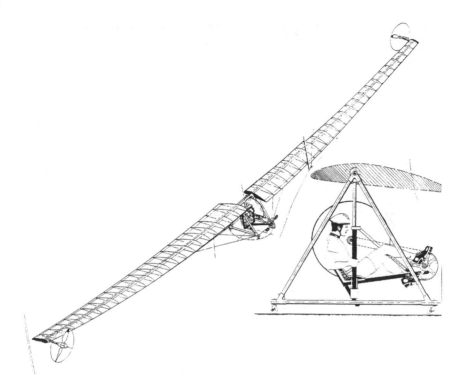

Figure 11.30
Da Vinci III drawing by Matai Kiraly.

chose a design similar to Naito's, except that the lower rotor would be below the pilot, in full ground effect, whereas the upper rotor would be above
the pilot, not having the benefit of ground effect but not producing the
buffeting interference. The rotors are 30 m in diameter, and, as on the
Omer submarine, their pitch is controlled by an on-board microprocessor
to provide stability to the craft and station keeping above the 10-m-square
launching pad. The craft uses carbon fiber as a structural skin, with structural foam and carbon as backing.

Human-powered blimp: The White Dwarf

Piloted by its designer, Bill Watson, the White Dwarf HP airship (figure 11.31),
conceived and owned by the popular comedian Gallagher, is filled with approximately 6,000 cu.ft. (170 cu.m.) of helium. The 1.6-m-diameter propeller,
mounted on a pylon behind the pilot, can have its thrust angle altered through
nearly 100 degrees to allow altitude control via a lever seen in the pilot's left
hand.... Two triangular tanks, under and behind the pilot, carry water ballast.
(Anonymous 1985)

Figure 11.31
Human-powered blimp White Dwarf. (Photo by Bryan Allen.)

The cross-channel pilot Bryan Allen, in a talk to the MIT Daedalus team, said that aircraft of the White Dwarf type seemed to be the only type of human-powered aircraft that could provide recreation and possibly utility with reasonable availability. (Successful flights by heavier-than-air human-powered airplanes are typically preceded by days, sometimes weeks or even months, of waiting for perfect conditions.) If one wants to travel by blimp to a chosen destination, a wind speed below that of the blimp is essential. Allen described a trip he took over the California countryside on a lazy summer day, requiring little power input, floating over communities in relative silence, and able to greet and talk with people below him (often to their considerable surprise).

This happy picture is a suitable one on which to close this short survey of alternatives for human power.

References

Anonymous. (1985). "The HP Airship White Dwarf." *Human Power* 3, no. 3 (Spring):1.

Furton, Doug. (2000). "Helios: A helicopter with legs." *Rotor & Wing* (April 2000). Published by *Aviation Today*.

McCullagh, James C., ed. (1977). *Pedal Power*. Emmaus, Penn.: Rodale Press.

Mellin, Bob. (1996). *Railbike: Cycling on Abandoned Railroads*. San Anselmo, Calif.: Balboa Publishing.

Naito, Akira. (1991). "Review of developments in human-powered helicopters." *Human Power* 9, no. 2 (Summer):1, 7–9.

Poole, P. K. (1993). "H.P. submarines: Design parameters." *Human Power* 10, no. 4 (Fall):24–26.

Roper, Chris. (1995). "History and present status of human-powered flight." In *Human-Powered Vehicles*, ed. Allan Abbott and David Wilson, 217–238. Champaign, Ill.: Human Kinetics.

Thiel, Philip. (1991). "Pedal-power on the French canals." *Human Power* 9, no. 1 (Spring):4.

Whitt, F. R., and D. G. Wilson. (1982). *Bicycling Science*, 2d ed. Cambridge: MIT Press.

12 Human-powered vehicles in the future

Introduction

To write about the future is, of course, risky. It is easy to review recent trends and to extrapolate. However, we will give some relevant data in this chapter on bicycle usage and manufacture, with appropriate cautions on extrapolating from them. We shall point out that, although we like to think of ourselves as free creatures, what we do is largely controlled by governmental actions, and that these actions are highly uncertain, even in democracies. We will look at developments in technologies affecting the different categories of bicycles and of other human-powered vehicles. We shall give our own "wish list" of technologies that we hope will be adopted and, in some cases, invented. Enthusiastic readers can also access publications that record the dreams of the future for cycles, such as *Encycleopedia* (McGurn and Davidson 2000), *Bike Culture*, and others produced by McGurn and Davidson at Open Road. (Open Road is now defunct, but one of its principals, Peter Eland, started VeloVision in 2001 to continue the tradition Open Road began.)

Government regulations and incentives

The author and the principal contributor of this book were born in Britain and now live in the United States. The very different nature of bicycling in the two countries could be taken to be representative of their different cultures. On the other hand, a major component of national behavior comes from laws and regulations and the degree to which these are enforced. While it could be stated that these laws and regulations in turn come from the people of their respective countries, the "law of unintended consequences" applies to laws themselves in addition to regulations and customs, and thereby laws shape communities in ways that those who originally proposed them or voted for them might not originally have foreseen.

For example, in the nineteenth century in the United States, the federal government saw an overwhelming need to connect the various parts of the country and to "open up the West," and it gave generous inducements to railroad companies to build westward lines. For this and many other reasons was born an era of "railroad barons" such as Cornelius Vanderbilt: people with great wealth and power. Oil was discovered, and "oil barons" such as John D. Rockefeller joined the ranks of America's multimillionaires. The Sherman Anti-Trust Act became law in 1890, and the Interstate

Commerce Commission (ICC) was given, by 1910, according Oscar and Mary Handlin,

extensive rate-fixing authority. The courts became battlegrounds across which lawyers sallied to establish the boundaries between licit combinations and conspiracies in restraint of trade ... litigation was a wholesome alternative to the overt violence and chicanery that had enlivened entrepreneurial contests in the 1870s.... While sometimes, as in the case of railroads and urban transit systems, those constraints [e.g., rate-fixing and rule-making] were so narrow as to stifle growth, in most industries entrepreneurs bore the burden lightly and even profited from it (Handlin and Handlin 1975).

The arrival of automobiles and the empires associated with them created conflicts. The ICC and other regulatory bodies seemed to have opposite effects on railroads and highways: railroads began losing money and merging or going out of business, while truckers began taking over freight hauling, even over long distances along the same routes covered apparently more efficiently by the railroads. Similarly, differential taxation and regulation made it far less expensive for a family and even an individual to drive an automobile or to take a bus between two cities than to take a train, and passenger railroads have died out in the United States except when highly subsidized.

Economists show that trucks and automobiles are also subsidized, in fact subsidized to a far greater extent than are the few persisting passenger railroads. However, the subsidies are of a totally different character. Subsidies for passenger railroads and subway systems are funded from tax monies that are handed over to the railroad managements. Subsidies for highway users are costs that are imposed on general taxpayers and on many others (for instance, the costs of highway maintenance, snow clearing, bridge repair, accident services, police, pollution, delay, and urban sprawl that are not charged directly to highway users). It is politically very difficult to correct this anomaly, because lobbyists connected with all the powerful groups that would be affected by such a change are very active in advancing legislation favorable to the industries they represent, and vice versa. There are virtually no lobbyists looking out for the interests of the weaker groups, including poor people, pedestrians, and bicyclists, who would benefit from the correction of these anomalies and the promotion of fairness.

In summary, users of automobiles in particular are highly subsidized in the United States, by an average, for the quantifiable costs alone, assessed by some economists as sixty-seven cents per mile in 2002 money. That this is an average value means that the subsidy in cities and at rush hours is very much higher than this, whereas the subsidy out in areas of low population density is negligible. Therefore the users of other forms of transportation, including urban bicyclists, are competing with this enormous motor-vehicle subsidy. In other countries with higher fuel and other taxes, the subsidies are lower than those in the United States, but they are still significant. And a fuel tax is a very crude method of recovering some of the external costs of using motor vehicles. To produce greater fairness in road use, three complementary forms of taxation are needed: electronically collected per-mile or per-kilometer road-use taxes and parking taxes, both varying with place and time of day, in addition to fuel taxes. (Preferably, proposers of taxation should also stipulate the destination of the monies collected. It is the author's opinion that these taxes should be deposited in a trust fund that is reduced to near zero each month by a uniform distribution to all [at least to all adult local] citizens through a "negative" income tax—i.e., a refund or rebate. In this way, poor people would receive a guaranteed small income. Rich people would receive the same rebate income, but their additional expenditures would be likely to be higher than this rebate if they used automobiles.) The author has been advocating this policy so stridently since the early 1970s that his friends have dubbed it "Wilsonomics." The necessity of such a policy is gradually gaining acceptance, even by economists. It has been picked up by Greenpeace Germany, and it may be incorporated into legislation there and possibly elsewhere. A different approach with similar consequences has been proposed recently by Barnes (2001).

The vital relevance of this policy to our mission in this book is that most bicycling occurs in urban and suburban areas, the same locations where there is increasing traffic accompanied by gridlock and "road rage," apparently all over the world. If there were a gradual introduction or increase of all three forms of taxation suggested above, to an extent appropriate for each urban area, there would be a gradual reduction in motor-vehicle use, starting with those people whose use of motor vehicles is a daily choice between two level-value alternatives and who would happily decide not to drive if it were made a little less attractive. Traffic obeys what is sometimes called the "cocktail-party equation." If one has the misfortune of attending a cocktail party and arrives early, conversation is possible without undue strain. People keep arriving, and the noise level increases gradually. The critical point is reached when one additional person suddenly causes everyone to have to shout to be heard by her companions (figure 12.1). From then on conversation is exceedingly difficult and unpleasant. Traffic behaves the same way: at a certain point, the introduction of one more vehicle produces a traffic jam. The converse is then true: at a particular point, with only a small reduction, traffic flows much more freely. Conditions become much pleasanter not only for bicyclists but for people taking buses, for instance. These can now also travel faster, and

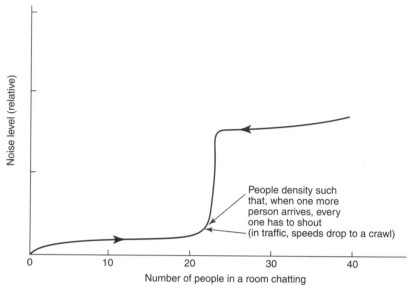

Figure 12.1
The "cocktail-party equation."

some of the motor-vehicle drivers will say to themselves, "The buses are quicker than I thought: I'll switch when it's possible to do so." And the buses can then become more frequent, reducing the waiting times and further increasing their appeal, and so forth.

There have been many movements in many countries to introduce road-use taxes, sometimes referred to as "congestion taxes," and there are now places where tolls on high-speed roads parallel to heavily used roads are collected electronically. However, the region-wide introduction of road-use taxes for a large nation or group of associated nations involves so great a complexity that it seems likely that they will be first introduced comprehensively in an island nation and, if successful, spread rapidly to others. A form of congestion charge was, however, introduced in London in 2003 and has won high praise for its early success.

In addition to taxing or tinkering with incentives, governments can also regulate. Motor vehicles can be prohibited in city centers, parks, and other recreational areas. Highways can be declared off limits for bicyclists. There have been several campaigns in Asian countries to banish rickshaws and to restrict bicycles, even in China. In democracies, motor-vehicle and oil-producer lobbies are very powerful, and bicyclists need to have lobbyists to counteract what would otherwise be absolute power, lest their interests be overriden entirely. "Power corrupts, and absolute power corrupts absolutely" (attributed to Lord Acton, nineteenth-century British historian). The comments of Andrew Oswald (2000) on what he perceives as ruination of Britain's universities by government measures are relevant as well to the situation regarding bicycles. "These measures were the work of outwardly rational and plausible politicians. As with most mistakes in life, they did not happen because of outright malice. They were made by honest men and women with the best of muddled intentions. The problem was sheer mental sloth, plus an eye on short-term exchequer advantage, rather than on any appraisal of long-term costs and benefits."

Forecasting the future use of bicycles and other human-powered vehicles is thus an impossible task, dependent on government actions that might be directed at one set of problems unrelated to bicycles and might yet have unintended effects on bicycle usage. "The price of liberty is eternal vigilance" (Thomas Jefferson, June 3, 1779, Yorktown).

Data on bicycle production and use

Past use of bicycles
We know of no comprehensive set of data of bicycle usage over time and regions. Hence we will give some statistics (where we have found them) as a guide for estimating missing data.

Bicycle sales in the United States
Perry (1995) gives estimated and collected data for bicycle sales in the United States since 1863. Sales of the new safety bicycles (our interpretation) rose from about 340,000 in 1890 to 2 million in 1897, the height of the bicycle boom (figure 12.2). Sales dropped to 250,000 by 1904 and oscillated from this level to occasionally twice this (being affected by war conditions) until 1932. Sales then began increasing, probably first because of the Great Depression, and later the Second World War, peaking at nearly 2 million again in 1941. Sales rose after the war, reaching nearly 3 million in 1947, presumably mainly because of purchases of bicycles for young people, and nearly 7.5 million in 1968. The (adult) bicycle boom of the early 1970s led to sales of 14–15 million per year from 1972–1974. Since then the sales figures have ranged between a low of 7 million and a high of 13 million, with 8 million of the bicycles sold having been domestically produced and 5 million imported.

Bicycle use in the United States
The Bicycle Federation of America reported that there were eighty-eight million bicyclists in the United States in 1988, twenty million of whom

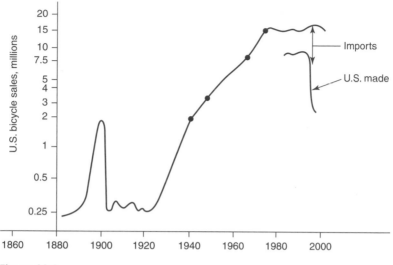

Figure 12.2
U.S. annual bicycle sales. (From various sources.)

were adults riding once per week, and estimated that this latter figure would increase 20 percent between 1988 and 1989 (Perry 1989). Adult commuters accounted for 2.7 million of these. There were 7.5 million users of all-terrain bicycles as of that year, million, and a 40 percent increase in that number was estimated for the next year. There were 180,000 cyclists involved in racing.

Future bicycle technology

Lightweight bicycles using composites

There have been many attempts to produce bicycles from plastics that, the designers hoped, would enable bicycles to be mass-produced at low cost. Both previous editions of this book have discussed some of these attempts. However, all of the plastic bicycles produced so far have lacked sufficient rigidity. Carbon-fiber composites are, however, at the other end of the scale, being extremely rigid, and have been used to make record-breaking but very-high-cost racing bikes (e.g., figure 12.3). Carbon composites have also been used to make lightweight low-aerodynamic-drag wheels and other bicycle components (figure 12.4). In general, these components have used variations of "wet layup" procedures incorporating a high proportion by volume of long carbon filaments in an epoxy matrix, procedures that do not lend themselves to mass production. As fiber length is reduced,

Figure 12.3
Carbon-fiber-reinforced road bike. (Courtesy Trek Corp.)

Figure 12.4
Lightweight low-drag composite bicycle wheels. (Courtesy Spinergy, Inc.)

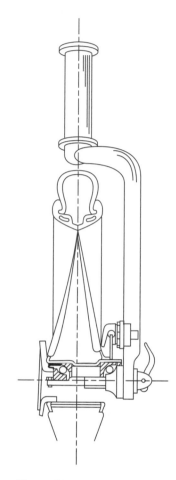

Figure 12.5
Monoblade forks.

the properties of the composite are degraded somewhat, but there comes a point at which the chopped-fiber-resin mix becomes fluid enough to flow into molds. Therefore there is promise that future bicycles could be injection molded to produce good stiffness and durability. The price of carbon fiber would have to fall considerably before such a process would lead to low-cost bicycles, however.

Tubeless tires
The increasing use of composite wheels is certain to foster the development of tubeless tires for bicycles. These should reduce weight. The uniformity

Figure 12.6
Flevobike, with wheels cantilevered on one side to permit the use of a single-fork design.

in roundness and width of composite wheels should also reduce rolling resistance.

Monoblade forks

Mike Burrows (1993, 2000) has incorporated single-blade forks into many of his advanced designs (figure 12.5), including his Windcheetah tricycles and many other highly original and fast bicycles. One-sided forks require wheels that are attached on one side (figure 12.6). These forks could bring all the advantages of automotive wheel-changing to bicycles: the transmission and brakes and hub would stay in place during a wheel change, and dealing with a flat would be just a question of replacing the lightweight rim

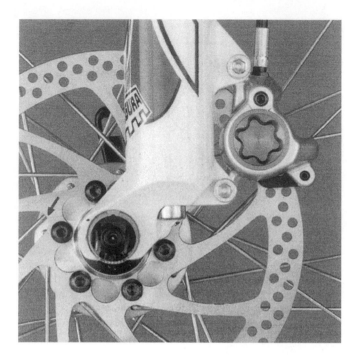

Figure 12.7
Magura Marta hydraulic disk brake. (Courtesy Magura-Germany.)

and its tire with a new one. The advantages of single-blade forks are so great that we hope and expect this form of construction to extend at least to mid-priced bicycles within a decade.

Disk brakes
Disk brakes apply their braking torque through the spokes of a bicycle wheel and can add stress to larger wheels (chapter 10). Nevertheless, they are being successfully used on all-terrain bicycles, which are promoting the development and use of these brakes (figure 12.7). They appear to be ide-ally matched to smaller (e.g., 406 mm, 20 inch) wheels and are likely to become much more widely, and more inexpensively, available than they currently are. Rim brakes on smaller wheels have two significant disad-vantages: braking heat is transferred to a smaller mass, so that on long downhills the rim can heat up to a high temperature and threaten the integrity of the tire and tube, and abrasive wear of the rim's braking sur-faces is more rapid on a smaller wheel, increasing the risk of explosive fail-ure. Disk brakes can consist of pad-disk combinations that have excellent wet-weather performance and involve a much longer period between the

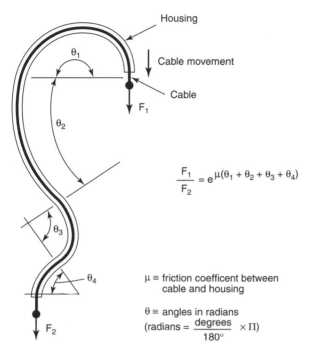

Housing

Cable movement

Cable

F_1

$$\frac{F_1}{F_2} = e^{\mu(\theta_1 + \theta_2 + \theta_3 + \theta_4)}$$

$\mu \equiv$ friction coefficent between
cable and housing

$\theta \equiv$ angles in radians
(radians $= \dfrac{\text{degrees}}{180°} \times \Pi$)

F_2

Figure 12.8
Attenuation in braking force when transmitted through cable bends.

need for pad replacement than for rim brakes and a much simpler procedure for doing so.

Hydraulic actuation
Cables with many bends produce a large and often unknown degree of
attenuation in the braking force applied at the wheels of a bicycle that
is different for each wheel and for each pattern of bicycle (figure 12.8).
Hydraulic action, in contrast, gives equal braking force regardless of distance or the tortuous nature of the pipe connecting hand lever to brake
path. Designers seldom produce bicycle cable systems that conform to the
guideline of cable manufacturers that cables should go around pulleys or
bends of diameters at least forty-two times the diameter of the cable itself
(figure 12.9). Consequently bicycle brake cables and gear-shift cables fail
frequently and without warning, producing, in the case of the brakes in
particular, extremely dangerous conditions. Hydraulic brakes are being
increasingly used on bicycles, and hydraulic gear shifters have also been
developed. These hydraulic systems should become universal, or else
greatly improved cable systems should be devised.

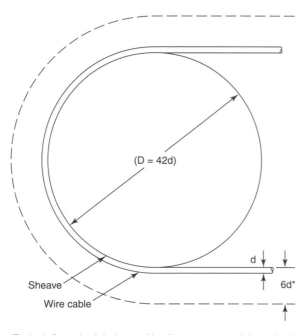

(D = 42d)

d

6d*

Sheave

Wire cable

*Typical of very bad design on bicycle components: (sheave/cable diameter ratio D/d should be a minimum of 42 for long fatigue life)

Figure 12.9
Bend limits for steel multistrand cables.

Elimination of exposed chains?

Many people are working to eliminate the exposed, oily, dirty, long chains on recumbents and other bicycles. Thomas Kretschmer (2000) is integrating through-the-hub pedal drive with a multispeed hub in a configuration (figure 9.19) that has a lot in common with the early Michaux-Lallement machines of the 1860s. Clemens Bucher (1998) is working with Flevobike to extend his development of a recumbent with a totally enclosed chain transmission (figure 12.10).

Derk Thijs produces Rowbikes in which the drive is a simple steel stranded cable passing around an aluminum spiral cone (a "fusee") needing no lubrication (see chapter 9 and in particular figure 9.23). (Thijs and his brother have patented a multispeed shifting mechanism for the fusee, and several records have been set and races won on their Rowbikes.)

It is standard practice in Europe to have the chain of a bicycle pass through PVC tubes in the long stretches between the chainwheel and the rear sprockets. These tubes add some chain friction but reduce the wear on chain and gears.

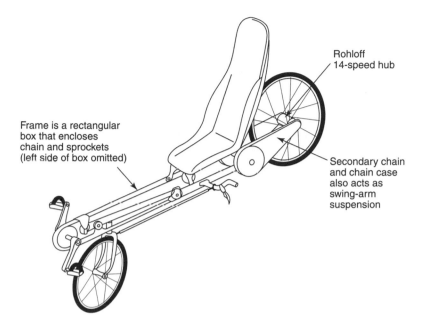

Rohloff
14-speed hub

Frame is a rectangular
box that encloses
chain and sprockets
(left side of box omitted)

Secondary chain
and chain case
also acts as
swing-arm
suspension

Figure 12.10
Totally enclosed chain transmission.

Electric drives, under development by Juerg Blatter and Andreas Fuchs (see chapter 9), would also eliminate the chain, would give optimum pedaling speeds, and would capture some braking energy for use in subsequent acceleration (see figures 9.25–26).

All-terrain bicycles

ATBs are still undergoing rapid development spurred by competition for a still-large market and by the desire on the part of component and bicycle manufacturers to win races, which will increase market share. The rapid developments in suspensions, gear-shifting mechanisms, disk brakes, wheels, and tires for ATBs have been beneficial to the rest of the bicycle industry, especially, perhaps, to the fledgling recumbent movement discussed below. Derailleur transmissions and exposed aluminum-alloy chainwheels are weak points in present ATBs, and it seems probable that some of the alternative transmissions described in chapter 9 will be adopted for these bicycles. Downhill racing is extraordinarily demanding and dangerous; it is like rock climbing on exposed high-altitude faces in that many technological aids that would reduce danger are spurned because overcoming danger by human qualities is a large part of the reason for the sport. Improved reliability of all components, especially, perhaps, tires and front forks (still prone to occasional batches of bad design or manufacture leading to horrible injuries), is a probable outcome of further competition and development. It seems from an observer's viewpoint that it would be easier to protect the rider by appropriate impact-absorbing clothing than by building more protection into the bike itself, but it may be difficult to encourage the sport's macho competitors to go beyond the helmets, gloves, and pads that they are already wearing.

Recumbent bicycles

Recumbent bicycles are those in which most of the rider's weight is carried in a seat, a little on the pedals, which are out in front of the body, and virtually none on the handlebars. (On a diamond-frame bicycle the weight is ideally divided equally among saddle, handlebars, and pedals). Enthusiasts for recumbents (the author is one) believe that this type of bicycle will increase in use for commuting, touring, and its own form of racing. (It will not, however, displace the all-terrain bike.)

At the time of publication (2004), recumbents are produced in many configurations. Some have the front wheel behind the cranks and bottom bracket, producing a short wheelbase (SWB) (figure 12.11a). Others have the front wheel ahead of the bottom bracket, resulting in a long wheelbase (LWB) (figure 12.11b). A growing proportion have the bottom bracket over the front wheel (giving a high pedaling positions), sometimes referred to as a "compact long wheelbase" (CLWB) (figure 12.11c).

Another variation among recumbents is in the position of the handlebars, either under the seat or above the seat. When above-seat steering is combined with LWB, the handlebars and stem of the bicycle can be quite long, and if the handlebars are directly mounted, there can be a pronounced "tiller effect": the handlebars must be swung widely to steer. An alternative arrangement is to use some form of gearing or universal joint at the head tube and a fixed position for the handlebar (a steering wheel has been used) (figure 12.12).

Recumbents can be unfaired, as in the examples above, or partially faired (figure 5.15) or fully faired (figure 5.14). They can be driven through the rear wheel, again as in the examples above, or through the front wheel, as in the Urieli bicycle of figure 9.11. They may have a rear wheel of conventional size and a smaller front wheel, as in figure 12.11, or two wheels of the same size, usually smaller than conventional wheels. Variations in the height of the bottom bracket are associated with variations in the angle of the seat back. Suspension of one or both wheels is increasingly incorporated into recumbents.

(a)

(b)

(c)

Figure 12.11
Recumbent-bicycle wheelbase types: (a) long; (b) short; and (c) compact long.
(Photo (a) by Ellen Wilson, (b) courtesy of Lighting Cycle Dynamics, and (c) by
Dave Wilson.)

Figure 12.12
Moller recumbent equipped with a steering wheel.

The profusion of recumbent-bicycle configurations is similar to that of early bicycles and of early automobiles. We would expect that favored arrangements will soon begin to appear.

Some of the advantages claimed for recumbent bicycles over the traditional diamond-frame pattern are the following.

1.   Greater safety because of the near impossibility of taking a "header" over the front wheel or of catching a pedal or foot on the ground when cornering.
2.   Far greater comfort in an almost complete absence of pain or trauma in the rider's hands and wrists, or back and neck, or crotch.
3.   Greatly improved braking, especially in LWB and CLWB models, because of a lower center of mass and a more favorable weight distribution.
4.   Better visibility forward and to the side for the rider compared with that for a diamond-framed road bike with dropped handlebars.
5.   Lower aerodynamic resistance for some unfaired configurations (particularly those with high bottom brackets) and for virtually all faired types compared with regular road bicycles.

6.   A large reduction in manufacturing and stocking costs for some configurations (particularly LWB) where one size fits all, compared with regular bicycles, which must be manufactured and stocked in three to six sizes.

Some disadvantages are the following.

1.   Rear vision is more difficult than for a conventional bicycle, so that good rearview mirrors are essential for safety.
2.   Recumbents are generally not good on rough terrain and in snow (because one cannot use changes in body position—"body English"—to improve balance).
3.   For the same reason, one cannot attack a hill, for instance, by jumping up on the pedals, which can give some muscle relief for riders of conventional bicycles. A wider range of gears is therefore desirable on recumbents.
4.   At the present stage of refinement, recumbents are generally heavier and much more expensive than their diamond-frame counterparts.

Sales figures, at least in the United States, of recumbent bicycles at the turn of the millennium exceeded those of tandems. However, most bicyclists in most regions of the world have not seen or even heard of one of these machines. Yet vigorous innovation in the design of recumbents is ongoing in several (mainly Western) countries.

There will be some who question the use of the word "innovation" in respect to ongoing work on recumbents. Bicycles and bicycling have a rich history, and it sometimes seems that a precedent can be found for every "new" development. The authors's involvement with recumbents started with his organization (1967–1969) of an international design competition in which he encouraged recumbency, entirely unaware (unbelievable but true) of the existence of earlier recumbents. Subsequently, friends constructed prototypes of five of my designs, and each one could later be said to bear at least some resemblance to earlier machines.

There is however, a fundamental difference between the enthusiasm for recumbents today and that of earlier periods: we now have technical publications and symposiums and the broadcast of information on the Internet, among other routes for disseminating information. These should ensure that future innovators will spend less time repeating earlier developments and more time "standing on the shoulders of giants" (to use the phrase employed by Isaac Newton—and it may not have been original with him—to describe the way progress is made in other technologies and sciences) to make advances.

Quantitative measures of recumbent-bicycle performance

Use of the word "advances" must also be considered carefully in regard to recumbents. In some respects—performance in races and in record setting—advances in recumbent bicycles can be measured, at least relatively. These are the same ways in which most sporting equipment, from conventional bicycles to tennis racquets, has been measured from the beginning of their sports. They have a significant disadvantage: athletic performance is mostly that of the athlete and only secondarily that of the equipment he uses. This disadvantage is compounded for recumbent bicycles, because top-category racers on upright bicycles do not want to compromise their muscle training by pedaling in a new position. It must be assumed that some new designs for recumbent bicycles have been eliminated from serious consideration because they have been seen to behave poorly in a race. Yet the reason for their poor performance may lie entirely in the low level of either skill or power or both of the rider, who is often the developer himself. Thus we honor racers Francis Faure, who rode Velocars in the 1930s to their place in history; "Fast Freddy" Markham, who has won many races and set many records for the Easy Racers company since the 1980s; and Sam Whittingham, a Canadian Category-2 bicycle racer, who holds the present (2003) world speed record (200-meter flying-start) at 36.2 m/s (130.4 km/h, 81.0 mile/h).

It is also possible nowadays to make quite accurate measurements of human power output in various positions, and of the energy losses arising from power transmissions, in aerodynamic drag, and in tire rolling friction, so that one could (in the best of all possible worlds) quote quantitative data on new recumbent bicycles. Yet there is a great deal about the characteristics of recumbent bicycles that is subjective. The machine may feel stiff or flexible, steady or "squirrelly" on the road; it may be very sensitive to sidewind gusts or may seem to ignore them; it may feel "steady as a rock" on high-speed descents or may suddenly degrade into shimmy in the steering; and so forth. Sometimes we lump these subjective characteristics together under the term "feel."

This term has different implications in different recumbent markets. We might identify three markets in different parts of the spectrum: that for the rider who wants to break records and to win races; that for the recreational rider who likes to go on rides with his club or family in the evenings or weekends and who wants to keep up with, or preferably go faster than, fellow riders and at the same time to feel comfortable and safe; and that for the commuter, who often needs to take a heavy briefcase or bag with lunch, rain gear, office clothes, etc., and who wants a comfortable and safe ride with the minimum of breakdowns, flat tires etc. Whereas in diamond-frame bicycles these three markets are supplied with very different machines, the recumbent is remarkable in that it is possible for someone to use a race-winning Gold Rush or Lightning (to choose two popular and successful U.S. machines) on which to go to work without too many compromises. However, normally a commuter would not be willing to spend a great deal more money than it would otherwise cost him for a bicycle that had lightweight disk wheels and thin-section lightweight tires, for instance, that would give only a marginal reduction in traveling time but a strong likelihood of sensitivity to side winds and to a propensity for more-frequent flats. Hence there is some differentiation among recumbent bicycles, just as there is for diamond-frame types. Only the first of the three markets identified above (racing) is served well by the use of race and record results as an indication of advances in the field.

Recent trends in recumbent design

Above-seat steering
Above-seat steering is gaining adherents over under-seat steering, partly because it gives more of a sense of familiarity to new riders. Seat adjustment is also likely to be simpler in above-seat models, and it is easy to mount a partial fairing (a wind and rain deflector) on the handlebar extension of such bicycles. Enthusiasts for under-seat steering still have reservations about above-seat steering for two reasons. One is that in a frontal crash, there is the possibility of the bars and other hardware giving facial injuries. The other is that in a really bad frontal impact in which one travels forward over the bike, one's automatic reactions might impel one to grip the handlebars instead of putting the hands out to safeguard one's head and spine in the fall.

Partial fairings
Fairings that can reduce aerodynamic drag somewhat (obviously less than for a total-enclosure full fairing), that do not render the bicycle dangerous in side winds, that protect the rider from rain and snow, and yet that are easy to get into and out of have been developed (figure 5.15). People write about their experiences using these partial fairings with something approaching rapture. If the recumbent movement maintains its present momentum, most manufacturers of recumbents will offer partial fairings. There is also a market for independent suppliers of partial fairings for recumbents, like that which exists for front fairings. It would be desirable to be able to buy a fairing built using umbrella technology, so that it could be deployed for longer trips and left stowed for shorter journeys.

20-20 CLWB and suspensions
The perhaps-mysterious symbol 20-20 CLWB is used to indicate that there is a trend for recumbent bikes to use similar midsize wheels, and the

nominal 20-inch (406-mm) size is usually chosen, rather than having a larger wheel (usually 700-mm or 27-inch) at the rear and a smaller (16-inch or 20-inch) at the front. There is also a trend toward the CLWB configuration with the crank axis over the front wheel (figure 12.11c). (However, at this book's publication, SWB recumbents are overwhelmingly popular in Europe and Britain, and there is approximately equal enthusiasm for SWB and LWB bicycles in the United States.) Recumbents are following the trend of all-terrain bikes in increasingly being fitted with suspension on one or both wheels.

Tricycles and quadricycles

Georg Rasmussen, a Danish physicist, has developed the Leitra enclosed tricycle (figure 12.13), which has hundreds of enthusiastic users, almost entirely in Europe. His pioneering work has been followed by the development of the Alleweder, the Twike, and the Cab-Bike (incorporating battery-electric power assist). Many similar (but generally cruder) carlike HPVs have been developed in the past. Some have flourished for a few years, for instance, in the years after World War I, and for a shorter period during the energy crises of the 1970s. It could be said that they were overcome by affluence: as people earned more money, they tended to "trade up" to vehicles of increasing size, power requirements, and speed.

The most recent wave of interest in HPVs also comes as a result of affluence. Too many people have too many motor vehicles and are getting too little exercise. City centers and beyond in many parts of the world are clogged with internal-combustion-engined vehicles that can move only slowly because of the congestion their sheer numbers create. The air is becoming dirtier. The planet is warming and the climate is changing, apparently usually for the worse. It would be logical, therefore, to predict that HPVs have a rosy future ahead of them. We hope they do! On the other hand, as noted earlier in the chapter, there are large flows of tax moneys and other funds aimed at developing battery-electric and fuel-cell vehicles and many other alternatives intended to contribute toward reducing the factors that bring about global warming (and toward not contributing to the growing obesity problem in many countries).

Transportation systems based on human-powered vehicles

The Mount Holley and Smithville Bicycle Railroad (figure 12.14) was opened in 1892 (Stockinger 1992). This and several other attempts at producing safer, faster, or more-enjoyable conditions for bicyclists are reviewed by Wilson (1992). Some of these systems provided a complete separate right-of-way or "guideway" that takes either special vehicles, as in figures

(a)

(b)

(c)

(d)

Figure 12.13
Enclosed commuting HPVs: (a) Leitra enclosed tricycle, with Mr. and Mrs. Georg Rasmussen; (b) enclosed recumbent bicycle; (c) cab-bike; (d) Alleweder semi-enclosed tricycle. (Photos by Dave Wilson.)

Figure 12.14
Mt. Holley and Smithfield Bicycle Railroad. (From Harter 1984.)

12.14 and 12.15, or regular bicycles, human-powered or totally externally powered or having powered assistance when needed. The combination of regular roads and railroads, streetcars, or buses (figure 12.16) is a special form of providing power assistance when needed. The author designed an electrically powered towing hook that can be fitted in a small trench in some roads where there are steep uphills (figure 12.17).

Off-vehicle power assist

The author has lived through the post–World War II period when bicycles could be fitted with small gasoline engines. Some of these were beautifully engineered into rear wheels, with the piston engine, the controls, and even the fuel tank all in the wheel hub, and the wheel thus equipped could

Figure 12.15
The suspended HP monorail. (From Harter 1984.)

Figure 12.16
Bicycles on a streetcar. (From Harter 1984.)

simply be substituted for the normal rear wheel. It seemed that it was only a short time before the users of these power-assist bicycles had switched to full motorcycles, including the Vincent HRD Rapide, guaranteed to be capable of 150 mile/h (240 km/h, 67 m/s) straight from the showroom. He has therefore always had a preference for power that could be provided externally. Previous editions of this book illustrated the Syracuse Crusway, in which regular bicycles could be connected to an overhead hook that would haul them up periodic steep inclines and then release them to coast down long, gradual descents. Other "guideway" proposals have been made (Kor 1992, 1998) to combine human power and electrical power in enclosed roadway tubes, in one case by arranging fans to blow strong breezes in the direction of travel. Another "modified highway" system was to have the bicycle's front wheel carried by a small powered truck along certain routes.

We shall repeat from the second edition of this book a concept that the author, at least, would like to see tried in hilly cities like Athens or San Francisco. A chosen bicycle route that goes over a steep and forbidding hill would be equipped, in the uphill direction, at the boundary between the pedestrian walkway and the road, with a moving handrail of the type used in escalators (figure 12.18). Parking on the road would be banned. Bicyclists could pedal over to the handrail and hold it with one hand in a level region

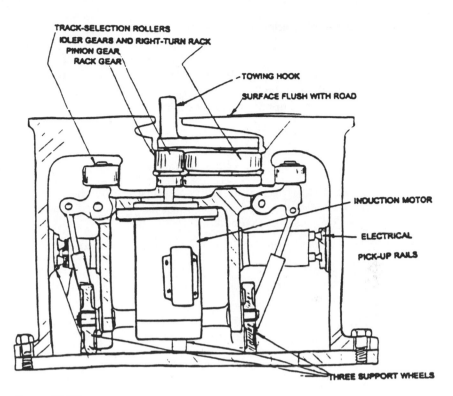

TRACK-SELECTION ROLLERS
IDLER GEARS AND RIGHT-TURN RACK
PINION GEAR
RACK GEAR

TOWING HOOK

SURFACE FLUSH WITH ROAD

INDUCTION MOTOR

ELECTRICAL
PICK-UP RAILS

THREE SUPPORT WHEELS

Figure 12.17
Electrically powered traction "rabbit." (From Wilson 1992.)

Figure 12.18
Powered handrail for uphill assistance.

before the uphill and would be pulled up at around walking speed. On the pedestrian side of the handrail, walkers and in-line skaters could also hold the handrail to be pulled uphill. It could be a lot of fun.

References

Barnes, Peter. (2001). *Who Owns the Sky?* Washington, D.C.: Island Press.

Bucher, Clemens. (1998). "Recumbent with encapsulated drive chain." In *Proceedings of the Third European Conference on Velomobile Design*, ed. Carl Georg Rasmussen, 184–188. Roskilde, Denmark.

Burrows, Mike. (1993). "Cantilever wheel hubs." *Human Power* 10, no. 2 (Fall–Winter):8–10.

Burrows, Mike. (2000). *Bicycle Design*. York, U.K.: Open Road, and Mukilteo, Wash.: AlpenBooks Press.

Handlin, Oscar, and Mary F. Handlin. (1975). *The Wealth of the American People*. New York: McGraw-Hill.

Harter, Jim. (1984). *Transportation: A Pictorial Archive from Nineteenth-Century Sources. 525 Copyright-Free Illustrations Selected by Jim Harter*. New York: Dover Publications.

Kor, Jim. (1992). "The Skyway project." In *Proceedings of the Fourth IHPVA Scientific Symposium*, ed. Chester R. Kyle, Jean A. Seay, and Joyce S. Kyle, 27. San Luis Obispo, Calif.: International Human Powered Vehicle Association.

Kor, Jim. (1998). "Solos micro metro." *Journal of Advanced Transportation* 32, no. 1 (Spring).

Kretschmer, Thomas. (2000). "Direct-drive (chainless) recumbent bicycles." *Human Power*, no. 49.

McGurn, Jim, and Alan Davidson. (2000). *Encycleopedia 2001*. York, U.K.: Open Road.

Oswald, Andrew. (2000). "A sorry state." *Economist* (July 1).

Perry, David B. (1989). "Bicycle use: More people riding for all reasons." *Pro Bike News* 9, no. 4 (April).

Perry, David B. (1995). *Bike cult*. New York: Four Walls, Eight Windows.

Stockinger, Herbert H. (1992). "The bicycle railroad." In *American Heritage of Invention & Technology*. New York: Forbes.

Wilson, David Gordon. (1992). "Transportation systems based on HPVs." In *Proceedings of the Fourth IHPVA Scientific Symposium*, ed. Chester R. Kyle, Jean A. Seay, and Joyce S. Kyle, 11–16. San Luis Obispo, Calif.: International Human Powered Vehicle Association.

Appendix

Notation

a	acceleration	(m/s^2)
A	area	(m^2)
C_A	coefficient of adhesion	
C_D	coefficient of drag	
$C_{D,SA}$	coefficient of drag based on surface area	
C_{LL}, C_{SL}, C_{SS}	abbreviated representations for equation coefficients multiplying lean or steer angular velocities; each is a complex combination of bicycle parameters and speed	
C_R	coefficient of rolling resistance	
d	diameter	(m)
d	for derivative notation	
D	outside diameter	(m)
E	modulus of elasticity	(Pa)
f_F	forced frequency	(Hz, s^{-1})
f_N	natural frequency	(Hz, s^{-1})
F	force	(N)
F_A	aerodynamic (resisting) force	(N)
F_{Acc}	acceleration force	(N)
F_B	bump (resisting) force	(N)
F_F	force of friction	(N)
F_P	propulsive force	(N)
F_R	force of rolling resistance	(N)
F_S	slope (resisting) force	(N)
$F_{V,f}$	vertical force from front wheel	(N)
$F_{V,r}$	vertical force from rear wheel	(N)
g	acceleration due to gravity	$(9.807\ m/s^2$ at earth's sea level)

g_c	constant in equation $F = ma/g_c$	(1 in SI; 32.17 lbm-ft/(lbf-s^2) in U.S. units)
h	hysteresis factor in urethane	
I	section moment of inertia	(m^4)
I_{xx}, I_{yy}, I_{xy}	moments and product of inertia relative to center of mass	(kg-m^2)
k	spring constant	(N/m)
K_A	aerodynamic-drag factor, $0.5C_DA\rho$	(kg/m)
K_C	convergence factor for solving power equation	
K_{CS}	Cornering stiffness of a pneumatic tire, the ratio of lateral force to the "drift slope" (i.e., ratio of lateral creep to forward motion)	(N)
$K_{H/V}$	ratio of wheel's angular momentum to its rolling velocity	(kg*m)
$K_{LL}, K_{LS}, K_{SL}, K_{SS}$	abbreviated representation for equation coefficients multiplying lean or steer angles: each is a complex combination of bicycle parameters, speed, and the acceleration of gravity	
L	length, as defined in text	(m)
L_{CL}	tire contact length ahead of the axle	(m)
L_{FO}	fork offset perpendicular to steering angle	(m)
L_{MT}	mechanical trail	(m)
L_{PT}	pneumatic trail	(m)
L_{TT}	tire-tread thickness	(m)
L_W	wheelbase	(m)
L_{WW}	wheel (tire) width	(m)
m	mass	(kg)
m_{eff}	effective mass of system (total mass plus effect of wheel rotation)	
$M_{LL}, M_{LS}, M_{SL}, M_{SS}$	abbreviated representation for equation coefficients multiplying lean or steer angular accelerations; each is a complex combination of bicycle parameters	

N	rotational speed	(rev./min)
p	pressure	(Pa, N/m^2, bar)
Q	heat energy	(J)
$\dot{Q}$	rate of heat transfer	(W)
r	radius, as defined in text	(m)
r_{eq}	radius of equivalent roller	(m)
r_T	radius of tire cross-section	(m)
r_1, r_2	roller radii	(m)
r_W	radius of wheel	(m)
R	gas constant	(for air: 286.96 J/kg-°K)
R_e	Reynolds number	$\equiv (\rho V d/\mu)$
R_T	radius of turn, from center of turn to rear contact	
s	slope, the tangent of the angle of rise, or the vertical climb divided by horizontal travel	
$s_\%$	slope given as a percentage	(s = s$_\%$/100)
S	stopping distance	(m)
t	time	(s, min, h)
t_h	wall thickness	(mm)
T	temperature	(°C)
T_K	absolute temperature	(K)
T_Q	torque	(N-m)
u	ratio: (speed on hill)/(speed on level) for fixed rider power (a measure of hill steepness)	
v	specific volume	(m^3/kg)
V	velocity	(m/s)
V_{DD}	velocity at which rolling drag doubles	(m/s)
V_{IN}	inversion speed	(m/s)
V_V	velocity of vehicle	(m/s)
V_W	headwind velocity	(m/s)

W	work energy	(J)
$\dot{W}$	power, rate of doing work	(W)
$\dot{W}_R$	rider power	(W)
$\dot{W}_W$	power delivered to driving wheel	(W)
x	coordinate in axial direction	(m)
x_{CM}	position of center of mass ahead of rear-wheel contact point	(m)
y	coordinate in vertical direction	(m)
y_{CM}	height of center of mass above ground	(m)
y_{WS}	wheel sinkage	(m)
Y	compressive yield strength of soil	(Pa)

Greek symbols

α_S downslope angle   (degrees)

η_m mechanical efficiency

λ tilt of steering axis back from vertical   (degrees)

μ coefficient of friction

π pi   (3.1416)

θ angle through which torque acts   (radians)

θ_C angle of crank   (radians)

θ_L rightward lean angle of frame from vertical; sometimes large (e.g., in fast turns) (radians)

θ_P downward pitch angle of frame around rear axle, relative to horizontal; nearly always small (radians)

θ_S leftward steer angle of handlebars from straight ahead; small except in slow, sharp turns (radians)

θ_W angle of wheel   (radians)

ρ density ($\equiv \nu^{-1}$)   (kg/m^3)

ν Poisson's ratio

Definitions and equalities

$=$ "equals"

$\equiv$ "is defined as"

$\propto$ "is proportional to"

$(\,\dot{}\,)$ derivative with respect to time of the bracketed quantity (i.e., a time rate of change or velocity)

$(\,\ddot{}\,)$ second derivative with respect to time (i.e., a rate of change of a velocity, in other words, an acceleration)

c amper constant (ratio between force and velocity) (N s/m or kg/s)

Conversion factors

Mass:	x lbm $= 0.4536x$ kg
Force:	x lbf $= 4.448x$ N
Length:	x in. $= 25.4x$ mm
	x ft $= 0.3048x$ m
	x miles $= 1.609x$ km
Area:	x ft$^2 = 0.0929x$ m^2
Volume:	x ft$^3 = 0.02832x$ m^3
Pressure, stress, modulus of elasticity:	x lbf/in.$^2 = 6,895$ Pa
	$(1 \text{ Pa} = 1 \text{ N/m}^2)$
	$= 6.895x$ kPa
	$(100 \text{ kPa} = 1 \text{ bar} = 14.503 \text{ lbf/inches}^2)$
Density:	x lbm/ft$^3 = 16.017x$ kg/m^3
Velocity:	x mph $= 0.447x$ m/s
	$= 1.609x$ km/h
	x knots $= 0.52x$ m/s
Torque:	x lbf-ft $= 1.356x$ N-m
Energy:	x ft-lbf $= 1.356x$ J
	x Btu $= 1,054.9x$ J
	x kcal $= 4,186.8x$ J
	x kWh $= (3.6 \times 10^6)x$ J $= 3.6$ MJ

Power: x hp $= 746x$ J/s $= 746x$ W
                                x kcal/min $= 69.78x$ W
                                x ft-lbf/s $= 1.356$ W

Specific heat: x Btu/lbm-°R $= 4{,}187x$ J/kg-°K

Heat flux: x Btu/ft²-h $= 3.154x$ W/m²
                                x kcal/m²-h $= 1.163$ W/m²

Derivations

Force (newtons) = Mass (kg) × Acceleration (m/s²)
Energy or Work (joules) = Force (N) × Distance (m)
Power (watts) = Work (joules) per unit Time (s)

Mass and weight

When we refer to the weight of (for instance) a bicycle or its rider, we are, strictly, giving the gravitational force acting on the bicycle or rider. The correct units would therefore be newtons or pounds force (lbf). If we were to take a bicycle to the moon, its weight would be about one-sixth of its weight on the earth. The mass would remain unchanged. Therefore, we have usually given the mass (in kilograms or in pounds mass [lbm]) when we have by common usage referred to the "weight." Weight is given by the relation

$$\frac{\text{Mass} \times \text{Gravitational acceleration}}{g_c},$$

where g_c is a constant that in the S.I. system equals unity and in English units equals 32.17 lbm-ft/lbf-s².

Properties of dry air at normal pressures

Temperature			Specific heat C_p	Thermal conductivity	Density[a]	Viscosity[a]
°K	°C	°F	(kJ/kg-°K)	k (kW/m-°K)	ρ (kg/m³)	(m²/s)
275	2	35.6	1.0038	2.428×10^{-5}	1.284	1.343×10^{-5}
300	27	80.6	1.0049	2.624×10^{-5}	1.177	1.567×10^{-5}
325	52	125.6	1.0063	2.816×10^{-5}	1.086	1.807×10^{-5}

[a] These properties are at 1 bar, atmospheric pressure.

Gear-speed conversion chart

Note: This chart is derived from one issued by the Tandem Club (U.K.). In continental Europe, the gear size is often specified as π × meters, or 3.1416 × meters, which gives the distance traveled for one revolution of the cranks.

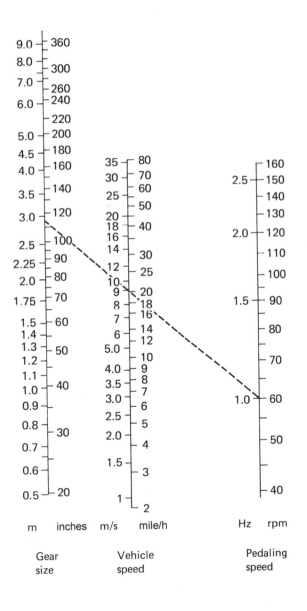

m	inches	m/s	mile/h	Hz	rpm

Gear		Vehicle		Pedaling	
size		speed		speed	

Timeline

Date	Historical events	Science and technology	Bicycle and tricycle developments	Diffusion of technology
Pre-1800		Cranks from 1500s. Iron tramways (1767). Canal era 1769+. Cast iron, wrought iron. Crucible steel (1740). Tin, copper, brass, bronze, silver, gold. Steam power. Coal gas.	No known bicycles of any type, despite myths. Skiing and ice-skating around 3000 B.C. Chinese invent wheelbarrow 300 B.C. as an aid to the infantryman, reported in Europe, 1792. War on skis between Finns and Danes, 1200. Roller skates on theatre stages, 1761. Outdoor demonstration of in-line roller skates (Netherlands), 1790.	Modified wheel-barrow spreads through Europe 1200–1300.
1787	U.S. Constitution.			
1785–1799	French Revolution.			
1801	Jefferson elected president.	Trevithick's steam road carriage.		
1802	Treaty of Amiens.			
1803	Britain declares war on France.			
1804	Lewis and Clark expedition.	Trevithick's steam railroad loco.		
1806	Battle of Trafalgar.			
1807		Fulton's steamboat. Davy discovers aluminum.		

1808	Mexican war for independence.			
1812	Madison declares war on Britain.			
1813	Battle of Leipzig.		Karl von Drais invents four-wheeled "driving-machine" with treadmill on rear axle.	
1815	Mt. Tambora, Indonesia, erupts. Waterloo: defeat of Napoleon.			
1816	"Year without summer": widespread crop failure. All-time high corn price.	Hall's wet-puddling process for wrought iron lowers price.	Karl von Drais reports four-wheeled "driving machine" with cranked back axle.	
1817	First good harvest in Germany.		Von Drais reports "running machine": first single-track two-wheeled balancing vehicle (French patent, 1818).	
1818			Denis Johnson, London, patents a copy of "Draisienne," which is very popular.	
1820	Prussia and other German duchies ban outdoor sports as paramilitary.		Gompertz adds hand drive to Draisienne. Increasing restrictions on velocipede use.	U.K. patents on inline roller skates.

Date	Historical events	Science and technology	Bicycle and tricycle developments	Diffusion of technology
1821–1827		1821: Faraday's electric motor; 1825: Stephenson's first loco pulls 450 people at 24 km/h; 1827: Niepce invents photography.	Von Drais builds "writing machine" and emigrates to Brazil.	Railways spread to continental Europe.
1830		Brunel's steamship *Great Western* crosses Atlantic 16.3 km/h.	Bramley and Parker patent a hand-and-foot-powered tandem tricycle.	
1831	Nat Turner's revolt.	Faraday's electric generator.		
1838	Afghan wars.	First electric telegraph; Nasmyth's steam hammer.	Bernard of Vienna patents "Draisienne" on railway track—a single-track railcar. Davies's lecture: first "bicycle science" (1837).	
1839	Opium wars.	Babbitt bearing patent.		
1840–1842		Brunel's SS Great Britain: first iron steamship; first screw-driven steamship; silent-chain drive.	A pedaled-lever direct-steered bicycle formerly credited to Kirkpatrick Macmillan in Scotland appears to be a myth.	
1841		Goodyear vulcanizes rubber.		
1843			Von Drais, back in Mannheim, builds treadmill-driven railcar on toothed racks.	

1844–1845	Irish potato famine starts.	Morse invents telegraph.						Thomson invents pneumatic tire (Scotland).
1851								Willard Sawyer starts making four-wheeled pedaled velocipedes with cranked axles.
1855		Deville's method for aluminum manufacture.						
1856		Bessemer process for making steel from cast iron without added fuel.						
1860		Lenoir invents coal-gas engine.						
1861	American Civil War starts.	Siemens demonstrates open-hearth steelmaking.						
1866		Nobel invents dynamite.						Lallement patent for a pedaled Draisienne.
1867	Paris Exhibition. United States purchases Alaska.							
1868	Meiji restoration. Cuban ten-year war with Spain.	Mushet invents Tungsten steel.						Pierre Michaux and the Olivier brothers form Michaux & Cie.; Eugene Meyer develops his steel-spoked tension wheel.
1869	Suez Canal finished.							Hydraulic tubing is used in Pickering & Davis machine.

Date	Historical events	Science and technology	Bicycle and tricycle developments	Diffusion of technology
1870	Franco-Prussian war results in unification of Germany & Italy.		Malleable iron is used for velocipedes. Oliviers uses drop-forged frames. Ball bearings and 1-m front wheel used; spring-steel rims, rubber (solid) tires, suspension spokes.	
1874			Starley patents tangent-spoked wheel.	Tangent-spoked wheels become widely used.
1876		Bell demonstrates his telephone. Otto invents four-stroke coal-gas engine.		
1877	Northeast passage is sailed.		Starley patents differential gear.	Universally used on automobiles.
1878			Rudge patents adjustable ball-bearing axle.	Ball-bearing development is greatly speeded by bicycle makers.
1880			Hans Renold patents bush roller chain.	Wide subsequent use in cars and motorcycles.
1881			Starley brings out Royal Salvo tricycle.	

1884		Daimler's first gasoline engine. Linotype machine.	(Approx.) Garrood introduces hollow forks.	
1885		Benz's first auto. First steel-framed building.	Two Starley safety bicycles are introduced. (Not the first, but the company sets the pattern.)	
1886		Hatfield makes manganese steel.	Road-Improvement Association is formed in the United Kingdom by CTC.	Major benefit to subsequent road users.
1887	Queen Victoria's Golden Jubilee.	Parsons pilots steam-turbine-powered ship "Turbinia" at 64 km/h.		
1888			Dunlop reinvents the pneumatic tire.	Widespread use on motor vehicles.
1892		Diesel demonstrates his universal engine.		
1895		X-rays are discovered.	Bevel-gear shaft drive introduced in France.	
1896	Klondike gold rush.	Titanium is first isolated.	Two-speed hub gear is developed; Challand patents first "modern" recumbent (Belgium).	Many auto firms are started by bicycle makers.
1897			Reynolds develops butted tubes.	
1899	U.S. Philippines war; Boer war.		Linley patents derailleur gear.	

Date	Historical events	Science and technology	Bicycle and tricycle developments	Diffusion of technology
1901		Invention of radio.	Raleigh introduces the Sturmey-Archer three-speed gear.	
1903		Wright brothers' flight.		
1913			Marcel Berthet rides Velo Torpille semi-faired bicycle.	
1914	First World War starts (–1918)		Oscar Egg sets hour record of 44.247 km; Berlin races for stream-lined bicycles; UCI bans aero-dynamic enclosures.	
1920			(Approx. date) J-Rad treadle-action recumbent, Stuttgart.	
1929	Great Depression starts.		(Approx. date) Charles Mochet makes pedaled Velocars, four wheels, three gears.	
1932–1934			Mochet produces two-wheeled (recumbent) version of Velocar. Many world records set, including 45.055 km/hr. UCI bans recumbents.	

Sources:

Compton's 1999 encyclopedia (CD-ROM).

Dodge, Pryor. (1996). *The Bicycle.* Paris and New York: Flammarion.

Lessing, Hans-Erhard. (2001). Direct contributions and edits.

McGurn, Jim. (1999). *On Your Bicycle: The Illustrated Story of Cycling.* York, U.K.: Open Road.

McNeil, Ian. (1990). *An Encyclopaedia of the History of Technology.* New York and London: Routledge.

Index